T0192241

This book, together with its companion volume *The science of crystallization: macroscopic phenomena and defect generation*, make up a complete course to teach an advanced student how to understand and analyze scientifically any of the phenomena that are observed during natural or technological crystallization from any medium and via any technique of crystallization. It is an advanced text that goes into considerable detail concerning the many elements of knowledge needed to understand both qualitatively and quantitatively a crystallization event.

This particular volume deals with the important atomistic-level processes occurring at the interface between a crystal and its nutrient. It also provides the necessary scientific background of both thermodynamics and kinetics needed for the understanding of crystallization for both bulk crystals and thin film formation. It focusses on the dual characteristics of terraces, ledges and kinks with respect to their role in both interface energetics and phase transition kinetics. In addition, it focusses on the role of these structural elements in many important morphological features of both bulk crystals and thin films. This book and its companion volume are much more broadly based and science oriented than other available books in this field, and are therefore more able to address any area of application, ranging from the production of dislocation-free single crystals in bulk or film form, at one extreme, to structurally sound large metal ingots, at the other.

This book, together with its companion, provide the basis for advanced courses on crystallization in departments of materials science, metallurgy, electrical engineering, geology, chemistry, chemical engineering and physics. In addition the books will be invaluable to scientists and engineers in the solid state electronics, optoelectronics, metallurgical and chemical industries involved in any form of crystallization and thin film formation.

The science of crystallization:
microscopic interfacial phenomena

The science of crystallization: microscopic interfacial phenomena

William A. Tiller
Department of Materials Science and Engineering
Stanford University

CAMBRIDGE UNIVERSITY PRESS
Cambridge, New York, Melbourne, Madrid, Cape Town, Singapore, São Paulo

Cambridge University Press
The Edinburgh Building, Cambridge CB2 2RU, UK

Published in the United States of America by Cambridge University Press, New York

www.cambridge.org
Information on this title: www.cambridge.org/9780521381383

© Cambridge University Press 1991

This publication is in copyright. Subject to statutory exception
and to the provisions of relevant collective licensing agreements,
no reproduction of any part may take place without
the written permission of Cambridge University Press.

First published 1991
Reprinted 1995

A catalogue record for this publication is available from the British Library

Library of Congress Cataloguing in Publication data
Tiller, William A.
The science of crystallization: microscopic interfacial phenomena /
William A. Tiller.
p. cm.
ISBN 0-521-38138-X (US) – ISBN 0-521-38827-9 (US: paperback)
1. Crystallization   I. Title
QD921.T52   1991
548′.5–dc20   89-48048   CIP

ISBN-13 978-0-521-38138-3 hardback
ISBN-10 0-521-38138-X hardback

ISBN-13 978-0-521-38827-6 paperback
ISBN-10 0-521-38827-9 paperback

Transferred to digital printing 2006

To
My Loving Wife
Jean

Contents

PREFACE

This book is one of two companion books on the science of crystallization and both are meant to be teaching vehicles rather than scientific treatises although some parts get close to the line. They are intended to develop a student's understanding of all the interwoven processes that are involved in either the natural geological formation of crystals or in the tailor-making of crystals in the laboratory for a specific use. Although the specific techniques used in the laboratory are many and extremely varied, the basic atomic processes involved are quite similar. Thus, it is felt that a basic description of the processes involved in any one specific technique allows ready extrapolation to the understanding of other techniques.

In this book, many of the experimental examples will relate to crystallization from the melt; however, examples and applications are also given to show how these same principles may be used to understand crystallization from all other nutrient media. In some cases, we will be more interested in the corollaries "Under what conditions does amorphous deposition occur?" and "What is the dissolution rate?". We shall see that these same principles and ways of thinking allow us to consider the more general case of "phase transformations," whether the path is to a crystalline or to an amorphous product (e.g., the formation of an SiO_2 film on Si during thermal oxidation).

With respect to materials, the point of focus in the examples is to Si, wherever possible, with data on compounds and metals to provide contrast and broadening. The material should therefore be of great interest to those involved in semiconductor processing and solid state electronics.

Although the text is fairly mathematical in nature, it is designed to lead the student first and foremost to a qualitative appreciation and understanding of the many factors and principles involved so that a good intuition might eventually develop concerning these matters and so that the student will then understand how to create novel new techniques for tailor-making specific crystals. The examples and problem sets have been designed to lead the student to a quantitative understanding of the field so that he or she may learn to make meaningful numerical appraisals of a particular situation. In this regard, the goal is to reveal mathematical shortcuts to an approximate result that still contains all the essential physics of the problem rather than to whittle down the physics of the real problem so that a closed-form mathematically beautiful solution of a highly idealized problem can be obtained.

After an introduction designed to illustrate the philosophy of approach and the scope of the treatment for both books, the text of this book is divided into two sections. The next four chapters deal with the principles and applications of the basic elements involved with microscopic interfacial phenomena present in almost every crystallization event that one may wish to consider. These chapters allow the student to learn how the background science and basic principles interact to explain the phenomena seen at this level. The last three chapters form a section, entitled "Background Science", which deals with the general thermodynamic and kinetic underpinning needed to effectively explore the territory of the first five chapters as well as the territory of the companion book. This Background Science section is placed at the end of this text to demonstrate that it can be bypassed by those students already possessing a qualitative understanding of that material. While this text deals with microscopic interfacial phenomena, the companion text deals with the macroscopic phenomena and defects. As such, it treats convection, heat transport, solute redistribution, interface morphology development and physical defect generation.

From the material presented in these two books, it is hoped that the student will notice that, from example to example, the essential features differ only in degree rather than in kind and that the same set of considerations are involved in the development of dislocation-free single crystals of silicon as in the development of structurally sound 100 ton steel ingots.

The present text and that of the companion book are more broad based and science oriented than alternative books in this field and, thus, should be texts from which university courses at the senior or master's degree level can be drawn by departments of Materials Science, Metallurgy, Electrical Engineering, Geology, Chemistry and Chemical Engineering. In addition, technical personnel in the solid state electronics, metallurgical and chemical industries involved in any form of crystallization or film formation should find them insightful and useful, if not indispensable.

I owe a great debt to many individuals for the completion of this work. First, to my PhD students over the past two and a half decades for helping me to refine and understand the concepts set forth here and in the companion text. To Irv Greenfield who provided me with the opportunity to make a good start on this version of the book (1980) after two earlier false starts (1964 and 1970). To my Stanford classroom students in this topic area who, over the years, have helped to correct my many inconsistencies with respect to symbols and logic. To my secretary, Miriam Peckler, who provided much support for the earlier versions. To Jerri Rudnick who has converted my handwritten manuscript to Tex. To my colleague and friend Guy Marshall Pound (deceased) for reading and helpfully criticizing the text. To Bruce Chalmers who first introduced me to this subject. To John Dodd who first made me want to learn. To my wife Jean for her loving patience in putting up with both the many years of my distraction to this task and the perpetually littered surfaces of our home with working notes and papers.

Stanford, California   William A. Tiller
September 18, 1989

Symbols

1. Arrangement

The most often used subscripts and superscripts are given immediately below and this is followed by the list of symbols. English symbols are listed first and Greek second with the symbols being arranged in alphabetical order. For each letter, printed capitals are listed first, script capitals second and lower case third.

2. Subscripts

a - adsorption site	β - phase	C - solute
$\hat{c}$ - crystal	c - critical value	E - excess
$\tilde{E}$ - eutectic	e - electronic	F - fusion
f - face	i - interface	i^* - intrinsic
∞ - far field value	K - kinetic	k - kink
L - liquid	ℓ - ledge	M - melting
N - nutrient	o - reference condition	P - pressure
Φ - electrostatic	S - solid	$\tilde{S}$ - solution
s - surface	s^* - sublimation	T - thermal, temperature
V - vapor	v^* - vaporization	

3. Superscripts

a - adsorbate	e - electron	j - species
M - major solute	m - minor solute	$\hat{v}$ - vacancy
$*$ - equilibrium value		

4.   English symbols

A - area

A_0 - area occupied by a growth unit in a crystal

A^* - activated state

$\mathcal{A}_0$ - area of surface with unit probability for a new nucleation event

a, a', a'' - lattice distances in different directions

$\hat{a}$ - thermodynamic activity

a_0 - molecular radius

$a_{\hat{c}}$ - crystal radius

a_d - droplet area

a^* - convection parameter $(a^* = 2\tilde{U}_\infty/\pi R)$

a_* - dimensionless suction parameter

$\tilde{a}$ - sphere diameter

$\bar{a}$ - nucleation parameter $(\bar{a} = \pi h \gamma_E^2/3\Delta S_F \mathcal{R}T)$

$\hat{B}$ - dynamic bond number $(\hat{B} = G_V/Re_M)$

$\mathcal{B}$ - non-planar field decay coefficient

$$\left(\mathcal{B} = \tfrac{1}{2}\left\{1 + \left[1 + \left(\tfrac{2D}{V}\omega\right)^2\right]^{1/2}\right\}\right)$$

$\mathcal{B}_*$ - value of $\mathcal{B}$ at the marginal stability condition

b - growth rate parameter $(\dot{R} = bt^{-1/2})$

$\tilde{b}$ - Henry's law constant

b_* - Burger's vector

$\hat{b}$ - diffusion parameter $(\hat{b} = a^2/4\alpha t_0)$

C - concentration

C_T - concentration of transition state

$\bar{C}$ - Laplace transform concentration

$\tilde{C}$ - electrolytic capacitance

$\hat{C}$ - Taylor constant

c - velocity of light $(2.998 \times 10^8$ m s$^{-1})$

c_p - specific heat at constant pressure

c_v - specific volume ratio $(c_v = v_0^j/v_0^0)$

$\tilde{c}$ - number of chemical elements in system

D - diffusion coefficient

$\tilde{D}_T$	- transitional diameter for a dendrite array
$\mathcal{D}$	- an effective diffusivity
d^f	- film thickness
E	- energy
$\bar{E}$	- partial molar internal energy
$E_{\bar{T}}$	- total energy
E_j^f	- energy of the jth type of fault
E_g	- band gap energy
E_F	- chemical binding energy at the Fermi level
E_v	- valence band energy
E_c	- conduction band energy
E_B	- boundary energy
E^*	- Young's modulus
$\hat{E}$	- electric field
E_i	- exponential integral function
Erf	- error function
Erfc	- complementary error function
ΔE_r	- energy change due to roughening
ΔE^f	- formation energy
ΔE_b	- excess energy per bond
$E_{\hat{L}}$	- electron lattice energy
F	- Helmholtz free energy
$\hat{F}$	- force between half spaces
F_*	- Cochran function
ΔF_r	- Helmholtz free energy change due to roughening
ΔF_{rec}	- Helmholtz free energy change due to surface reconstruction
f	- Helmholtz free energy density
$\tilde{f}, \tilde{f}'$	- unspecified general mathematical functions
$\hat{f}$	- fraction of active sites
$\bar{f}_{mn}$	- surface stress tensor
f_*	- correlation factor
$\tilde{f}_*$	- degrees of freedom
$\vec{f}$	- volume force
G	- Gibbs free energy
G^*	- Gibbs free energy of the activated state
G_*	- Cochran function

$\tilde{G}$	- shear modulus
G_+	- Green's function
ΔG	- Gibbs free energy driving force
ΔG_{sv}	- driving force consumed by the thermodynamic state variable change ($\infty-i$)
ΔG_N	- free energy change per unit volume transformed
ΔG_A	- activation energy
ΔG_p^f	- pseudo kink formation free energy
ΔG_m	- free energy of mixing
$\widetilde{\Delta G_0^j}$	- free energy change of j due to an interface field
$\Delta \bar{G}_+, \Delta \bar{G}_-$	- free energy change for transitions to right or to left
ΔG_S	- free energy of separation
ΔG_0	- bulk free energy change
ΔG^{uh}	- free energy change for unhindered rotation
δG_{C^*}	- free energy change for surface ledge creation
δG_{A^*}	- free energy change for surface ledge annihilation
δG_{I^*}	- free energy change for surface ledge interaction
$\mathcal{G}$	- temperature gradient
$\mathcal{G}_n$	- temperature gradient in the normal direction
g	- Gibbs free energy density
$\tilde{g}, \tilde{g}^1$	- unspecified general mathematical functions
$\hat{g}$	- geometrical factor
g_*	- fraction solidified
H	- enthalpy
$\bar{H}$	- partial molar enthalpy
H_*	- Cochran function
$H_{\hat{B}}$	- Biot number
$\tilde{H}$	- height of cavity
$\hat{H}$	- magnetic field strength
ΔH	- heat content change
$\Delta H'$	- heat change per molecule
ΔH_m	- heat of mixing
$\Delta H''$	- thermodynamic state with two degrees of freedom factored out
ΔH_g	- gram molar heat of fusion
h_ℓ	- ledge height
h_c	- transport coefficient
h^f	- facet height

$h', \bar{h}'$	- vector distances
$\tilde{h}$	- heat transfer coefficient
h^*	- Planck's constant
I	- nucleation frequency
$\tilde{I}$	- moment of inertia of a cluster
$\hat{I}$	- rate of elementary process
$\mathcal{I}$	- kinetic amplification factor in spinodal decomposition
i	- number of molecules in a cluster
i^*	- number of molecules in a critical size cluster
J	- flux of heat or matter
$\mathcal{J}$	- Bessel function
j	- species
K	- thermal conductivity
$\hat{K}$	- equilibrium constant
$\hat{K}_R$	- reaction constant
K_*	- transmission coefficient
K_f	- drag coefficient
$\bar{K}_0$	- gross equilibrium solute partition coefficient for a polycomponent system
$\tilde{K}$	- surface curvature parameter ($\tilde{K} = (\gamma/\mathcal{G}\Delta S_F)^{1/2}$)
$\mathcal{K}$	- surface curvature
$k_0, \tilde{k}_0$	- phase diagram solute partition coefficient ($C_S/C_L, X_S/X_L$)
k_i	- interface solute partition coefficient
$\hat{k}_i$	- net interface solute partition coefficient
k	- effective solute partition coefficient
$k_{\hat{E}}$	- transformed solute distribution coefficient
$\bar{k}$	- rate constant
$\bar{k}_p$	- parabolic rate constant
$\bar{k}_\ell$	- linear rate constant
$\bar{k}_+, \bar{k}_-$	- rate of transition to right or to left
$\bar{L}$	- dimensionless length ($\bar{L} = Vt/a_{\hat{c}}$)

L_* - crucible size
$\tilde{L}$ - $\tilde{H}/\ell$ - dimensionless height
$\hat{L}_i$ - a thermodynamic potential component

ℓ - distance
ℓ_{tw} - length of twin plane

M - resistive torque
$\hat{M}$ - atomic mobilty
$\hat{M}_{\hat{E}}$ - effective ionic mobility
$\hat{M}_{ij}$ - a thermodynamic potential component
M_a - Marangoni number

m - mass
m_r - reduced mass
m_L - liquidus slope
$\bar{m}_L$ - average liquidus slope component
$\hat{m}_L$ - effective liquidus slope
$\tilde{m}$ - degree of a kink
m_* - log spiral convection parameter
$m^\wedge$ - slope in stirring/field strength plot

N^j - number of j species
N_a - number of adsorption sites/cm^2
$\tilde{N}$ - number of nearest neighbor sites in a crystal
N_k - number of heterogeneous nuclei/cm^3
N_* - transport number
$\hat{N}_{ij}$ - thermodynamic potential function component

$\mathcal{N}_A$ - Avogadro's number

n^j - number of moles of j
n_j^f - number of j-type faults
$\hat{n}$ - distance in the interface normal direction
$\hat{n}_x, \hat{n}_y$ - projections of the unit normal vector on x and y
$\tilde{n}_1$ - number of nearest neighbor sites in the interface
n_{bf} - number of broken bonds in the face
$n_{\hat{r}}$ - number of rearrangements
n_n - average number of foreign nuclei
n_D - number of dislocations

$n_{\hat{v}}^*$	- equilibrium concentration of vacancies
n_I^*	- equilibrium concentration of interstitials
$n_{I\alpha}$	- number of interstitial sites per unit cell
P	- pressure
$\tilde{P}$	- dimensionless pressure ($\tilde{P} = P/\rho U_\infty^2$)
$\hat{P}$	- Péclet number ($\hat{P}_T = VR/2D_T, \hat{P}_C = VR/2D_C$)
Pr	- Prandlt number ($Pr = \nu/D_T$)
P_*	- Cochran function
p, p_+, p_-, p_0	- probabilities
p_f	- stacking fault formation probability
p_{sp}	- spiral dislocation formation probability
$\tilde{p}$	- number of coexisting phases
$\bar{Q}$	- total energy content
$\tilde{Q}$	- centrifugal fan volume
$\dot{Q}_*$	- non-conservative system effect on solute content
$\hat{Q}$	- parameter
q	- charge
$\bar{q}$	- energy distribution
$\tilde{q}_\omega$	- heat flux
q_*	- rate factor for non-conservative system
$\hat{q}$	- canonical partition function
R	- radius
R_*	- crucible radius
R^*	- critical nucleus radius
$\tilde{R}_1, \tilde{R}_2$	- principal radii of curvature
$\hat{R}_0, \hat{R}$	- metallic radii
$\bar{R}_n^2$	- mean square displacement
Re	- Reynold's number ($Re = U\ell/\nu$)
Re_M	- Marangoni Reynold's number
Ra	- Rayleigh number
$\mathcal{R}$	- gas constant per g mole, 8.134 JK^{-1} mole $^{-1}$, 1.987 cal deg^{-1} mole $^{-1}$
r	- radial coordinate

$\hat{r}$	- surface roughness
r_0	- equilibrium separation of atoms
r_D^c	- dislocation core radius
r^*	- critical radius of two dimensional embryo

S	- entropy
$\bar{S}$	- partial molar entropy
S_b	- entropy per bond
$\tilde{S}_{ij}$	- compliance
S_*	- sticking coefficient
$\hat{S}$	- Laplace transform variable
Sc	- Schmidt number $(Sc = \nu/D_C)$
S_p^q	- charged species parameter

ΔS	- entropy change
$\Delta S''$	- entropy change for states with two degrees of freedom factored out
ΔS_r	- entropy change on roughening

$\mathcal{S}$	- stability function
$\mathcal{S}_*$	- value of $\mathcal{S}$ at the marginal stability condition

s	- surface entropy density, surface coordinate
$\hat{s}$	- interface shape
$\hat{s}_*$	- optimum surface shape
s_{c^*}, s_{A^*}	- coordinates for surface creation, annihilation regions

T	- temperature
$\dot{T}$	- cooling rate
T_L	- liquidus line, liquidus surface
T_s	- heat source temperature
$T_{s'}$	- heat sink temperature
$\tilde{T}_s$	- substrate temperature
T_g	- glass transition temperature
T_t	- transition temperature
T_w	- wall temperature
$\hat{T}$	- Taylor number
ΔT	- thermal driving force (temperature potential difference)
Δ_T	- superheat

$\Delta T^{*}, \Delta T^{**}, \Delta T^{***}$ - special transition driving forces for attachment kinetics

ΔT_{k}^{fr} - driving force for attachment at the root of a facet

ΔT_{NP} - driving force due to non-planar isotherms

ΔT_{c} - critical supercooling

δT - temperature oscillation amplitude

t - time

t_{k} - relaxation time to enter a kink

$U, u, \vec{u}$ - fluid velocity

$\hat{U}, \hat{u}$ - intermolecular potential

$\hat{U}'_{dd}$ - potential for interacting layers

U_{*} - potential driving a fluid

U_{M} - Marangoni velocity

U_{g} - buoyancy driving velocity

$\bar{U}_{R}$ - return flow velocity

$\bar{U}_{\infty}$ - uniform velocity at ∞

U_{N} - natural convection velocity

U_{D} - axial forced flow velocity

$\hat{u}_{j}$ - multiplicity factor

$\vec{u}_{R}$ - relative velocity

$\tilde{u}_{j}$ - dimensionless velocity in the jth coordinate direction

V - growth velocity

V_{*} - optimum velocity

V_{y} - velocity of surface in the y-direction

V_{ℓ} - ledge velocity

$\tilde{V}$ - Volta potential, solid velocity

$\bar{V}_{c}$ - critical velocity for loss of transport communication

$\bar{V}_{c}^{*}$ - critical velocity for loss of interface equilibrium

$V_{\hat{E}}$ - transformed frame velocity

V_{max}^{CSC} - maximum velocity without formation of constitutional supercooling

V_{0} - climb velocity

$\underline{V}$ - macropotential, voltage

$\mathcal{V}$ - nucleation velocity limit

v - volume

$\tilde{v}$	- partial molar quantity
v_m	- molar volume of a material
dv/v_0	- fraction of volume transformed
Δv_t	- volume transformed
$\hat{v}$	- vacancies
W	- facet width
$\hat{W}$	- a general potential
W_a	- work of atomic adsorption rearrangement
W_e	- work of electron rearrangement
$W_{\alpha\beta}$	- work of separation of two phases
$\bar{w}$	- a parameter
$\tilde{w}$	- possible number of complexions
X	- mole fraction
X_t	- fraction transformed
X^*	- optimum density of roughened states
$\bar{X}_k$	- average kink spacing in a'' units
X_d	- fraction of droplets
$\bar{X}_{\hat{s}}$	- mean surface diffusion distance in a' units
X_{ox}	- oxide thickness
$\hat{X}$	- position of the interface
x_j	- experimental coordinate of the j-type
$\tilde{x}$	- coordinate on face relative to a crystal corner
Y	- Young's modulus ($Y = E^*/(1 - \sigma^*)$)
$\bar{Y}_\ell$	- average ledge spacing
$\hat{Y}_{\ell m}$	- harmonic test function
$\tilde{Y}_j$	- dimensionless coordinate (y_j/e)
y	- coordinate direction
Z_3	- three-body interaction parameter
$\tilde{Z}$	- Zeldovich factor
$Z_{\hat{s}}$	- z-coordinate variation on the surface shape
z	- distance
z_*	- number of equivalent jumps

$\tilde{z}_1$ - number of nearest neighbors
$\hat{z}^j$ - valence of the j species

5.   Greek symbols

α - Jackson α-parameter
α - phase
$\alpha_1, \alpha_2, \alpha_3$ - parameters
α^*, α^{A_j} - thermodynamic α-factors
α_k - vaporization fraction
α' - monomer in cluster formation process
$\tilde{\alpha}$ - field width at the interface
$\bar{\alpha}$ - growth rate parameter ($V = \bar{\alpha}t^{-1/2}$)
$\bar{\alpha}^*$ - thermal expansion coefficient, interface attachment
         parameter at the onset of instability
α_D - dislocation parameter
α_β - interface breakdown parameter

β - phase
$\beta_0, \beta_1, \beta_2$ - parameters used in the interface attachment process
$\tilde{\beta}$ - surface energy per unit area of low index projected face
β' - embryo in cluster formation process
β_* - fraction dissociated to tetramer state
β_* - reaction rate parameter ($\beta_* = -\bar{k}_+ C_A/\kappa T$)
β^* - volume expansion coefficient
$\bar{\beta}$ - magnitude of field strength at the interface

Γ^j - surface exess density of j
$\tilde{\Gamma}$ - surface capillarity parameter ($\tilde{\Gamma} = \gamma/\Delta S_f$)

γ - surface Gibbs free energy, phase
$\hat{\gamma}$ - thermodynamic activity coefficient
γ_ℓ - ledge free energy
γ_f^0 - face free energy relative to a standard condition
γ_e - electronic contribution to the surface free energy
γ_{tw} - twin boundary energy
γ_0'' - surface torque term ($\partial^2\gamma/\partial n_x^2$)
γ_c - quasi-chemical contribution
γ_t - transitional diffuseness contribution
γ_d - dislocation/strain contribution
γ_a - adsorption contribution

γ' - interfacial free energy per molecule

γ_g - gram molar surface tension

$\Delta\gamma^{\hat{R}}$ - free energy change in forming a ridge

δ - boundary layer thickness, off-stoichiometry fraction, phase, strain

$\tilde{\delta}$ - amplitude of surface undulation

$\dot{\tilde{\delta}}$ - time rate of change of amplitude

δ_* - half-width of activation barrier

$\tilde{\Delta}$ - surface creation parameter $(\lambda''/\omega^2\delta)^{1/2}$

ε - an effective bond energy

ε_n - pair bonds to the nth neighbour

ε_a - adsorption energy per bond

ε_0 - binding energy

$\hat{\varepsilon}_j$ - parameter of the jth type

$\tilde{\varepsilon}_{mn}$ - strain tensor

ε_* - dielectric permeability

$\tilde{\varepsilon}$ - lattice parameter strain

$\bar{\varepsilon}$ - emissivity

$\Delta\varepsilon$ - energy fluctuation

ζ - Fermi energy

$\tilde{\zeta}$ - a dimensionless coordinate $\left(z\left(\dfrac{a^*+\omega}{v}\right)^{1/2}\right)$

$\eta, \tilde{\eta}$ - expanded electrochemical potential

$\bar{\eta}$ - viscosity

η^* - a parameter

$\hat{\eta}$ - a quantity

$\tilde{\eta}$ - dimensionless coordinate (y/δ_m)

$\tilde{\eta}_*$ - flow field variable $(z/2(\nu t)^{1/2})$

θ - angle

$\hat{\theta}$ - surface coverage

$\hat{\theta}^a$ - surface coverage of adatoms

$\hat{\theta}^{\hat{v}}$ - surface coverage of vacancies

$\tilde{\theta}$ - dimensionless temperature $((T-T_a)/(T_M-T_a))$

θ_* - diffusion parameter $((Vx/D)^{1/2})$

$\Delta\theta$ - non-dimensional supercooling or Stefan number $(c\Delta T_\infty / \Delta H_F)$

κ - Boltzmann's constant
κ_* - a drag coefficient
κ_D - inverse Debye length (λ_D^{-1})
$\hat{\kappa}_0, \hat{\kappa}_1, \hat{\kappa}_2$ - potential function components

λ - grid spacing
$\hat{\lambda}$ - $\gamma_{max}/\gamma_{min}$
λ_D - Debye length
λ_D^e - Debye length for electrons
λ_* - de Broglie wavelength
$\tilde{\lambda}$ - width of interface field
λ^* - optimum perturbation wavelength
$\Delta\lambda/\lambda$ - lattice parameter change

μ - chemical potential
μ_0 - standard state chemical potential
μ_0^*, μ_* - magnetic permeabilities
$\Delta\mu_b$ - excess chemical potential per bond
$\bar{\mu}$ - viscosity

ν - kinematic viscosity $(\nu = \bar{\eta}/\rho)$
$\hat{\nu}$ - field frequency
$\hat{\nu}$ - vibrational frequency
$\bar{\nu}$ - a surface roughness parameter $(2\varepsilon_1/\kappa T^*)$
$\tilde{\nu}_r$ - normal frequency of reaction

ξ - distance coordinate in a moving frame of reference
$\hat{\xi}$ - a surface roughening parameter

ρ - density
ρ_m - mass density

σ - stress
$\hat{\sigma}$ - supersaturation
σ^* - Poisson's ratio
σ_* - electrical conductivity
$\bar{\sigma}_*$ - electrical charge density
$\tilde{\sigma}$ - Stefan–Boltzmann constant
$\bar{\sigma}$ - supersaturation number

Φ - electrostatic potential, Galvani potential

ϕ - angle
$\hat{\phi}$ - dimensionless velocity potential
$\tilde{\phi}$ - dissipation function
$\bar{\phi}$ - dimensionless concentration (C/C_∞)
ϕ_* - Fourier spectrum of perturbations

τ - time constant
τ_k - average time for two adjacent kinks to grow together
τ_{R_1} - time for a surface pill box disc to grow to radius R_1
τ_1^a - jumping time of adsorbate
$\hat{\tau}_\ell$ - ledge overgrowth time
$\hat{\tau}_f$ - face overgrowth time
$\bar{\tau}_0$ - modulus parameter
$\tau_{\hat{s}}$ - adatom lifetime on surface
τ_σ - shear stress
τ_{diff} - effective solute redistribution time
τ_{rise} - effective rise time to achieve a steady state solute distribution

$\tilde{\Psi}$ - electron work function

ψ - angle
$\hat{\psi}$ - nucleation time lag factor

Ω - molecular volume
$\hat{\Omega}$ - angular velocity

ω - frequency, angular velocity
ω_u - frequency of configurational rearrangements
ω_D - frequency of diffusive transport
$\omega_{\bar{\eta}}$ - frequency of viscous flow events
ω_t - frequency of transition
ω_{cr} - crucible rotation rate

χ - electron affinity
$\tilde{\chi}_{\hat{s}}$ - surface potential
χ_E^*, χ_M^* - electric and magnetic susceptibilities

1

Introduction and philosophy

We have all been delighted and fascinated by the variety of crystal forms that abound in nature and please the eye. Each pattern in this vast panorama appears unique. The variations seen in the growth forms presented in Fig. 1.1 give us some idea of how the dendritic growth form can change in appearance as we change from material to material or from one crystal growth process to another. Even for a single material and a single process we often find great variability as illustrated by Nakaya's[1] eight common snow-flake patterns given in Fig 1.2. Such variability and uniqueness of form caused earlier scientists to speculate on and argue about the relative importance of heredity factors versus environment factors in the ultimate form of such crystals.

A relatively recent experiment by Mason[2] tended to illuminate this issue and strongly favor the environmental factor control point of view. He placed a nylon fiber in the water vapor diffusion chamber illustrated in Fig. 1.3 and observed the form of the ice crystals that developed at the different temperature locations along the fiber. This great range of forms is listed in Table 1.1. After these forms were growing stably in the chamber, he lifted the fiber some distance so that those crystals growing between temperatures T_1 and T_2 on the fiber were shifted to another temperature range $T_3 - T_4$ characteristic of a different crystal form on the earlier result. As he watched, the old crystal form began to change as the crystal grew larger and eventually developed the form

Fig. 1.1. Crystal growth forms: (center) Bi hopper crystal grown from the melt; (upper left) Cr dendrite grown from the vapor; (upper right) GaP grown from an alloy melt at low temperature; (lower left) Ag dendrite grown by electro deposition; and (lower right) spherulite grown from a blend of 40% isotactic polypropylene and 60% atactic polypropylene.

characteristic of the new temperature zone $T_3 - T_4$ that was observed in the initial experiment.

All along the fiber the ice seed crystals, having the form characteristic of their initial temperature environment, grew to develop the forms characteristic of the final temperature environment. Clearly, heredity factors were not relevant in this experiment. We shall see in what follows that small changes in the state variables of the environment play the complete role in changing a crystal's morphology in almost all cases. Likewise, for fixed state variables of the environment, small changes in the material parameters as we go from system to system play the complete role in changing the crystal morphology for the different material

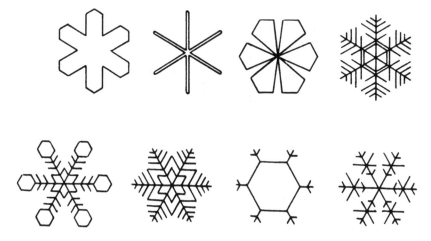

Fig. 1.2. Nakaya's eight snow crystal types.

system. However, in a few isolated cases, we will see that heredity factors play a crucial role in a crystal's morphology.

1.1   Pathways to understanding

There are two basic ways of approaching crystal growth phenomena. The first is the time-honored method of inquiry which treats the phenomenon under study as a "black box" whose internal characteristics are unknown but are amenable to probing and analysis. Such a situation always occurs in the early stages of development of a particular field of knowledge and is illustrated in Fig. 1.4(a).[3] We apply some input stimulus (IS) to the box and determine some output response (OR). By correlating the OR with the IS, we deduce information about the most probable behavior of the box for this degree of stimulus. We then speculate on models that would reproduce such a spectrum of responses and design critical tests for discriminating between acceptable models.

Our first steps toward determining the behavior of the black box of Fig. 1.4(a) is to characterize it in the following form:

$$\frac{\text{OR}}{\text{IS}} = \tilde{f}(\hat{\varepsilon}_1, \hat{\varepsilon}_2, \ldots, \hat{\varepsilon}_n; x_1, x_2, \ldots, x_n) \qquad (1.1a)$$

$$\approx \tilde{f}'(\hat{\varepsilon}_1, \ldots, \hat{\varepsilon}_j; x_1, \ldots, x_k) \quad \text{with} \quad \hat{\varepsilon}_j^* \lessgtr \hat{\varepsilon}_j \lessgtr \hat{\varepsilon}_j^{**}$$

$$\text{and} \quad x_k^* \lessgtr x_k \lessgtr x_k^{**} \qquad (1.1b)$$

for all indices $j, k = 1, 2, \ldots, m$. In Eq. (1.1a), $\tilde{f}$ represents the true functional relationship between all the possible material parameters $\hat{\varepsilon}_j$

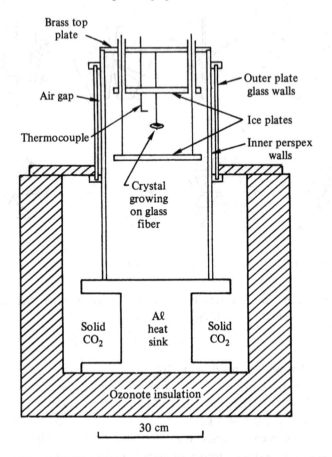

Fig. 1.3. The diffusion cloud chamber used by Mason and Hallett for growing ice crystals at various temperatures and supersaturations.

and the experimental variables x_k of the system where unlimited range is allowed for these parameters and variables. In Eq. (1.1*b*), the observed functional relationship $\tilde{f}'$ between a limited but seemingly sufficient number of the parameters and variables, for a given degree of reliability, is indicated for bounded ranges of the parameters and variables. Since the number of experiments, Q_1, needed to be performed to map OR/IS uniformly in this $j + k$ space is

$$Q_1 = (d/\lambda)^{j+k} \tag{1.2}$$

where λ is the average grid spacing for measurement along any coordinate axis and d is the total experimental range encompassed along that coordinate direction, we endeavor to reduce $j + k$ and d/λ as much as

Table 1.1. *Variation of ice crystal habit with temperature (Hallett and Mason)*

0 to $-3°$C	Thin hexagonal plates
-3 to $-5°$C	Needles
-5 to $-8°$C	Hollow prisms
-8 to $-12°$C	Hexagonal plates
-12 to $-16°$C	Dendritic crystals
-16 to $-25°$C	Plates
-25 to $-50°$C	Hollow Prisms

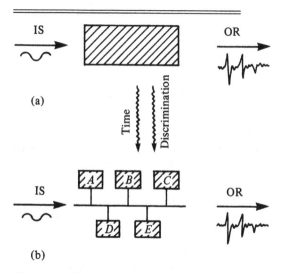

Fig. 1.4. Schematic representation of (a) the "black box" approach and (b) the "system" approach for studying nature (IS = input stimulus; OR = output response).

is practical. Since the time and cost involved in an experimental study are proportional to Q_1, practicality dictates that we try to reduce Q_1.

To obtain enough data along a given coordinate direction for the testing of a theory or for the fitting to a mathematical function, $d/\lambda \sim 10$. We thus find that a typical well-defined physics problem, for which $j + k \sim 3 - 4$, requires $\sim 10^3 - 10^4$ experiments. At $1 - 10$ data points collected per day, this is a typical thesis problem of ~ 3 man-years duration. For a typical engineering problem, $j + k$ increases to ~ 10 and, by using dimensionless groups, this may be reduced to 5–6 which requires the numbers of man-years of effort to increase for comparable reliability with the typical physics problem. For a typical crystal growth problem, $j + k \sim 20$ so that d/λ must be reduced to $\sim 1 - 2$ for a comparable number of man-years investment. Because of this, one is confined to a "recipe" or "art" mode to obtain a successful result and the technology resulting from such efforts must be characterized as an "Art-Based Technology."

With the passage of time and experimentation we are led to the second way of approaching crystal growth phenomena. We begin to discriminate the various unique phenomena and processes clustered together in the black box of Fig 1.4(a) and choose to describe it as an interacting system as illustrated schematically in Fig 1.4(b) and mathematically via the following equations

$$\frac{\text{OR}}{\text{IS}} = \tilde{g}[\tilde{f}_1, \ldots, \tilde{f}_k] \tag{1.3a}$$

$$\approx \tilde{g}'[\tilde{f}_1', \ldots, \tilde{f}_j'] \tag{1.3b}$$

where $\tilde{f}_i$ and the $\tilde{f}_i'$ are of the form represented by Eqs. (1.1a) and (1.1b) and where $\tilde{g}$ and $\tilde{g}'$ represent the exact and approximate functional relationships between the various $\tilde{f}_i$ and $\tilde{f}_i'$. For such clustered phenomena, in this representation the $\tilde{f}_i'$ can be treated as elementary parts or subsets in the overall system or ensemble.

In order to develop a science of events that conforms to Eqs. (1.3a) and (1.3b) for a considerable range of variations of the $\hat{\varepsilon}_j$ and x_k, it is necessary to develop methods of systems analysis for analyzing the events (the phenomena). The steps to be taken appear to be: (i) identify the critical and individual phenomena included in the single black box that encompasses the clustered event, i.e., boxes A through E of Fig. 1.4(b); (ii) gain an understanding of the $\tilde{f}_i$ for each box in isolation so that a quantitative response spectrum can be determined for a quantitative input stimulus; (iii) gain an understanding of the $\tilde{f}_i$ as they interact with various other $\tilde{f}_j$ in pairs, triplets, etc. and (iv) partition the total potential for the clustered event into the partial potentials consumed by the various elemental $\tilde{f}_i$ as they interact with each other; and (v) given the total available input potential, assess the kinetics of change of the coupled system and the stability of the mode of change.

For a clustered event with n discriminated subsystems and $j + k$ total coordinate axes, the average number of mapping experiments, Q_2, is

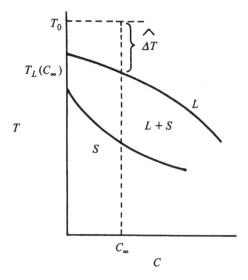

Fig. 1.5. Binary phase diagram for alloy melt of composition C_∞ that is initially superheated an amount $\widehat{\Delta T}$.

given by

$$Q_2 = n \left(\frac{d}{\lambda}\right)^{\left(\frac{j+k}{n}\right)} + \frac{n}{2}(n-1)\left(\frac{d}{\lambda}\right)^{2\left(\frac{j+k}{n}\right)} + \cdots \qquad (1.4)$$

where the second term arises from pair interactions. If one takes $j+k \sim 20$ and $n = 5$ with $d/\lambda \sim 10$, then $Q_1 \sim 10^{20}$ which is an impossibly large number while $Q_2 \sim 5 \times 10^4 + 10^9$ which is still large but appreciably smaller. If $j+k$ can be reduced to ~ 10 by forming dimensionless groups, then $Q_2 \sim 10^5$ which becomes a much more manageable problem and we can expect reliable scientific studies to be forthcoming from this approach with a reasonable amount of effort. Such an approach leads to a "Science-Based Technology." In this approach, careful basic experiments on each of the subsystems in isolation can be performed so that the first term of Eq. (1.4) is readily known in any clustered event. This allows the OR/IS to be determined for the pair and higher order interactions.

As an example, suppose we want to predict the solid structure (grain size and shape, degree of macro- and micro-segregation, etc.) that arises from freezing a volume v of binary alloy liquid having a solute content, C_∞, initially superheated an amount $\widehat{\Delta T}$ and then cooled at its outer surface at a given rate $\dot{T}$ per unit area of surface (see Fig. 1.5). Even

Table 1.2. *Crystallization variables and parameters*

Areas of study	Boundary value problems	Material parameters	Interface variables	Macroscopic variables	Con-straints
Phase equilibria		$\Delta H, T_0,$ k_0, m_L		$T_L(C_\infty)$	
Nucleation		$N_h, \Delta T_c$		t	
Solute partitioning	Diffusion eq. (C)	D_{CS}, D_{CL}, k_i	$C_i, T_L(C_i)$ $V, \hat{s}$	C_∞	
Fluid motion	Hydrodynamic eq. (u)	ν	δ_C	u_∞	C_∞
Excess solid free energy		$\gamma, \Delta S,$ E_j^f, n_j^f	T_E		$\dot{T}$
Interface attachment kinetics		β_1, β_2			
Heat transport	Heat eq. (T)	$K_S, K_L,$ D_{TS}, D_{TL}	T_i	T_∞	
Interface morphology	Perturbation response and coupling eqs.				
Defect generation	Stress eq.				

to begin to make such a prediction, it is necessary to discriminate at least nine separate $\tilde{f}_i$ which are indicated in Table 1.2; moreover, at least twenty material parameters, at least seven interface variables which control the processes going on at the interface between the crystals and the liquid, and at least five major field equations must be considered. If one relied solely on the philosophy represented by Eqs. (1.1a) and (1.1b), it would be relatively impossible to predict the behavior of one alloy system on the basis of the performance of another, since the variation of any one of the parameters or variables leads to large variability in the morphology of the growing crystals and thus in the resulting structure of the solid. This approach has been the dominant one in the casting industry until very recently.

When one uses the approach of Eqs. (1.3a) and (1.3b), the problem begins to become manageable and one can partition the total excess free energy driving the total reaction at any time into a set of partial excess free energies consumed by the various elemental $\tilde{f}_i$ in the system as a function of time. Following this approach, the basic multiparameter, multivariable metallurgical problem ($j+k \sim 20-40$) has been reduced to the simultaneous solution of nine interrelated physics problems ($j + k \sim 3 - 4$).

1.2 Identifying the $\tilde{f}_i$ in a systems problem

During the growth of a crystal from an alloy melt, two unique modes of growth can be distinguished. The first is called *constrained growth* wherein the temperature of the crystal is lower than that of the melt from which it is growing (the normal single crystal pulling mode). The second is called *unconstrained growth* wherein the temperature of the crystal is higher than that of the melt from which it is freezing (the normal isothermal quench mode). Since the latter is the growth mode involved in certain phases of the casting example given earlier, let us delineate the $\tilde{f}_i$ and identify the key components of the total bulk driving force, ΔG_∞, acting in that example.

First, the bulk driving force for solid formation, ΔG_∞, is given by

$$\Delta G_\infty = \tilde{f}_1[\widehat{\Delta T}, \dot{T}, \Delta H_F, T_L(C_\infty), t] \tag{1.5a}$$

where ΔH_F is the heat of fusion per unit volume, t is time and $T_L(C_\infty)$ is the liquidus temperature for the liquid concentration C_∞. Here, $\tilde{f}_1$ represents the proper functional relationship between these quantities. Clearly, phase equilibria information for this system is a prerequisite. As the driving force grows in magnitude with time, a condition is reached wherein heterogeneous catalysts begin to nucleate separated crystals of solid at a frequency I given by

$$I = \tilde{f}_2[N_h, \Delta T_c, \dot{T}, C_\infty, t] \tag{1.5b}$$

N_h is the number of heterogeneous nuclei per unit volume and ΔT_c is the critical undercooling at which such nuclei are effective in forming solid. We shall consider the actual undercooling at which these nuclei form to be ΔT_∞.

Considering the growth of any one crystal, the phase diagram of the alloy tells us that the solubility of the solute in the solid, C_S, is different from that in the liquid, C_L ($C_S(T_L) = k_0 C_L$ where $k_0 \overset{<}{>} 1$ at the liquidus temperature T_L), and a partitioning of solute will occur at the interface leading to an interface liquid concentration, C_i, given by

$$C_i/C_\infty = \tilde{f}_3[V, k_0, D_C, \hat{s}, \delta_C, t] \tag{1.5c}$$

where $\hat{s}$ is the shape of the crystal growing at velocity V while D_C is

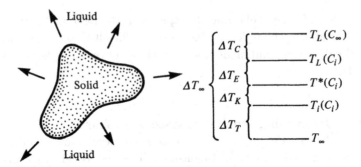

Fig. 1.6. Important temperatures involved in the growth process for the crystal on the left; relative magnitudes of the ΔTs indicate the degree of solute diffusion, capillarity, attachment or heat transport control.

the solute diffusion coefficient in the liquid and δ_C is the solute boundary layer thickness at the interface which is strongly influenced by fluid convection, i.e.,

$$\delta_C = \tilde{f}_4[V, \hat{s}, D_C, \nu, U_\infty, t] \qquad (1.5d)$$

In Eq. (1.5d), ν is the kinematic viscosity of the fluid and U_∞ is the convective stream velocity of the bulk fluid relative to the crystal. Because of this solute partitioning, a portion, ΔG_C, of the total bulk driving force, ΔG_∞, is used up in the matter transport process. This can be translated into a portion, ΔT_C, of the total undercooling ΔT_∞ (see Fig. 1.6), i.e.,

$$\Delta T_C = T_L(C_\infty) - T_L(C_i) = -\bar{m}_L(C_i - C_\infty) \qquad (1.5e)$$

where $\bar{m}_L$ is the average liquidus slope between C_i and C_∞.

Next, because the crystal interface is curved and the crystal may contain non-equilibrium defects, a portion ΔG_E of the total driving force is used to provide the excess free energy of solid above that involved in the phase diagram which we have used as our standard state. Thus we have

$$\Delta T_E = \Delta G_E/\Delta S_F = T_L(C_i) - T^*(C_i) \qquad (1.5f)$$
$$= \tilde{f}_5[\hat{s}, \Delta S_F, \gamma, \sum_j E_j^f, \sum_j n_j^f] \qquad (1.5g)$$

where T^* is the equilibrium temperature, ΔS_F is the entropy of fusion per unit volume, γ is the solid–liquid interfacial free energy while n_j^f and E_j^f are the number and energy of the faults of type j. Further, because the crystal is growing, there must exist a departure from equilibrium,

ΔG_K, at the interface to drive the crystal attachment process for the atoms leading to

$$\Delta T_K = T^*(C_i) - T_i = \tilde{f}_6[\hat{s}, V, \beta_1, \beta_2, t] \qquad (1.5h)$$

Here, T_i is the actual interface temperature while β_1 and β_2 are two parameters used to define the dominant attachment mechanism. Finally, because the crystal is evolving latent heat as it grows, a portion ΔG_T of the total ΔG_∞ is needed to drive this heat transfer process,

$$\Delta T_T = T_i - T_\infty = \tilde{f}_7[K, D_T, \Delta H_F, V, \hat{s}, t] \qquad (1.5i)$$

In Eq. (1.5i), T_∞ is the bath temperature, K represents the thermal conductivities of the liquid and solid while D_T represents the corresponding thermal diffusivities.

The coupling equation which defines the partitioning of the total bulk undercooling, $T_L(C_\infty) - T_\infty$, into its component parts is

$$\Delta T_\infty = \Delta T_C + \Delta T_E + \Delta T_K + \Delta T_T \qquad (1.5j)$$

which is illustrated in Figs. 1.6 and 1.7. The actual growth velocity, V, of the crystal is given by

$$V = \tilde{f}_8[\hat{s}, \Delta T_C, \Delta T_E, \Delta T_K, \Delta T_T, t] \qquad (1.5k)$$

so that it is a function of crystal shape and everything else. If the shape is known, V can be calculated; however, to determine the instantaneous shape one must consider the stability of the interface to small amplitude perturbations so that one finds the optimum velocity V^*, given in terms of the optimum shape $\hat{s}^*$. To make meaningful statements concerning the types of defects generated, one must incorporate modulus features and local stress effects and generate $\tilde{f}_9$ for each specific defect.

In the previous pages, both the ΔGs and the ΔTs have been casually referred to as driving forces for the crystallization process. However, the exact choice of thermodynamic potential to use in a particular situation varies: for example, (i) the Gibbs free energy G is the correct thermodynamic potential to use for a system of constant temperature and pressure, (ii) the Helmholtz free energy, F, is the proper choice for a system of constant temperature and volume, while (iii) the entropy, S, is the proper choice for a system of constant energy and volume. Of course, at the low pressures involved for most crystallization events there is not much difference between G and F. Further, the use of ΔT as a surrogate for ΔG is purely as a convenience which builds on its usefulness as a potential for heat transport.

The goal of the mathematical transport analyses for this type of situation is primarily to provide a translation from the bulk thermodynamic

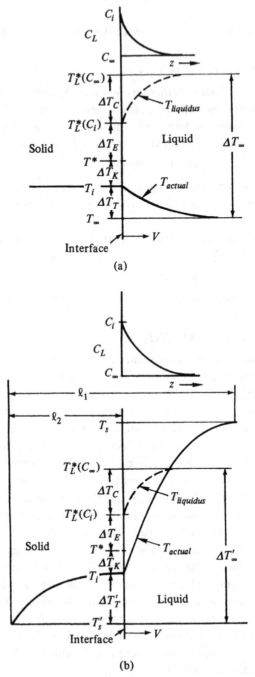

Fig. 1.7. Solute and temperature distributions plus key temperatures for (a) unconstrained crystallization and (b) constrained crystallization.

state variables (C_∞, T_∞), to the interface state variables (C_i, T_i). The former allow one to calculate the bulk driving force, ΔG_∞ while the latter allow the interface driving force, ΔG_i, to be calculated. From Eqs. (1.5g) and (1.5h) plus Figs. 1.6 and 1.7(a),

$$\Delta G_i = \Delta G_E + \Delta G_K \qquad (1.6a)$$

so that

$$\Delta G_\infty = \Delta G_{sv} + \Delta G_i \qquad (1.6b)$$

where ΔG_{sv} is the free energy difference in the liquid associated with changing the thermodynamic state variables from (C_∞, T_∞) to (C_i, T_i). From the attachment kinetic portion of Eq. (1.6a) and Eq. (1.5h), a microscopic growth velocity may be obtained

$$V_{micro} = \tilde{f}_6''(\Delta G_K) \qquad (1.7a)$$

From Eqs. (1.5c) and (1.5i), a macroscopic growth velocity due to transport is obtained

$$V_{macro} = \tilde{f}_3''(\Delta G_C) = \tilde{f}_7''(\Delta G_T) \qquad (1.7b)$$

These two velocities must, of course, be equal so that these conditions allow a complete specification of all the interface variables for a given interface shape. We should note that the coupling equation for ΔG_∞ (Eq. (1.6b)) is different than that for ΔT_∞ (Eq. (1.5j)) because G_S and G_L are generally complex functions of C and T.

1.3   Extensions to the constrained growth mode

We can readily extend the schematic illustration of the coupling equation given by Fig 1.7(a), to the constrained growth case. By defining ℓ_1 as the distance between the heat source (T_s) and the heat sink $(T_{s'})$ and ℓ_2 as the distance between the interface and the heat sink in a typical constrained growth situation, the key temperatures and partial undercoolings may be given as in Fig. 1.7(b). The choice of $T_{s'}$ and thus $\Delta T_T'$ is quite arbitrary; however, the other factors are identical to those used in the unconstrained growth example (Fig. 1.7(a)). The important point to note is that all considerations apply equally well to either growth mode.

1.4   Generalization

The principles described above apply to all types of crystallization – melt growth, solution or flux growth, vapor growth, growth by electrodeposition, growth from gels or growth by solid-state transformation. It is only the manner of description, the points of emphasis and the scale of the numbers that differ from system to system.

For the most general type of transformation, we must characterize both the constrained and the unconstrained modes of growth in terms of the driving force differences, $\Delta \hat{G}$, between different points in the nutrient phase, N. The most dramatic difference is between the interface and the far field location; i.e.,

$$\Delta G_{sv} = \Delta \hat{G}_{i\infty} = \Delta G_{\infty} - \Delta G_i \qquad (1.8)$$

Compared to our earlier definition, we would say that, if $\Delta G_{\infty} > \Delta G_i$, the system is macroscopically unconstrained while, if $\Delta G_{\infty} < \Delta G_i$, the system is macroscopically constrained. Since we are really interested in the effect upon crystal growth, we need a condition that applies at the growing interface and, consistent with the above, we choose

(a) constrained growth: $(\partial \Delta \hat{G} / \partial n)_{\hat{s}} < 0$ (1.9a)

(b) unconstrained growth: $(\partial \Delta \hat{G} / \partial n)_{\hat{s}} > 0$ (1.9b)

in terms of the normal vector n to the surface $\hat{s}$ extending into the nutrient phase. Of course, for crystallization to proceed, $\Delta G_i = G_{Ni} - G_{\hat{c}i} > 0$ where N and $\hat{c}$ refer to nutrient and crystal respectively. In the next chapter we shall deal more fully with the evaluation of ΔG_i for the various cases of interest.

In many cases of interest, and especially for geological systems, not only are C_i and T_i different than their bulk values C_{∞} and T_{∞} but we have a different interface pressure, P_i, compared to the bulk value, P_{∞}. This arises when the density of the crystal is different than the nutrient and the crystallization rate is rapid compared to the viscous flow processes occurring at the interface. Likewise, if we wish to deal with crystallization from an ionic nutrient phase, the interface electrostatic potential, Φ_i, will generally be different from that of the bulk, Φ_{∞}. Since the driving force for such systems depends upon these thermodynamic state variables, to obtain ΔG_{sv} in Eq. (1.6b), we must make the transition from the state $(C_{\infty}, T_{\infty}, P_{\infty}, \Phi_{\infty})$ to the state (C_i, T_i, P_i, Φ_i) which adds an involvement with the stress relaxation equations and Poisson's equation as well as heat and matter transport. Thus, instead of going from $T_L(C_{\infty})$ to $T_L(C_i)$ on a simple liquidus line via an average liquidus slope, $\bar{m}_L$, one would go from $T_L(C_{\infty}, P_{\infty}, \Phi_{\infty})$ to $T_L(C_i, P_i, \Phi_i)$ through a liquidus volume via an average slope $\bar{m}$ given by

$$\bar{m} = (\bar{m}_C^2 + \bar{m}_P^2 + \bar{m}_\Phi^2)^{1/2} \qquad (1.10)$$

where the subscripts C, P, and Φ refer to the composition, pressure and electrostatic potential axes respectively. This also requires that, in Eq. (1.5j), T_C is replaced by T_{sv} given by

$$\Delta T_{sv} = \left\{ [\bar{m}_C(C_i - C_\infty)]^2 + [\bar{m}_P(P_i - P_\infty)]^2 \right.$$
$$\left. + [\bar{m}_\Phi(\Phi_i - \Phi_\infty)]^2 \right\}^{1/2} \tag{1.11}$$

However, through all of this, Eqs. (1.6) still hold.

The foregoing applies to a binary system whereas, in many cases of interest, one is interested in the crystallization of multicomponent systems such as a compound crystal plus dopants. The main new feature to be introduced for this case can be demonstrated by considering a ternary alloy consisting of solvent, major solute (M) and minor solute (m) where P and Φ remain constant throughout the system. For this system we have gone from a (C, T) case to a (C^M, C^m, T) case so that we just proceed as before and alter Eq. (1.5j) wherein ΔT_C is replaced by ΔT_{sv} given by

$$\Delta T_{sv} = \left\{ [\bar{m}_M(C_i^M - C_\infty^M)]^2 + [\bar{m}_m(C_i^m - C_\infty^m)]^2 \right\}^{1/2} \tag{1.12a}$$
$$\approx \bar{m}_M(C_i^M - C_\infty^M)$$
$$+ \frac{1}{2}\bar{m}_m(C_i^m - C_\infty^m)\left[\frac{\bar{m}_m(C_i^m - C_\infty^m)}{\bar{m}_M(C_i^M - C_\infty^M)} \right] \tag{1.12b}$$

Although the presence of the third constituent may influence the magnitude of the other contributions in Eq. (1.5j), the basic philosophy and format is unchanged. If the third constituent is changed and the other two are not, then ΔT_{sv} contains this third term which involves $\bar{m}_m(C_i^m - C_\infty^m)$. From this, the student should readily see how to extend this approach to the most general case.

Although Eqs. (1.6) are correct in all cases and Eq. (1.5j) is a very useful representation when one is dealing with crystallization from the melt, Eq. (1.5j) is not appropriate when dealing with growth from solution or from the vapor phase. For these cases, we can also generate useful coupling equations using their driving force surrogates ΔC and ΔP. This follows from the small value expansions of the appropriate ΔG representation, on a per cubic centimeter basis, i.e.,

(a) melt-solid:

$$\Delta G \approx \Delta S_F \Delta T = -\Delta H_F \left[\frac{(T - T_M)}{T_M} \right]; \qquad \frac{\Delta T}{T_M} \ll 1, \tag{1.13a}$$

(b) solution-solid:

$$\Delta G = \frac{\mathcal{R}}{v_m} T \ln \left(\frac{\hat{\gamma}_C C}{\hat{\gamma}_{C^*} C^*} \right) \approx \frac{\mathcal{R}}{v_m} T \left[\frac{(C - C^*)}{C^*} \right]; \qquad \frac{\Delta C}{C^*} \ll 1, \tag{1.13b}$$

(c) vapor-solid:

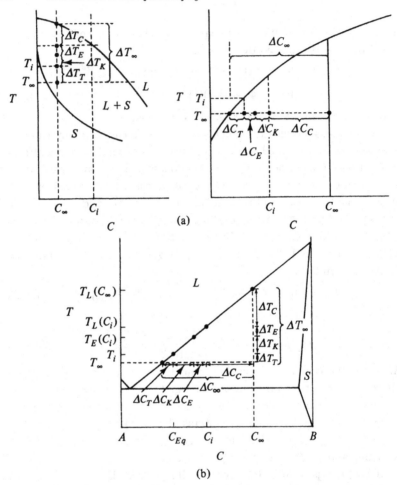

Fig. 1.8. (a) Typical phase diagram illustrations of the coupling equations for ΔT and ΔC. (b) General eutectic diagram illustrating the direct relationship between ΔT and ΔC.

$$\Delta G = \frac{\mathcal{R}}{v_m}T \ln\left(\frac{P}{P^*}\right) \approx \frac{\mathcal{R}}{v_m}T\left[\frac{(P - P^*)}{P^*}\right]; \quad \frac{\Delta P}{P^*} \ll 1, \qquad (1.13c)$$

where the $\hat{\gamma}$s are thermodynamic activity coefficients. In Eq. (1.13a), this approximation assumes that the heat capacity difference between liquid and solid is zero. In Eq. (1.13b), this applies only for a dilute solution otherwise another ΔG contribution is needed for the solvent. From Eqs. (1.13) we can see that, for small departures from equilibrium, $\Delta T, \Delta C$ and ΔP are all proportional to ΔG so that, corresponding to Eq. (1.5j), we can write

$$\Delta C_\infty = \Delta C_C + \Delta C_E + \Delta C_K + \Delta C_T \qquad (1.14a)$$

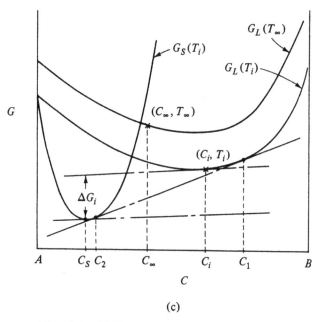

(c)

Fig. 1.8. (c) Free energy curves and T_∞ and T_i with the solid concentration, C_S, yielding maximum interface driving force, ΔG_i, for an interface liquid concentration C_i.

and

$$\Delta P_\infty = \Delta P_C + \Delta P_E + \Delta P_K + \Delta P_T \qquad (1.14b)$$

and use these coupling equations in the appropriate situations. In Fig. 1.8, we can see the companion representations on the relevant phase diagram as we go from (C_∞, T_∞) to (C_i, T_i) using the ΔT_∞ or the ΔC_∞ coupling equation approach. In Fig. 1.8(b), we see the direct correspondence for the case of crystal growth from solution.

In Fig. 1.8(c), the general situation is illustrated via free energy curves where (C_∞, T_∞) and (C_i, T_i) represent the far field point and the interface point, respectively, on the liquid free energy curve, G_L. C_1 and C_2 represent the equilibrium liquidus and solidus concentrations, respectively, for the temperature T_i. As the situation departs from equilibrium, C_i departs from C_1, in the direction shown, and the solid concentration, C_S, departs from C_2. Using the parallel tangent procedure one sees that C_S is the solid concentration yielding the maximum interface driving force, ΔG_i. One also sees that ΔG_i increases as $C_1 - C_i$ increases

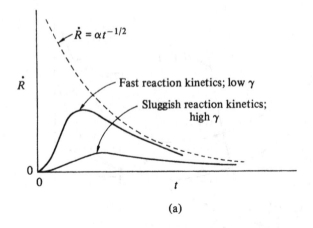

(a)

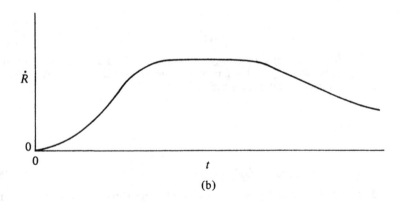

(b)

Fig. 1.9. Schematic representation of particle growth velocity, $\dot{R}$ versus time, t: illustrating (a) the effect of interface reaction kinetics plus the asymptotic approach to the transport dominance regime; and (b) the case of very sluggish interface reaction with rapid transport.

and the liquid departs further from equilibrium. The parallel tangent applies when the attachment kinetic coefficient, β_1, is the same for both the A and B species. When $\beta_1^A \neq \beta_1^B$, the tangents will not be parallel since $\Delta\mu_K^A \neq \Delta\mu_K^B$ for the same V.

1.5   Subprocess dominance

In the past there has been the tendency to try and over-simplify a particular growth situation by selecting just one or two of the subprocesses in the appropriate coupling equation as being dominant and then totally neglecting the other factors. For example, in growth from the melt, ΔT_T is often selected; in growth from solution, ΔC_C is often selected; in growth from the vapor, ΔP_K is often selected, etc. This procedure has obvious advantages but also serious disadvantages because many key factors that might be used to modify favorably or explain a given laboratory result have been eliminated. In addition, as we shall see later, some of the minor ΔG_j contributions participate in particular phenomena in such a sensitive way that, even though their relative magnitude compared to the major term is small, small variations in this factor can create very large variations in the overall result. This arises because of the importance of a specific pair event interaction in Fig. 1.4(b). Let us consider some examples to illustrate the point.

1. Unconstrained crystallization Perhaps an even stronger way to make the general point, that one should not neglect consideration of any contribution in the appropriate coupling equation until one has clearly evaluated its insignificance to the problem or phenomenon under consideration, is to consider the time-coordinate for the growth of any crystal. For the general case of a particle of solid nucleating in a large volume of supercooled liquid and growing spherically outwards, the growth velocity, V, is given schematically as a function of time by Fig. 1.9. Here, we see that, at zero time, all of the driving force for solid formation goes into the nucleation event so that $\Delta G_\infty = \Delta G_E$ and $V = 0$. At very long times, almost all of the driving force for solid formation goes into transport processes so that $\Delta G_\infty \approx \Delta G_{sv}$ and V is proportional to $t^{-1/2}$. At intermediate times, the driving force is consumed by all four subprocesses to varying degrees depending upon the actual time.

Since this time-shifting selection of the dominant driving force, or a growth process with a time-varying driving force, may be unfamiliar considerations to many students, let us expand the idea further by considering the growth of a spherical crystal from a molten solution with a phase diagram like that given in Fig. 1.8(b). Let the critical nucleus form at time $t = 0$ at radius $R^* = 2\gamma/\Delta G_\infty$ which comes from homogeneous nucleation theory. We see that this is just the Gibbs–Thomson equation result of $\Delta T_\infty = 2\gamma/R^*\Delta S_F$ where the crystallite is in equilibrium with the melt and all the available driving force/atom is stored in the excess free energy of surface creation. Thus, since there is no other driving force available for the other processes, V must equal zero at $t = 0$. As the crystal begins to grow slowly, the surface/volume ratio per atom decreases so a portion of the driving force becomes available for the other processes. When $R = 2R^*$, half of the ΔG_∞ becomes available for other processes and, when $R \sim 10R^*$, almost all of the bulk driving force can be used for other processes. In most cases throughout this book the ΔGs are given in terms of energy per cubic centimeter.

To gain a quantitative appreciation for the shift in dominant process with crystal radius, let us arbitrarily decide that, if a particular sub-process consumes more than half of ΔG_∞, then it will be selected as the dominant process. To evaluate the important regimes for this example, we have

$$\Delta G_\infty = \Delta G_E + \Delta G_K + \Delta G_{sv} \qquad (1.15a)$$

If we neglect the heat flow, assume that no excess defects are being formed in the volume of the growing crystal and take a simple Laplace solution to the matter transport, Eq. (1.15a) becomes

$$\frac{2\gamma}{R^*}\left(1 - \frac{R^*}{R}\right) + \frac{\mathcal{R}T}{v_m}\ell n\left[1 - \frac{\Omega^{-1}}{C_\infty}\left(\frac{R\dot{R}}{D_C}\right)\right] + \frac{\mathcal{R}T}{v_m}\ell n(1 - \beta_1\dot{R}) = 0$$

$$(1.15b)$$

In Eq. (1.15b), the first term is $(\Delta G_\infty - \Delta G_E)$, the second term is $-\Delta G_{sv}$ while the third term is $-\Delta G_K$. Here, R is the radius of the crystal, $\dot{R}$ is its growth rate $(\dot{R} \equiv V)$, Ω is the molecular volume in the nutrient, β_1 is a lumped attachment kinetic parameter from Eq. (1.15a), while $\mathcal{R}$ is the gas constant and v_m is the molar volume. The larger is β_1, the more sluggish is the interface reaction kinetics. Rearranging Eq. (1.15b) by combining the second and third terms leads to

$$\left[1 - \alpha_1\left(\frac{\dot{R}}{R^*}\right)\right]\left[1 - \alpha_2\frac{R}{R^*}\left(\frac{\dot{R}}{R^*}\right)\right] = \exp\left[-\alpha_3\left(1 - \frac{R^*}{R}\right)\right] \quad (1.15c)$$

where

$$\alpha_1 = \beta_1 R^*; \quad \alpha_2 = \frac{\Omega^{-1}R^{*2}}{C_\infty D_C}; \quad \alpha_3 = \frac{2\gamma v_m}{R^*\mathcal{R}T} \qquad (1.15d)$$

This is a quadratic equation with the general solution

$$\frac{\dot{R}}{R^*}\left[2\alpha_1\alpha_2\left(\frac{R}{R^*}\right)\right] = \left[\alpha_1 + \alpha_2\left(\frac{R}{R^*}\right)\right]$$

$$- \left(\left[\alpha_1 + \alpha_2\left(\frac{R}{R^*}\right)\right]^2 - 4\alpha_1\alpha_2\left(\frac{R}{R^*}\right)\right)$$

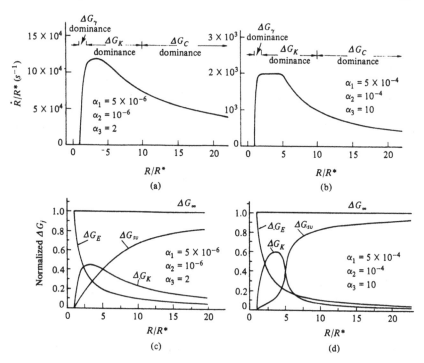

Fig. 1.10. Relative crystal velocity versus relative crystal size plots: (a) plot of $\dot{R}/R$ versus R/R^* for a specific $(\alpha_1, \alpha_2, \alpha_3)$ set; (b) same as (a) but with more sluggish reaction kinetics and faster transport; (c) normalized ΔG_j via the use of ΔG_∞ for the same parameter set as (a); and (d) same as (c) but for the parameter set of (b).

$$\left\{ 1 - \exp\left[-\alpha_3 \left(1 - \frac{R^*}{R} \right) \right] \right\}^{1/2} \qquad (1.15e)$$

We note from Eq. (1.15c) that $\dot{R}/R \to 0$ as $R/R^* \to 1$ and that $R\dot{R}/R^{*2} \to \alpha_2^{-1}[1 - \exp(-\alpha_3)]$ as $R/R^* \to \infty$. Thus, the long time (large R) behavior gives $\dot{R} \propto t^{-1/2}$ as expected. Eq. (1.15e) has been plotted in Figs. 1.10(a) and (b) for two sets of $(\alpha_1, \alpha_2, \alpha_3)$ values to illustrate the shape changes that can occur as these parameters are varied. The particular ΔG-dominance regimes can be seen most clearly by considering the following expressions

$$\frac{\Delta G_E}{\Delta G_\infty} = \frac{R^*}{R} \qquad (1.16a)$$

$$\frac{\Delta G_K}{\Delta G_\infty} = -\alpha_3^{-1} \ell n \left[1 - \alpha_1 \left(\frac{\dot{R}}{R^*} \right) \right] \qquad (1.16b)$$

and

$$\frac{\Delta G_C}{\Delta G_\infty} = -\alpha_3^{-1}\ell n\left[1 - \alpha_2\left(\frac{R}{R^*}\right)\left(\frac{\dot{R}}{R^*}\right)\right] \qquad (1.16c)$$

These normalized components have been plotted in Figs. 1.10(c) and (d) for the particular cases listed in Figs. 1.10(a) and (b), respectively. As one can see from this "spectrum analysis" type of approach, clear dominance is never really developed by the ΔG_K factor for either case so that, in general, one would do better to call this middle regime a "mixed" dominance regime. Inverting Eq. (1.16*b*) leads to

$$\frac{\dot{R}}{R} = \alpha_1^{-1}\left[1 - \exp\left(-\alpha_3\frac{\Delta G_K}{\Delta G_\infty}\right)\right] \qquad (1.16d)$$

Thus, when α_3 is large ($\alpha_3 = 10$, say) and when $\Delta G_K/\Delta G_\infty$ reaches $0.3 - 0.4$, the exponential shrinks to ~ 0 and one has

$$\frac{\dot{R}}{R^*} \approx \alpha_1^{-1} \qquad (1.16e)$$

which gives us the plateau in Fig. 1.10(b). From the magnitude of such a plateau, β_1 can be readily obtained. The reduced amplitude of $(\dot{R}/R^*)_{max}$ between Figs. 1.10(a) and 1.10(b) is due to the increased value of β_1 for the latter case.

When a multiplicity of crystallites nucleate and grow in a bulk fluid, the volume fraction transformed, $X(t)$, follows an overall Avrami-type of time law given by

$$X(t) = 1 - \exp(-t/\tau)^n \qquad (1.17)$$

and yields a typical "S"-shaped curve for X versus t. Here, n depends upon both the time law for new crystal formation and the time law for the individual crystal growth rate (see Table 8.4). Because of the variable time laws for individual particles demonstrated by Figs. 1.9 and 1.10, an experimental determination of an integral or half-integral value for n has only a small probability and a clear interpretation of the meaning for n_{expt} is difficult.

2. ΔG_E components in solid/solid precipitation Si crystals grown by the Czochralski (*CZ*) process contain both O ($C_\infty^O \sim 3 \times 10^{18}/\text{cm}^3$) and C ($C_\infty^C \sim 10^{17}/\text{cm}^3$), the latter being disposed on substitutional sites while the former are located on interstitial sites. During cooling, a driving force exists in the Si crystal for the precipitation of both SiO$_2$ particles and SiC particles. Because of the molecular volume differences, Ω, between Si(20 Å^3), SiO$_2$(45 Å^3) and SiC(10 Å^3), point defect formation and stresses are involved in the particle growth process for each case. Thus, ΔG_E for each of these precipitation processes has several important components worthy of consideration.[4]

Thermal annealing of CZ Si in the 400 °C – 1200 °C range is expected to cause interstitial oxygen, O_I, species to come out of supersaturated solution, vacancy, V_{Si}, species to be consumed and self interstitials or interstitialcy, Si_I, species to be created according to the following general formula

$$(1+Y)\text{Si(s)} + 2O_I + XV_{Si} \rightleftarrows \text{SiO}_2(\text{s}) + Y\text{Si}_I + \text{stress}, \sigma_1 \quad (1.18a)$$

with

$$K_1 = \frac{[\text{SiO}_2][\text{Si}_I]^Y}{[O_I]^2[V_{Si}]^X} \quad (1.18b)$$

Here, if $X = 0$ and $Y = 0$, the stress, σ_1, is extremely high. If $Y = 0$ and only the equilibrium concentration of vacancies is available to relieve the compressive stress, σ_1 is still high. To reduce σ_1, the system generates Frenkel defects at the interface and provides additional free volume via the V_{Si} created. This additional reaction has the form

$$\text{Si(s)} \rightleftarrows V_{Si} + \text{Si}_I \quad (1.18c)$$

with

$$K_2 = [V_{Si}][\text{Si}_I] \quad (1.18d)$$

Thus, vacancies and oxygen diffuse inwards towards a growing SiO_2 particle while self interstitials diffuse outwards. One expects X, Y and σ_1 to adjust to produce the maximum driving force for this precipitation reaction. The effect of stress, σ_1, can be incorporated into Eq. (1.18b) via its effect on K_1.

Including the effect of C is to add a vacancy source to the system via the reaction

$$\text{Si(s)} + \text{C(s)} \rightleftarrows \text{SiC(s)} + \frac{3}{2}V_{Si} + \text{stress}, \sigma_2 \quad (1.18e)$$

with

$$K_3 = \frac{[\text{SiC}][V_{Si}]^{3/2}}{[\text{C}]} \quad (1.18f)$$

Here, the 3/2 is needed for volume conservation. Again, the effect of the stress, σ_2 can be incorporated into Eq. (1.18f) via its effect on K_3. Clearly, a synergism exists between these three reactions with the formation of SiC providing the free volume to feed the SiO_2 nucleation and initial growth stages. Likewise, the formation of SiO_2 provides the Si_I species to drive Eq. (1.18e) to the right via consumption of V_{Si} species through Eq. (1.18c). Since much more O_I is present than C, the combined reactions eventually lead to a net production of Si_I species.

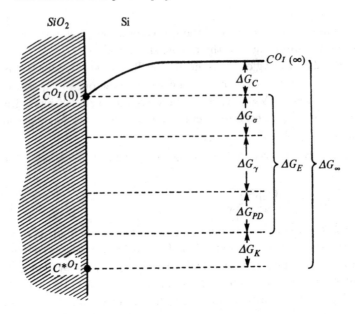

Fig. 1.11. Partitioning of the total driving force for precipitate formation, ΔG_∞, into that needed for the five important subprocesses. C^{O_I} is the concentration of O_I in the Si.

For our purposes here, there are three important contributions to ΔG_E for the growth of an SiO_2 particle: (1) the interfacial energy, ΔG_γ, (2) the stored strain energy, ΔG_σ, and (3) the point defect formation energy, ΔG_{PD}. Thus, the coupling equation for this particle growth will be given by

$$\Delta G_\infty = \Delta G_C + \Delta G_K + \Delta G_E \qquad (1.19a)$$

where

$$\Delta G_E = \Delta G_\gamma + \Delta G_\sigma + \Delta G_{PD} \qquad (1.19b)$$

Fig. 1.11 illustrates this partitioning process where each of the partial driving forces is a function of particle shape and particle growth rate. The qualitative variations of some of these quantities with precipitate shape aspect ratio, y/R, are illustrated in Fig. 1.12. Thus, although $\Delta G_C, \Delta G_{PD}$ and ΔG_γ favor the sphere morphology, ΔG_σ favors the disc morphology. At long times, $\Delta G_\gamma \to 0$ as the absolute size of the particle becomes larger; however, ΔG_σ and ΔG_{PD} do not go to zero as long as the particle growth velocity, V, is greater than zero.

In the past, the majority of investigators have assumed that, if a particle's growth rate is proportional to $t^{-1/2}$, the interface concentration

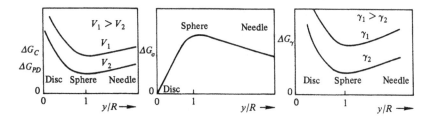

Fig. 1.12. Qualitative variations of $\Delta G_C, \Delta G_{PD}, \Delta G_{\sigma}$ and ΔG_{γ} with aspect ratio, y/R, for precipitates of different shape.

must be very close to C^* and thus that $\Delta G_C \approx \Delta G_{\infty}$ in Fig. 1.11; i.e., a transport dominated process develops. This is a very common error in the field of phase transformations and it comes about largely from the neglect of the ΔG_E contribution. It is not sufficiently recognized that a $t^{-1/2}$ law for particle growth requires only that $\Delta G_C \approx$ constant in the regime. Thus, provided $\Delta G_E \sim$ constant and ΔG_K is quite small at long times, a $t^{-1/2}$ time law will be observed even with $C_i \gg C^*$; i.e., the reaction is proceeding with interface conditions far from equilibrium. This is illustrated in Fig. 1.13 for particle growth under two different stress conditions with $\sigma_2 > \sigma_1$. In both cases, the long time behavior gives $V \propto t^{-1/2}$ with $V(\sigma_1) > V(\sigma_2)$. The interface is not at equilibrium but C_i is out of equilibrium by the amount

$$\frac{\mathcal{R}T}{v_m} \ell n \left(\frac{C_i}{C^*} \right) = \Delta G_{\sigma} + \Delta G_{PD} \qquad (1.20)$$

Calculations for spherical SiO_2 particle growth in carbon-free Si indicate that $\Delta G_{PD} > \Delta G_{\sigma}$ and that $C_i/C^* \gg 1$ because $\Delta G_E/\Delta G_{\infty} \sim 0.8$ at long times. Only when the diffusion fields of adjacent crystallites overlap and lower $C_{\infty}(t)$ to the solubility value, $C^*(T)$, does equilibrium obtain. Since this requires a reduction in $\Delta G_C, R(t)$ will increase as t^n where $n < 1/2$ during this overlap phase. Thus, one must first observe a decreasing n before it is possible to say that C_{∞} is starting to approach C^*. For this example, one must also be assured that ΔG_{PD} and ΔG_{σ} have decreased to zero before one can be confident that the concentration of O_I left in solution truly represents C^*. The mathematical analysis for the SiO_2 particle growth rate follows the procedure of Eq. (1.15) but with the inclusion of the $\Delta G_{\sigma} + \Delta G_{PD}$ terms in ΔG_E. This merely adds an additional contribution, $+(\Delta G_{\sigma} + \Delta G_{PD})v_m/\mathcal{R}T$, to the exponential containing α_3 in Eqs. (1.15c) and (1.15e).

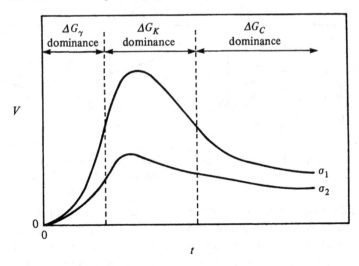

Fig. 1.13. Growth velocity of a spherical precipitate as a function of time for two stress states $\sigma_2 > \sigma_1$. Subprocess dominance regimes are shown.

3. ΔG_C *contribution in a pure system* During the crystallization of an elemental material, a volume change generally occurs at the interface. For a fluid system, this expansion ($\Delta v/v > 0$) or contraction ($\Delta v/v < 0$) generates an interface strain that is quickly dissipated by fluid flow. However, in a nutrient solid or a very viscous fluid, expansion of the forming crystal leads to a compressive stress and the formation of self interstitials or self interstitialcies to lower that stress. Likewise, contraction of the forming crystal leads to a tensile stress and the formation of vacancies to lower that stress. This latter case applies to solid phase epitaxial regrowth (SPER) of Si where an amorphous layer of pure Si transforms to crystalline Si.

The important point for us here is that all of the bulk driving force is not available for the interface reaction process but some must be used to drive the V_{Si} transport and some may be involved in excess energy storage in the crystalline Si; i.e., again Eq. (1.15a) holds where

$$\Delta G_{sv} = \Delta G_C = \frac{RT}{v_m} \ell n \left(\frac{C_i^{V_{Si}}}{C^{*V_{Si}}} \right) \qquad (1.21a)$$

and the velocity of the transforming interface will be given by

$$V = \beta_1^{-1}\{1 - \exp[-(\Delta G_\infty - \Delta G_C - \Delta G_E)]\} \qquad (1.21b)$$

from Eq. (1.15b). The V_{Si} creation is treated here as a ΔG_C term rather than as a ΔG_E term in order to allow ΔG_E to represent any additional factors present in the example. This point defect supersaturation at the interface and in the region ahead of the interface leads to the nucleation of stacking fault loops near the interface, intrinsic faults when the interface is a vacancy source and extrinsic faults when the interface is an interstitial source. An array of such loops forming parallel to the interface will act as an effective sink for excess point defects by expansion of the loops. Such a uniform climb process will lead to a uniform collapse of the nutrient phase towards the interface when $\Delta v/v < 0$ and a uniform expansion of the nutrient phase away from the interface when $\Delta v/v > 0$. The expanding loops tend to annihilate each other while new loops form within the old when the point defect supersaturation becomes large enough. Since the loop specific line-length energy is a function of distance from the interface, the new loops form at some optimum distance ahead of the interface and at some optimum supersaturation. This is what controls the magnitude of C_i and thus V via Eq. (1.21).

In this example, which appears at first sight to yield $\Delta G_K = \Delta G_\infty$, one sees the manifestation of an unsuspected contribution in $\Delta G_C = \Delta G_{PD}$ which acts as a throttle to control the magnitude of ΔG_K at the interface; i.e., C_i can become sufficiently large that $\Delta G_C \approx \Delta G_\infty$ except at $V \to 0$.

4. SiO$_2$ film formation on Si An important example illustrating various subprocess dominances is the formation of an oxide film on Si via the thermal oxidation process. Although this film is amorphous rather than crystalline, the structural aspects are not an important part of the overall thermokinetic process so the example has practical as well as pedagogical value.

The initial approach of experimenters was to try and fit the oxide thickness data to an equation of the form $X_0^n = \alpha t$ where n is a function of temperature, gas chemistry, time, etc. However, this did not work except in a very narrow range of the variable space and a better approach was needed. This was supplied in 1964 by Deal and Grove.[5] Following the approach of Deal and Grove, one assumes a pseudo-steady state growth model controlled by equal O fluxes at (i) the gas/oxide interface (J_1), (ii) across the oxide film (J_2) and (iii) at the Si/SiO$_2$ interface (J_3). This is illustrated in Fig. 1.14 where it is assumed that activity coefficients can be neglected. Mathematically, we require that

$$J_1 = J_2 = J_3 = J = V/\Omega_{SiO_2} \qquad (1.22a)$$

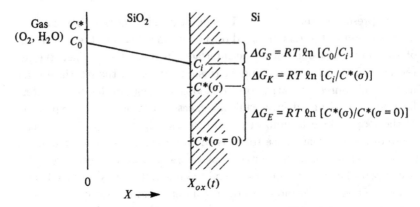

Fig. 1.14. Schematic representation of key concentrations, concentration profile in SiO_2 and subprocess driving forces during steady state thermal oxidation of Si.

where

$$J_1 = h_c(C^* - C_0) \tag{1.22b}$$

$$J_2 = D_{eff}(C_0 - C_i)/X_0 \tag{1.22c}$$

$$J_3 = \bar{k}\hat{a}_i = \bar{k}\hat{\gamma}_i C_i = \bar{k}' C_i \tag{1.22d}$$

In Eqs. (1.22), Ω_{SiO_2} is the molecular volume of SiO_2, the fluxes refer to O_2 species, h_c is the transfer coefficient at the gas/SiO_2 interface, X_0 is the oxide film thickness, $\bar{k}$ is the reaction rate constant at the Si/SiO_2 interface, $\hat{a}_i$ is the activity and $\hat{\gamma}_i$ is the activity coefficient at the Si/SiO_2 interface. In this process, it is known that the O_2 diffuses across the film to react at the Si/SiO_2 interface and that stress σ develops at the interface so that $\hat{\gamma}_i \neq 1$. It is common practice with this system to assume that the interface is far enough away from equilibrium that the back-reaction rate is negligible compared to the forward-reaction rate so Eq. (1.22d) is used.

The solution of Eqs. (1.22) leads to the following growth law

$$\frac{X_0^2}{B} + \frac{X_0}{B/A} = t + \tau \tag{1.23a}$$

where

$$B = 2D_{eff}C^*\Omega_{SiO_2}; \quad A = 2D_{eff}(\bar{k}'^{-1} + h_c^{-1}) \tag{1.23b}$$

$$\frac{B}{A} = \frac{\bar{k}'h_c}{\bar{k}' + h_c}(C^*\Omega_{SiO_2}); \quad \tau = \frac{(X_i^2 + AX_i)}{B} \tag{1.23c}$$

which leads to

$$X_0 \approx (B/A)(t + \tau) \quad \text{for } t \ll A^2/4B \tag{1.23d}$$

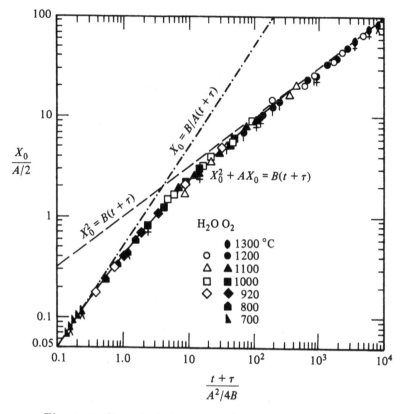

Fig. 1.15. General relationship for thermal oxidation of Si. The solid line represents the general relationship, the dashed and chain lines its two limiting forms. The values of τ correspond to $X_i = 0$ and 200 Å for wet and dry O respectively. (From Deal and Grove.[5])

$$X_0^2 \approx Bt \quad \text{for } t \gg A^2/4B \text{ and } t \gg \tau \qquad (1.23e)$$

Except for a very short time when the oxidation rate is almost a decaying exponential with time, this linear/parabolic time law is well matched by the experimental data as seen in Fig. 1.15.

In this process we again see the $t^{1/2}$ law for the growth rate at long times (ΔG_C dominance) and a constant rate at intermediate–short times (ΔG_K dominance). Experimentally, one finds a very high rate at very short times (unknown dominance). A large ΔG_E effect appears to develop during this very short time regime and remains constant throughout the remainder of the oxide growth. For oxide growth in wet O_2 at temperatures above the glass transition temperature ($T_g \sim 950$ °C), the high stress that was built in at the interface relaxes by viscous flow so that the resulting oxide is relatively stress free. For film growth at temperatures below 950 °C, most of this interface generated stress is still retained in the oxide film.

In this example, it has been generally thought that the $t^{1/2}$ growth law means a transport-limited process and that interface equilibrium obtains for the reaction. However, in this case we find that $\Delta G_\infty \sim 150 - 200$ kcal mole^{-1} while ΔG_C is calculated from the experimental data to be less than 10 kcal mole^{-1} .[6] Because the system obeys a linear/parabolic growth law, we can assume that $\Delta G_K + \Delta G_C \sim 10$ kcal mole^{-1} with the total value being utilized at early times by the interface reaction process and at later times by the transport process. We thus see that this time law behavior requires only that $\Delta G_K + \Delta G_C = $ constant for all time even if $(\Delta G_K + \Delta G_C)/\Delta G_\infty \ll 1$. The thermodynamic assessment provides the additional information that ΔG_E must be very large $(\Delta G_E = \Delta G_\infty - (\Delta G_K + \Delta G_C))$ and that this excess energy storage process developed at the interface is what dominates the reaction. If one wishes significantly to modify the oxide film forming reaction, it is this subprocess that should be the focus of attention. Of course, one again sees that the existence of a $t^{-1/2}$ law for growth rate does not automatically mean that interface equilibrium obtains during the reaction.

5. Solution or vapor growth A simple but fairly general example in this category is to consider a planar wafer source that might be a binary compound or a binary alloy at temperature T_1 separated by a distance d from a wafer substrate at temperature $T_2 < T_1$. The intervening medium may be either a vapor or a liquid solution. Thus, in this example, the source species at T_1 evaporates/dissolves in the intervening medium to approach the local equilibrium concentration, C_{i1}. However, at the substrate, a similar process is taking place to approach its local equilibrium, C_{i2}. In general, $C_{i1} > C_{i2}$ because $T_1 > T_2$ so transport keeps occurring across the intervening medium which upsets the local equilibria at the two surfaces. This generally leads to evaporation/dissolution at the source material $(T = T_1)$ and deposition at the sink material $(T = T_2)$. The total driving force, ΔG_∞, is well defined in this case since it just depends on $\Delta T = T_1 - T_2$ provided this is a closed system. In this case, the thermodynamic coupling equation is given by

$$\Delta G_\infty(\Delta T) = \Delta G_{i1} + \Delta G_{sv} + \Delta G_{i2} \qquad (1.24)$$

which is illustrated diagramatically in Fig. 1.16 as a series/parallel electrical equivalent circuit; i.e., potential is dropped across or consumed by

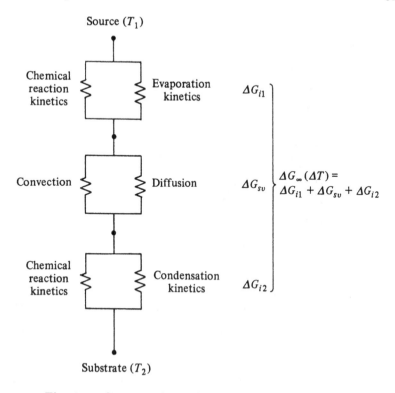

Source (T_1)

Chemical reaction kinetics

Evaporation kinetics

ΔG_{i1}

Convection

Diffusion

ΔG_{sv}

Chemical reaction kinetics

Condensation kinetics

ΔG_{i2}

$$\Delta G_\infty (\Delta T) = \Delta G_{i1} + \Delta G_{sv} + \Delta G_{i2}$$

Substrate (T_2)

Fig. 1.16. Sequence of series/parallel processes involved in the growth of crystals from the vapor.

each of these three circuit elements in order to produce the matter flux from the source to the substrate. In many cases, it is not just a matter of simple elemental evaporation and condensation but rather it is a matter of fairly complex reactions occurring at each surface in parallel with the evaporation and condensation processes.

6. Electrodeposition This technique has long been utilized for the growth of metallic and semiconductor crystals and, most recently, has been applied to the growth of special molecular crystals of the organic conductor type. The apparatus for electrocrystallization generally comprises a two-compartment cell with a permeable membrane separating the compartments (see Fig. 1.17). Generally Pt electrodes are used for the growth of the molecular crystals which often involves the oxidation of a donor molecule $(D^0 \rightarrow D^+)$ at the working electrode, in the presence of an acceptor ion, A^-, in the solvent. If potential control of the working

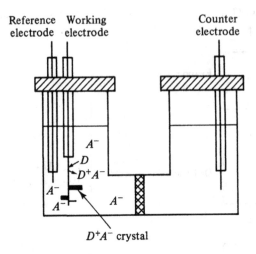

Reference Working
electrode electrode

Counter
electrode

D^+A^- crystal

Fig. 1.17. Schematic illustration of a conventional H-cell used for electrocrystallization. In this example, donor molecules are oxidized at the anode in the presence of A^- resulting in the growth of a D^+A^- crystal at the electrode. A glass frit is commonly used to prevent contamination of the desired compound with products of the counterelectrode reaction.

electrode is required, an appropriate reference electrode is also required. Electrolysis is simply performed under certain voltage and current conditions for an appropriate length of time and the crystals are harvested from the electrode on completion.

Our understanding of the crystal growth process for a D^+A^- crystal is as follows. The solvent contains both A^- and C^+ species from dissociation of an AC compound ($C_\infty^{A^-} = C_\infty^{C^+}$) and, in addition, it contains the D species (C_∞^D). Applying a positive voltage to the working electrode beyond the redox potential value for D causes D to convert to D^+ at the working electrode at a rate proportional to the voltage difference for small differences. This leads to D^+ diffusing away from the electrode, and both A^- and D diffusing towards the electrode so we have $C^D(z,t), C^{D^+}(z,t)$ and $C^{A^-}(z,t)$. Thus, in the region of small z near the working electrode, a potential exists for the formation of the D^+A^- species either homogeneously in the solvent ($C^{D^+A^-}(z,t)$) or heterogeneously on the electrode ($C_i^{D^+A^-}(t)$). When the D^+A^- concentration exceeds the solubility limit sufficiently, nucleation and crystal growth of the D^+A^- solid will form on the working electrode according to the thermodynamic procedure illustrated in Fig. 1.8(c). However, in this case,

$C_i^{D^+A^-}$ is a growing function of time and the interface driving force, $\Delta G_i(t)$, keeps increasing until the initial stages of nucleation occur.

During the crystal growth stage, since the D^+A^- species may be being produced either near the electrode or only at the electrode, the concentration profile in the vicinity of the interface can be very different depending upon the relative rates of species formation by ion supply and reaction and species loss via crystallization. In this case, the free energy budget balance focuses on ΔG_i rather than on ΔG_∞ since the latter doesn't seem to have much relevance ($C_\infty^{D^+} = 0$) here. One also expects some advantages to be gained from the use of constrained growth techniques rather than unconstrained growth techniques for such materials.

1.6 General path for resolving laboratory crystal growth problems

From basic studies under well-controlled conditions, it is necessary to determine the crystal's characteristics or behavior as a consequence of growing the crystal under a specific set of interface variables (C_i, T_i, P_i, Φ_i). This involves gaining a knowledge of some of the $\tilde{f}_i$ in Eqs. (1.3) and (1.5). Such information may come from basic studies on other systems that are somewhat similar in materials characteristics. It is next necessary to determine the relationship between the interface variables and the far field variables $(C_\infty(t), T_\infty(t), P_\infty(t), \Phi_\infty(t))$; i.e., the remainder of the $\tilde{f}_i$. Here, one is involved with the exact or approximate solution to specific mathematical boundary-value problems. The goal is to know that, when the far field variables are programmed to change in a specific way, the interface variables will change in a known fashion so that the interface morphology and the crystal characteristics will be determined. In practice, one is generally trying to obtain a desired set of crystal characteristics. Thus, with the foregoing knowledge in hand, one can write the proper program for adjusting the far field variables on the crystal growth device being utilized so as ultimately to yield the desired crystal characteristics. When one goes through this exercise, it is often found that the crystal preparation device being utilized does not have sufficient flexibility or good enough control to manifest the needed program so that one knows a new technique must be designed and one generally knows how to go about designing such a new technique.

In any technological problem that one is trying to resolve there is available only a fraction of the pertinent information needed to make the critical decision; thus, the total path to gaining a successful experimental solution to such a systems problem consists of two segments: (i) a scientific trajectory based upon the available information at hand and (ii) an empirical study of the system. Both segments are of vital importance to the success of the overall endeavor. With the first segment, one hopes that one's assumptions and scientific understanding will get one into the proximity of a successful solution. With the second segment, one learns something about the validity of those initial assumptions. An iteration of these two steps provides the most rapid path to the success point.

The greater is one's scientific understanding of the crystallization process, the more accurate will be the scientific trajectory segment of the path and the smaller and less costly will be the time-consuming empirical study (Eqs. (1.1) and (1.3) approach) needed to reach the success point. That is a major justification for this book and its companion.[7]

Problems

1. Think about the total number of parameters and variables involved in the understanding of the many professional activities with which you are somewhat familiar – physics, general engineering, materials science, geology, medicine, business, psychology, sociology, etc. Let your intuition soar and make an estimate of the total number of variables and parameters involved in understanding each of these activities. If a constant annual manpower of 10^7 technical people throughout the world addressed their research attention to each of the areas and, if 10^7 man-years of effort has been spent to date in bringing physics to its present level of predictive reliability, how many years will it be before each of these fields of understanding has achieved the same level of predictive reliability as present-day physics?

2. Consider the thin film crystal growth activities of molecular beam epitaxy (MBE), chemical vapor deposition (CVD) and liquid phase epitaxy (LPE). Delineate the specific $\tilde{f}_j$ involved in each of these activities, define the appropriate ΔG_∞ driving the film formation process and draw a picture for each of the systems events analogous to Fig. 1.4(b). At the simplest level of modeling, list the important parameters and variables involved in each of the $\tilde{f}_j$.

3. In question 2, neglecting the LPE case, how might the situation be changed if you directed a beam of photons at the surface during deposition? Assume that you have a variable frequency, variable power photon source available. (Some insights relative to this question can be gained from reference 8.)

4. You are growing a pure crystal isothermally from a melt by transferring latent heat, $\Delta H_F = 1000$ cal cm^{-3}, across a thermal boundary layer, $\delta_T = 0.01$ cm at $T_M = 1000$ K. The interface is flat, the thermal conductivity of the liquid is $K_L = 10^{-2}$ cal $°C^{-1}$ cm^{-1} s^{-1} and attachment kinetics gives $V \approx 10^{-2}\Delta T_K$. For values of $V = 10^{-4}, 10^{-2}$ and 1 cm s^{-1}, what is the total driving force, ΔG_∞, needed for growth at each velocity and is it limited predominantly by heat transport or by attachment kinetics?

5. A lamellar, α/β phase eutectic crystal of periodic spacing $\lambda = 1$ μm has $\gamma_{\alpha\beta} = 400$ erg cm^{-2} and is being grown at velocity V. If the average entropy of fusion is $\Delta S_F = 1$ cal $°C$ cm^{-3}, what magnitude of average kinetic coefficient, $\bar{\beta} \approx V/\overline{\Delta T}_K$, is needed to give $\overline{\Delta T}_K = \overline{\Delta T}_E$ at $V = 10^{-3}$ cm s^{-1} and 1 cm s^{-1}?

6. A Si crystal is being grown isothermally from a Sn + Si melt at $V \approx 10^{-3}$ cm s^{-1}. Given that linear attachment kinetics are operating with $\beta = 10^{-3}$ cm $°C^{-1}$ s$^{-1}, \Delta T_T = 1$ K, $\delta_C/\delta_T = 10^{-1}$ and $\Delta T_\infty = 32$ K, what is the magnitude of ΔT_E? (Use the following data: $K_L = 10^{-2}$ cal $°C^{-1}$ cm^{-1} s$^{-1}, D_C = 10^{-4}$ cm^2 s$^{-1}, \Delta H_F = 1000$ cal cm$^{-3}, \Delta S_F = 1.5$ cal $°C^{-1}$ cm$^{-3}, \Omega_{S_n} = 30$ Å$^3, \Omega_{Si} = 20$ Å^3 and $\bar{m}_L = 20$ $°C$/At%).

7. You are growing an SiO_2 layer on Si at 1000 $°C$ in 1 atm of dry O_2 where $\Delta G_\infty = 154$ kcal mole^{-1}. From O_2 solubility and oxide growth rate data, it is known that $C_\infty^{O_2} = 5 \times 10^{16}/$cm^3 and $C_i^{O_2} = 3.6 \times 10^{15}/$cm^3 for a 1 μm thick oxide. $C_i^{O_2}$ is less for a thinner oxide. For these conditions, what is the value of ΔG_i? The oxide growth rate for 1 μm thickness is given by $V \propto t^{-1/2}$ but would you say it is diffusion-limited?

8. For a perfect spherical crystal of radius R growing from a solution at concentration C_∞, the crystal concentration $C_S \gg C_i$, heat transport can be neglected and R is determined by Eqs. (1.15). For $\alpha_3 = 5$, find the domain of α_1 and α_2 where a plateau like that seen in Fig. 1.9(b) occurs in the $\dot{R}/R^*$ versus R/R^* plot. For such a case, one could assume $\Delta G_K \approx$ constant

and could thus extract the interface reaction rate constant, β_1, from experimental measurement of such plateaus.

9. Suppose that the growing crystal of question 8 is in a stiff viscoelastic solution (geological medium) and that an atomic volume difference exists between the solute in the solution and in the crystal. The strain energy developed as a result of this crystallite growth adds a constant excess energy per crystallized atom, $\Delta\mu_\sigma$, independent of crystal size but dependent on elastic moduli of the two media. Incorporate this additional factor into Eqs. (1.15), being careful to maintain the proper units of all the terms.

10. Using the pseudo-steady state approximation for the growth of an SiO_2 film on Si by thermal oxidation (Eqs. (1.22)), show that the resulting growth law is given by Eqs. (1.23).

2

Interface energetics and molecular attachment kinetics

This chapter is dedicated to developing the understanding of interface energetics and molecular attachment kinetics and this can be carried out most effectively by considering a completely pure material (all ΔG_C effects are eliminated). Both crystal growth by condensation from a pure vapor and crystal growth from a pure melt are suitable vehicles for pedagogical description. However, since the latter has more technological history than the former at the moment (relatively pure Si boules), it will be used as the major example. Since this application utilizes a seed crystal, we will not deal here with the initial stage of nucleation of a crystal. (Nucleation is dealt with in Chapter 8 (Background Science).)

Crystal growth of bulk Si is largely via the CZ technique illustrated in Fig. 2.1. The melt is generally contained in a pure SiO_2 crucible and either induction heated via a graphite susceptor (as seen here) or resistance heated via a graphite heater. The crystal is rotated at some angular frequency to produce axial thermal symmetry and forced convection in the melt.

To initiate the crystal, the melt is stabilized at a temperature slightly above the melting point of Si while a rotating seed crystal, held in a chuck, is lowered into intimate physical contact with the melt and a portion of the seed melted back. Then the seed is slowly withdrawn and new solid begins to crystallize on the seed. The melt temperature is programmed to be slowly lowered in synchronization with the seed withdrawal rate so that the crystal diameter will smoothly broaden to the desired growth diameter. If during this period a defect or stray

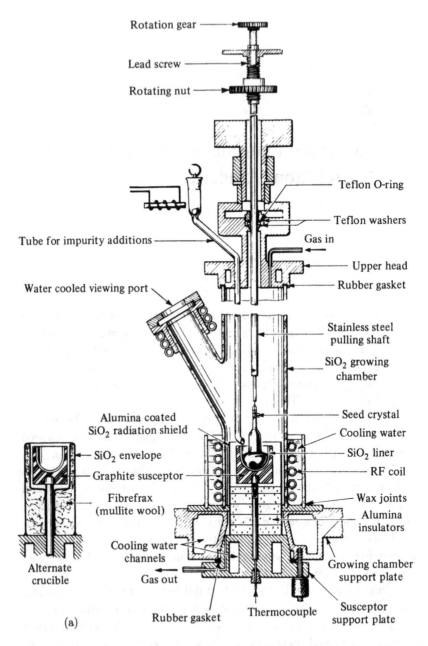

Rotation gear

Lead screw

Rotating nut

Teflon O-ring

Teflon washers

Tube for impurity additions

Gas in

Upper head

Rubber gasket

Water cooled viewing port

Stainless steel pulling shaft

SiO_2 growing chamber

Alumina coated SiO_2 radiation shield

Seed crystal

Cooling water

SiO_2 envelope

SiO_2 liner

Graphite susceptor

RF coil

Fibrefrax (mullite wool)

Wax joints

Alumina insulators

Cooling water channels

Alternate crucible

Gas out

Growing chamber support plate

(a)

Rubber gasket

Thermocouple

Susceptor support plate

Fig. 2.1. (a) Detailed diagram of a *CZ* crystal pulling apparatus. Many systems use a graphic resistance heater rather than an RF coil for heating.

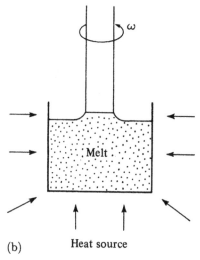

(b)   Heat source

Fig. 2.1. (b) Schematic illustration of the melt/crystal region in (a).

crystal is observed on the surface of the turning crystal, the pull rate and heat rate are reversed and the defect portion is melted off before one once again initiates the cooling and growing cycle. In Fig. 2.2, we see an example of the 1.5 cm diameter, 15 cm long type of Si crystal that used to be grown in the 1950s while it is now more commonplace to grow 10 cm diameter crystals several feet long. Present scale-up is to 15 cm diameter crystals by most commercial sources (with 25 cm diameter being assessed) while some are seriously studying continuous Si pulling at these diameters using a fine Si particle feed stock.

In Fig. 2.3, the macroscopic interface shape and interface undercooling $\Delta T_i = T_M - T_i$ are schematically represented for the case of a growing crystal interface that is slightly convex to the liquid. Because of crystal rotation, fluid is sucked up along the crucible axis and spun out laterally in a radial direction near the upper liquid surface. Because of the balance of surface tension forces at the point of contact of the liquid and the crystal edge, a liquid meniscus is formed at the contact region. This meniscus shape is often used for crystal diameter control (light reflection technique). We shall see later that the crystal interface shape in the region near the meniscus plays a very important role in the overall crystal growth features.

Since we are dealing with a pure system, free energy curves for the liquid and solid are straight lines in the vicinity of T_M, as illustrated in Fig. 2.4, and the connection between ΔT_i and ΔG_i may be readily seen ($\Delta G_i = \Delta S_F \Delta T_i$).

Fig. 2.2. Comparison of CZ-pulled, Si single crystal ingots produced in 1978 (top) and in 1960 (bottom). The 1978 ingot weighs approximately 17 500 g, is 90 cm long and 10 cm in diameter. The 1960 ingot weighs 50 g, is 15 cm long and 1.5 cm in diameter.

2.1   Interface energetics

A general overview of the energy contributions involved in the formation of an interface is presented in Chapter 7. In this chapter, we will concentrate on only two of these components, the quasi chemical component and the roughening component.

In Fig. 2.5, one can see various atomic states that are possible for a Si atom on a region of crystal that contains the more macroscopic features of faces, ledges, corners and edges. These are unique entropic states, each characterized by a certain excess energy due to lost interactions from the removed material in the upper half-space. Although the excess energy really comes from lost two-body, three-body, etc., interactions and inclusion of the many-body effect (any angular dependence in a potential energy function is an expression of the many-body effect) is absolutely necessary for meaningful quantitative assessment, and sometimes even for proper qualitative assessment of the excess energy, the two-body-only approximation is useful for initial description because it provides helpful insight. We shall use the two-body-only approximation in most of what is to follow and will characterize the excess energy in terms of broken pair bonds. Lost nearest neighbor (NN) pair bonds will contribute an excess energy of ε_1; second, third and higher neighbor

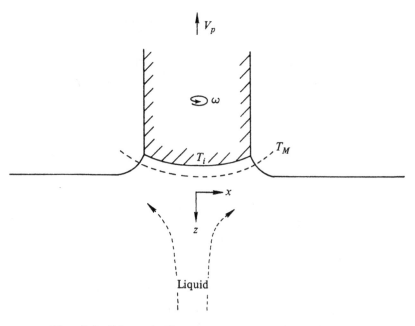

Fig. 2.3. Schematic illustration of a growing interface position (T_i) relative to the freezing point isotherm (T_M) for a crystal pulled at a rate V_p and rotated at a rate ω.

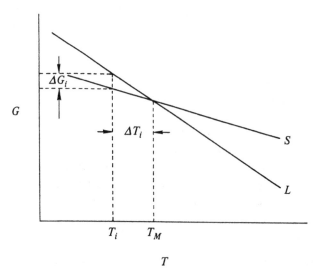

Fig. 2.4. Temperature variation of free energy, G, as a function of T for a pure solid (S) and its liquid (L) to illustrate how ΔG_i is determined.

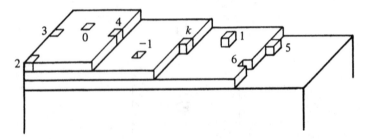

Fig. 2.5. Representation of the various, energetically unique, surface or interface sites for molecules.

lost pair bonds will contribute $\varepsilon_2, \varepsilon_3, \ldots, \varepsilon_n$, respectively. Depending on the kind of electronic interaction between the atoms under consideration, one usually makes the approximation to truncate the interaction beyond the ith bond. For simplicity one often uses a NN–only approximation (ε_1). A better approximation is an $(\varepsilon_1, \varepsilon_2, \varepsilon_3)$ choice while, for metals, one really needs to consider $(\varepsilon_1, \varepsilon_2, \ldots, \varepsilon_n)$ where n is quite large.

The average bond strength can be readily related to the heat of transformation of the material. For example, the heat of vaporization, ΔH_V, of Si is ~ 100 kcal mole^{-1} ($\Delta H_{s^*} = 4.67$ eV) and each Si atom has four NNs. Thus, with a NN–only approximation, four half-bonds are associated with each Si atom and, separating 1 mole of atoms from the crystal state to the vapor state, leads to $4\mathcal{N}_A$ ($\mathcal{N}_A$ = Avogadro's number) of these NN half-bonds so that

$$\varepsilon_1 = \frac{\Delta H_{s^*}}{\tilde{N}\mathcal{N}_A} \tag{2.1}$$

where $\tilde{N}$ is the number of nearest neighbors in the crystal. For unrelaxed Si, we thus find that $\varepsilon_1 \approx 1.93 \times 10^{-12}$ erg so the excess surface energy per Si atom on the (111) is 0.96×10^{-12} erg or 0.53 eV (only half of a half-bond or broken bond per atom on average in the puckered plane) while the excess energy per Si atom on the (100) is 3.86×10^{-12} erg (two broken bonds per atom). Of course, there will be some excess entropy, s_1, per broken bond for each of these faces as well so that, on a per atom basis, the quasi chemical face excess free energy, γ_f^0, is given by

$$\gamma_f^0 = n_{bf}(\varepsilon_1 - Ts_1) \tag{2.2}$$

where n_{bf} is the average number of first NN broken bonds per atom in the face. Using this approach, the excess energy, E_f^0, for the three major Si faces based on the NN–only approximation is given in Table 2.1 and we see that $E_f^0 \sim 2000$ erg cm^{-2}. In Table 2.1, the surface excess

Table 2.1. *Calculated surface and ledge energies for Si*

Plane/Ledge	n_b	E_1	E_f^0	E_f^0 (with many-body effect)	
		(eV)	(ergs cm^{-2})	Unrelaxed (ergs cm^{-2})	Relaxed (ergs cm^{-2})
$\{111\}$ unreconstructed	0.5	0.58	1465	1225	1019
$\{110\}$ unreconstructed	1.0	1.17	1794	1601	1468
$\{100\}$ unreconstructed	2.0	2.34	2537	2310	2220
$\{100\} - (2 \times 1)$				2537	1434
$\{100\} - (2 \times 2)$				2537	1243
	Δn_b	(eV)		(eV)	(eV)
$\{111\}/[2\bar{1}\bar{1}]_U$	0.5	1.168		1.119	0.050
$\{111\}/[2\bar{1}\bar{1}]_L$	0.5	1.168		1.249	0.815
$\{111\}/[\bar{2}11]_U$	1.5	3.503		3.211	~ 0
$\{111\}/[\bar{2}11]_L$	0.5	1.168		1.106	0.710
$\{100\}/[001] \ (2 \times 1)$	0	0		0	0.188
$\{100\}/[110] \ (2 \times 1)$	0	0		0	$-0.110 \ (S_A)$

energy calculated using up to five NNs and including the many-body effect both before and after relaxation is also given and we see that a significant reduction in E_f^0 is obtained by relaxation. In an analogous way, the quasi chemical edge excess free energy can be obtained for important ledges on the (111) and (100) planes.

The pair-bond-only approximation can be applied to the crystal–melt interface; however, an additional approximation is needed to simplify the excess energy computation. For Si, on melting, the volume contracts $\sim 11\%$ and the bonding character becomes metallic so the liquid is quite unlike the solid. The usual approximation that is made here is the "lattice-liquid" approximation where the structure of the liquid is thought to be the same or very similar to the structure of the crystal. Then, ε_1 for the crystal–melt interface will be given by Eq. (2.1) but with ΔH_V replaced by ΔH_F, the heat of fusion. This approximation, which may be reasonable for many materials, will obviously not be a very good approximation for Si. However, applying it to Si leads to $\varepsilon_1 \approx 2.11 \times 10^{-13}$ erg so that the crystal/melt excess energy will be reduced by a factor ~ 10 compared to the crystal–vapor interface.

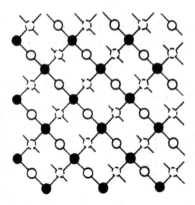

Top view

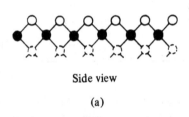

Side view

(a)

Fig. 2.6. Top and side views of Si surfaces and ledges before and after relaxation: (a) Si(100) surface unrelaxed.

2.1.1   Relaxation effects

The foregoing discussion applies to the smooth, unrelaxed faces and ledges of an ideal interface where the interactions are only by pair-bonds. However, for a *real* interface, two major atomic relaxation modes can occur to alter the interfacial energetics: (1) surface reconstruction and (2) surface roughening. For open crystal structures such as one finds with semiconductors, ceramics, etc., the former energy change is much larger than the latter. In surface reconstruction, the top few layers of atoms cooperatively take on an altered crystallographic geometry compared to the bulk crystal and this lowers the interfacial free energy. Fig. 2.6 shows the Si(100) surface and the $[2\bar{1}\bar{1}]$ ledge of the (111) plane before and after reconstruction. For both cases, Si dimer formation occurs to lower the energy as indicated in Table 2.1. Movement of the two ledges on the Si(111) requires sequential passage through both upper and lower states for a single atom-wide ledge so that reconstruction rates and ledge motion rates may interfere to produce defects at high

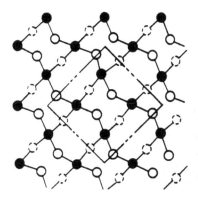

Top view

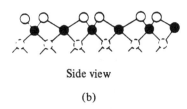

Side view

(b)

Fig. 2.6. (b) Si(100) surface showing dimerization and a (2×2) reconstruction pattern.

growth velocities. In surface roughening, single atom displacement occurs from the 0-state to the 1-state or from the 4-state to the 5-state in Fig. 2.5. The formation of these adatom states creates vacancy states in the face or ledge respectively. Such single atom roughening events can also play an important role in the cooperative atom event of surface reconstruction. This is thought to be the case for Si(111) reconstruction which forms both (1×1) and (7×7) surface patterns.

When these two relaxation factors plus the allowance for electron relaxation (see Chapter 7) are included, the excess free energies for the crystal face, γ_f, and the crystal ledge, γ_ℓ, are given by

$$\gamma_f = \gamma_f^o + \Delta F_r^* + \Delta F_{rec} + \gamma_e \tag{2.3a}$$

$$\gamma_\ell = \gamma_\ell^o + \Delta F_r'^* + \Delta F_{rec}' + \gamma_e' \tag{2.3b}$$

Here, ΔF_r^* is the Helmholtz free energy change due to roughening (negative), ΔF_{rec} is the Helmholtz free energy change due to reconstruction

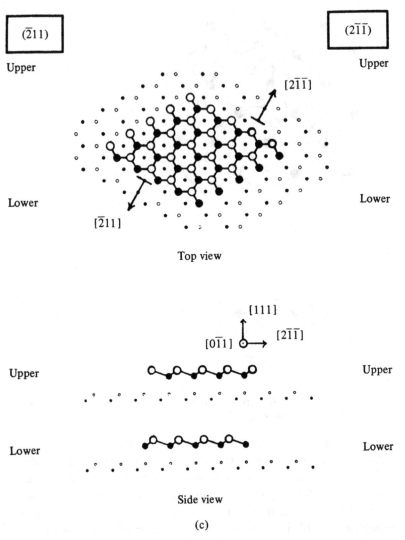

Fig. 2.6(c). Unrelaxed structures of the four basic single atom high ledges on Si(111), labelled according to their outward normal direction (either $[\bar{2}11]$-type or $[2\bar{1}\bar{1}]$-type). Movement in the $[2\bar{1}\bar{1}]$ requires passage through both upper and lower states for single-atom-wide kinks on the ledge but not for two-atom-wide kinks.

(negative) and γ_e is the free energy change due to space charge formation (negative) all for the face while the primed quantities refer to the excess value at the layer edge. Although quantitative assessment of ΔF_{rec} is beyond the scope of this book, a qualitative assessment can be given together with a straightforward approximation for ΔF_r^*.

For both Si(111) and GaAs(00$\bar{1}$), the free surface is found to exhibit a strongly compressive stress tensor that dominates all surface reconstruction processes.[1,2] Ledges of the $\langle 211 \rangle$-type act as a dilatational stress and thus tend to relax the net surface compressive stress. This leads to low values of γ_ℓ depending on the degree of reconstruction possible at the particular member of the $\langle 211 \rangle$ family. It is interesting to note that [$2\bar{1}\bar{1}$] ledges on Si(111) and both $\langle 110 \rangle$ and $\langle 100 \rangle$ ledges on GaAs(00$\bar{1}$) interact with each other at separation distances ~ 50 Å. Since this is far beyond the range of intermolecular potentials in each case, the interaction is clearly occurring via the long-range stress tensor effect. It is also interesting to note that, when $\langle 211 \rangle$ ledges on Si(111) reconstruct greatly to generate a small value of γ_ℓ, the kink formation energy is high and vice versa. In addition, the Si adatom–ledge interaction energy is low when γ_ℓ is low and vice versa. The point here is that, when the ledge is unable to lower the surface stress significantly, the formation of kinks on the ledge or adatoms near the ledge seems to be able to alter the surface reconstruction and lower the surface stress to a significant degree.

One of the consequences of the above result is that the hexagonal symmetry of Si(111) changes to trigonal symmetry. The [$2\bar{1}\bar{1}$]-type ledges form lowest-energy triangular clusters and thus will yield a trigonal equilibrium habit. The [$\bar{2}11$]-type ledges form highest kink density ledges and thus yield the trigonal rapid growth directions.

One final point to note is that, for Si(111) and [$2\bar{1}\bar{1}$] ledges, $\gamma_\ell(\lambda_\ell)$ decreases as λ_ℓ increases.[1] However, for GaAs(00$\bar{1}$) and either $\langle 110 \rangle$-type or $\langle 100 \rangle$-type ledges, $\gamma_\ell(\lambda_\ell)$ increases as λ_ℓ increases.[2] In the latter case, the ledges will tend to bunch while, in the former case, the ledges will tend to separate. These results are shown in Fig. 5.7.

Returning to a simple dangling bond picture, the various numbered states in Fig. 2.5 are unique entropic states, each characterized by a certain excess energy due to broken bonds parallel to the interface which are involved in their formation. It is the decrease in free energy of the system due to the configurational entropy gain that drives their formation, and it is the excess enthalpy due to their presence that limits

(d)

the process. To illustrate the computation procedure and the further approximations needed to find the optimum density of roughened states, X^*, and the roughening free energy change, $\Delta F_r^*(X^*)$, we consider the (100) plane of the simple cubic lattice with NN-only bonding and limit the roughening to 3-states, $0, +1$ and -1 (see Fig. 2.5). Calling the fraction of face states of the jth type, X_j, conservation requires

$$X_0 + X_{+1} + X_{-1} = 1 \tag{2.4a}$$

Assuming that symmetry applies for the $+1$ and -1 states, we have

$$X_{+1} = X_{-1} = \frac{1}{2}(1 - X_0) = X \tag{2.4b}$$

Thus, since each $+1$ state has four half-bonds parallel to the face, the energy increase, ΔE_r, due to roughening is[3]

$$\Delta E_r = 8\varepsilon_1 \, X(1 - X) + \cdots \tag{2.4c}$$

Here, the second term is due to dimer formation at the $+1$ and -1 levels which leads to a reduction of one half-bond per adatom or advacancy when one of their four neighboring sites is also occupied. We shall neglect higher order clustering effects.

The configurational entropy change due to these roughened states, ΔS_r, is given in terms of the number of possible complexions, $\tilde{w}$, by

$$\Delta S_r = \frac{\kappa}{n} \ell n \, \tilde{w} \tag{2.4d}$$

where κ is Boltzmann's constant and n is the number of surface atoms under consideration. If we make the further approximation that these configurations are arranged randomly in the three layers (Bragg–Williams (BW) approximation), then Eq. (2.4d) becomes

$$\Delta S_r = -\frac{\kappa}{n} \ell n \left\{ \frac{n!}{(nX)!^2 [n(1 - 2X)]!} \right\} \tag{2.4e}$$

$$= - \kappa [2X \ell n X + (1 - 2X) \, \ell n(1 - 2X)]$$

At the low pressures involved in these events, there is not much difference between the Gibbs and the Helmholtz thermodynamic potentials and the Helmholtz free energy change for this process, ΔF_r, is given by

$$\Delta F_r = \Delta E_r - T_i \Delta S_r \tag{2.4f}$$

Maximizing ΔF_r with respect to variation of X leads to X^* and ΔF_r^*

Fig. 2.6(d). Relaxed structures of the two basic $[2\bar{1}\bar{1}]$ and $[\bar{2}11]$ single-atom-high ledges.

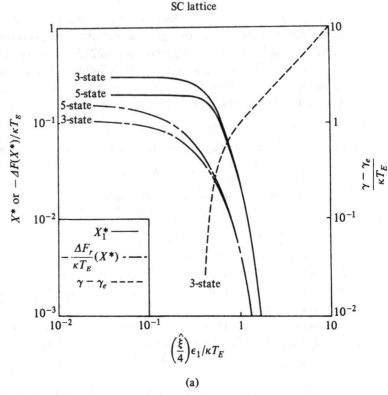

(a)

Fig. 2.7. (a) Plot of (i) equilibrium number of +1 states, X^*, for the 3-state and 5-state model, (ii) the decrease in free energy, $F(X^*)$, due to roughening for the models and (iii) the interfacial energy γ, for the 3–level model, all as a function of $\epsilon_1/_K T_E$ ($\hat{\xi}$ is a universal factor).

given by

$$\frac{X^*}{1 - 2X^*} = \exp\left[-\frac{4\varepsilon_1}{\kappa T_i}(1 - 2X^*)\right] \tag{2.4g}$$

$$\frac{\Delta F_r^*}{\kappa T_i} = \frac{8\varepsilon_1}{\kappa T_i}[X^*(1 - X^*)] + 2X^*\ell n\, X^* + (1 - 2X^*)\,\ell n(1 - 2X^*) \tag{2.4h}$$

This ignores a very small negative contribution from the change in the entropy of mixing. In Fig. 2.7(a), calculated plots have been given of X^* and $-\Delta F_r^*/\kappa T_E$ as a function of $\varepsilon_1/\kappa T_i$ for the (100) face of the simple cubic lattice ($\hat{\xi} = 4$).[4] Both 3-state and 5-state roughening have been considered. Here, we see that, as $\varepsilon_1/\kappa T_i \to 1$, roughening decreases very strongly but, as $\varepsilon_1/\kappa T_i \to 0.25$, maximal roughening occurs. We see also that, at small values of $\varepsilon_1/\kappa T_i$, the free energy is further lowered by going from 3-state roughening to 5-state roughening (addition of + 2

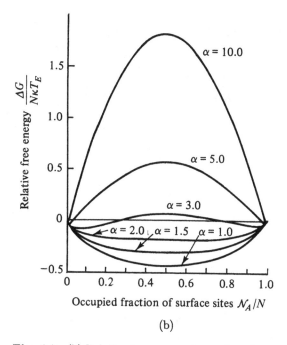

Fig. 2.7. (b) Relative free energy change for a 2-state, NN-only model versus +1 state population, $\mathcal{N}_A/N$, for various calues of Jackson's α-parameter.

and -2 states); thus, the system would automatically proceed from two states to five states to seven states, etc. so long as the free energy continues to decrease incrementally by doing so. It is interesting to note that $-\Delta F_r^* \tilde{<} \kappa T_E$ from Fig. 2.7(a) whereas, from Table 2.1, $\Delta F_{rec} \gg \kappa T_E$. The parameter $\hat{\xi}$ allows one to use this model for other crystal faces in the simple cubic (SC) system, extending the range of pair interactions, going to other crystal structures and correcting for the various approximations used in the model ($\hat{\xi}$ is the effective number of NN bonds in the plane). In Table 2.2 values of $\hat{\xi}$ for different conditions in the SC structure are given.[4] Because of the many approximations made in the computation, we can expect only qualitative match with experiment for the crystal/vapor case. For the crystal/melt case, the approximations are even less valid quantitatively so again only qualitative match can be expected. Since $\varepsilon_1/\kappa T_i$ is about an order of magnitude smaller for the crystal/melt case than for the crystal/vapor case at constant T_i, the former interface will be orders of magnitude rougher than the latter. It is further found that X^* changes little with departures from equilibrium so long as $\Delta T_i/T_E \ll 1$.

Table 2.2. *Universal scaling factor* $\hat{\xi}$ *for the SC lattice*[†]

Plane	(100)			(110)			(111)		
Bonds	1st	2nd	3rd	1st	2nd	3rd	1st	2nd	3rd
# of bonds above plane	1	4	4	2	5	2	3	3	4
# in plane	4	4	0	2	2	4	0	6	0
# below plane	1	4	4	2	5	2	3	3	4
$\hat{\xi}_1$		4			2			0	
$\hat{\xi}_{123}$		4.5			2.4			0.75	

[†] $(\varepsilon_2 = \varepsilon_1/8; \varepsilon_3 = \varepsilon_1/27)$

Using a 2-level, lattice liquid, pair potential only, NN only, BW approximation model, Jackson[5] defined an α-parameter to describe the roughening of a crystal–melt interface where

$$\alpha = \frac{\Delta H_F}{\kappa T} \left(\frac{\tilde{n}_1}{\tilde{N}} \right) \qquad (2.5)$$

and where $\tilde{n}_1$ is the number of NN crystal sites at the interface while $\tilde{N}$ is the total number of NN lattice sites in the crystal. For this case, he neglected the existence of -1 states and considered only 0 and $+1$ states to be present. He calculated the free energy of the interface as a function of roughness for a range of α finding the result shown in Fig. 2.7(b). Here, we see that, for $\alpha < 2.0$, the interface maximally roughens whereas, for $\alpha > 2.0$, only fractional roughening ($\theta_A < 0.5$) occurs. From Eqs. (2.1) and (2.5) we see that $\alpha/\tilde{n}_1 = \varepsilon_1/\kappa T$ so that Figs. 2.7(a) and (b) lead to similar conclusions. Because of the approximations used (see Chapter 7), one cannot expect such conclusions on $\alpha \approx 2$ to hold tightly and an error bar of a factor 2–5 is a more reasonable expectation, i.e., $\alpha \tilde{<} 0.5 - 1.0$ is probably where maximal roughening occurs. In practice, if the interface tends to roughen maximally for a 2-state model, it will proceed to form a 3-state, then a 5-state, etc., roughening condition.

Let us next consider the formation of analogous roughened states at the layer edge. The enthalpy increase of the system is just due to bonds broken parallel to the ledge. Thus, for our SC example with a (010) or

(001) ledge, the fraction of roughened sites on the ledge, X'^*, is given by

$$\frac{X'^*}{1 - 2X'^*} = \exp\left[-\frac{2\varepsilon_1}{\kappa T_i}(1 - 2X'^*)\right] \tag{2.6a}$$

and

$$\frac{\Delta F_r'^*}{\kappa T_i} = \frac{4\varepsilon_1}{\kappa T_i}[X'^*(1 - X'^*)] + 2X'^* \ell n\, X'^* + (1 - 2X'^*)\, \ell n(1 - 2X'^*) \tag{2.6b}$$

Comparing Eq. (2.6a) with Eq. (2.4g), we see that $X'^* \approx (X^*)^{1/2}$ for small values of X'^*. Thus, the ledges generally roughen much more than the face because the excess enthalpy of adatom formation is generally smaller for ledges than for the face. Correspondingly, we expect that $|\Delta F_r'^*|/\gamma_\ell^0 > |\Delta F_r^*|/\gamma_f^0$.

For practical utilization of these concepts, it is important to relate ε_1 properly to some physically measurable quantity so that some of the theoretical uncertainties are eliminated. The use of Eq. (2.1) provides one estimate for ε_1 while Eq. (2.2) provides another. In view of possible interface reconstruction or electronic effects, and the relative importance of 3-body versus 2-body effects at interfaces, Eq. (2.2) should give the better estimate for ε_1 if we replace γ_f^0 by γ_{exp}. For the crystal/melt case, the nucleation experiments of Turnbull[6] give $\gamma_{exp} \approx 0.5\Delta H_F$ for metallic systems so that $\varepsilon_1/\kappa T^*$ via Eq. (2.2) is related to ΔH_F as indicated in Eq. (2.1).

Using Si as our example and expecting it to behave like Ge, Turnbull's data gives $\gamma_{exp} \approx 0.35\Delta H_F$. Equation (2.1) gives $\varepsilon_1/\kappa T_M = \Delta H_F/4\kappa T_M \approx 1.6$ ($\Delta S_F = 6.5\kappa$) while Eq. (2.2) gives $\varepsilon_1/\kappa T_M = 0.35\Delta H_F/\kappa T_M \approx 2.25$ for the (111) plane since $n_b = 0.5$ for the puckered plane being treated as the planar element. Since $\alpha = \tilde{n}_1(\varepsilon_1/\kappa T_M)$ via Eq. (2.5), we see that $\alpha_{(111)}^{Si} \approx 4.8$ or 6.75 via Eq. (2.1) or Eq. (2.2), respectively. In both cases one would expect a low interface roughness. Since $\hat{\xi}$ in Fig. 2.7(a) is just the number of NN in the plane for a first NN-only model, $\hat{\xi} = 3$ for Si(111) so $X^* \approx 8.3 \times 10^{-3}$ or 1.2×10^{-3} depending upon whether Eq. (2.1) or Eq. (2.2) is used.

Computer calculations using the semiempirical Si potential energy function (PEF) that led to Fig. 2.6 have been used to evaluate point defect energies on a flat (111) vacuum interface (no ledges). Because of the strongly compressive surface stress tensor, point defects that relax the stress sufficiently are expected to form spontaneously on the surface. Because of this factor, surface Frenkel defect formation at large separa-

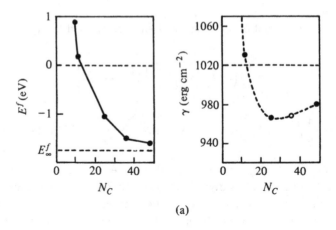

(a)

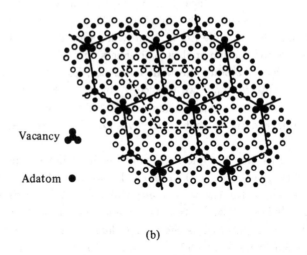

Vacancy

Adatom •

(b)

Fig. 2.8. Calculated point defect energies and geometries on a Si(111) surface: (a) vacancy formation energy, E^f, plus the surface excess free energy, γ, versus the number of surface primitive unit cells per vacancy, N_C; and (b) a (5×5) reconstruction pattern consisting of an array of surface Frenkel pairs.

tions has a formation energy of -2.1 eV. Of this, -0.25 eV comes from the Si adatom in the "hole" configuration (located over the hexagonal holes on the (111) plane) and -1.85 eV for the Si vacancy. The vacancy stress relaxation effects are seen to be much larger than those due to the adatom so that $\varepsilon_{1v} \neq \varepsilon_{1a}$ as assumed in the simple theory and ε_1

is not positive at dilute concentrations as also assumed in the simple theory. The changes in formation energy E_f as well as γ are shown in Fig. 2.8(a) for the case of only vacancies being present on the surface and we see that the optimum vacancy concentration is about 1 vacancy per 30 surface primitive unit cells (between a (5×5) and a (6×6) array) and γ has been reduced about 6%. A Frenkel defect array of the (5×5) type minimizes γ at a value around 940 erg cm^{-2}, an 8.1% reduction from the relaxed ideal surface value. This array is shown in Fig. 2.8(b). The use of a more complex Si PEF yields the experimentally observed (7×7) reconstruction pattern. The point of all this is that, although the simple theory can be expected to apply qualitatively for some systems, it can be in significant error quantitatively for others. Only by performing computer calculations with a good PEF can we be fairly sure of the quantitative situation in a particular case. Even though the PEF used for the above calculations is not perfect and the exact numbers are therefore somewhat in error, the qualitative findings can be expected to hold.

In spite of the shortcomings of the simple dangling bond theory, there is pedagogical value in extending it to solutions, metallic, aqueous or otherwise. Using the cubic lattice-liquid model and, assuming a highly solvent-rich solution with no interface segregation, Eq. (2.1) may be used but with the heat of solution, $\Delta H_{\tilde{S}}$, substituted for ΔH_V. Since the heat of solution may be expressed as

$$\Delta H_{\tilde{S}} = \Delta H_F + \Delta H_{mix}$$

where ΔH_{mix} is the heat of mixing. For a solution showing negative deviations from ideality, ΔH_{mix} is negative, the solubility in the solvent is higher and thus ε_1 will be reduced below the ideal value ($\Delta H_{mix} = 0$). In a solution showing positive deviations from ideality, $\Delta H_{mix} > 0$ and the crystal face will have a larger value of ε_1 than the ideal value. It thus follows that, for a given crystal face growing from solution in a series of solvents, the value of ε_1 may be expected to change depending upon the nature of the solute–solvent interactions with ε_1 decreasing and thus X^* increasing as these interactions become stronger. Thus, if we wish to grow Si crystals from molten solvents like Sn, Pb, Bi, Sb, Ga, we recognize that $\varepsilon_1/\kappa T$ will be increased compared to its melt value at $T = T_M$ as the solution temperature is reduced to $T = T_{\tilde{S}}$ and as the Si solubility is reduced in the solvent.

For strongly negative heats of mixing, $\Delta H_{\tilde{S}}$ can become negative, strong interface segregation of solvent is expected and the calculated

Table 2.3. *Thermodynamic data and α-values*
1. HMT crystals

Solvent	X_S	ΔH (kcal mole^{-1})	α	Growth rate
H_2O	0.1	-4.0		Fastest
C_2H_5OH	0.01	3.9	3.2	Moderate
Vapor		18	15	Slow

2. Glyceryl tristearate crystals

Solvent	X_S	ΔH (kcal mole^{-1})	α	Growth rate
Ideal	6×10^{-4}	48.5		
Trioleate	6.4×10^{-4}		4.5	Slow
$CC\ell_4$	0.04		2.8	Much faster

3. Succinic acid crystals

Solvent	X_S	$\alpha_{(010)}$	$\alpha_{(001)}$	Growth rate
Ideal	0.023			
H_2O	0.010	7.8	5.8	Faster in both cases
IPA	0.028	8.0	6.0	Slower in both cases

value of X^* can become very large. To illustrate this point, Bourne and Davey[7] studied the growth of hexamethylene tetramine (HMT) from the vapor as well as from aqueous and ethanolic solutions. Their thermodynamic data is presented in Table 2.3 where we see that the solubility of HMT, $X_{\bar{S}}$, is an order of magnitude greater in the aqueous than in the ethanolic solution and that $\alpha < 0$ via Eq. (2.1) for the aqueous solvent. They also found the crystal growth rate to be much faster in the aqueous solvent than in the ethanolic solvent than from the vapor at fixed supersaturation and under stirring conditions such that there was no transport limitation. This is consistent with an increasing magnitude of X^* as we go from vapor to ethanolic solution to aqueous solution. These authors have also found similar results for the growth of glyceryl tristearate crystals from glyceryl trioleate and $CC\ell_4$ solutions. They found the crystal growth rate to occur primarily in the [010] and

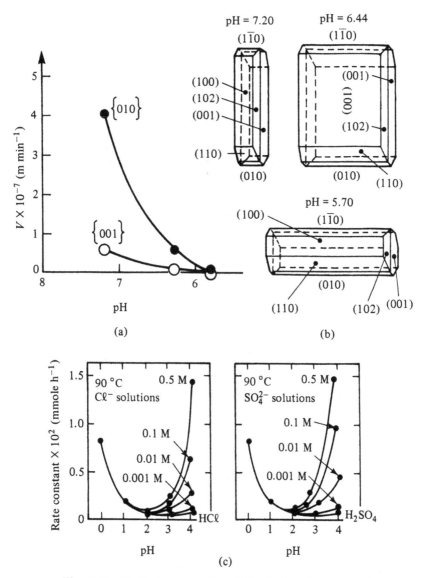

Fig. 2.9. pH effects on growth and dissolution of crystals: (a) crystallization velocity, V; (b) crystal habit changes for cytosine monohydrate; and (c) quartz dissolution rate constants as a function of electrolyte content.

to be much faster in the $CC\ell_4$ solution than in the trioleate solution which is consistent with their solubility data of Table 2.3. However, the

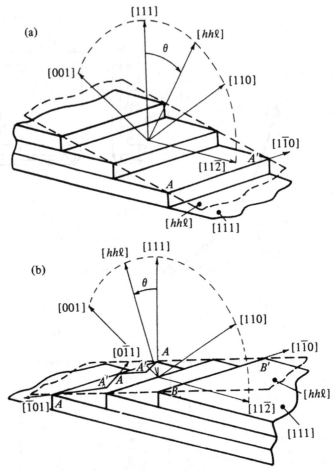

Fig. 2.10. Surface deviations from Si(111) plane illustrating ledge structure: (a) [1$\bar{1}$0] ledges for deviations in the directions of [110]: (b) [1$\bar{1}$0] ledges for deviations in the direction of [100]. F. Spaepan and D. Turnbull in *Laser Annealing of Semiconductors*, Eds. J. M. Poate and J. W. Mayer (Academic Press, New York, 1982) p. 30.

data on succinic acid crystal growth from H_2O and IPA solvents is not consistent with the calculated values of α.

An additional factor that must be considered in aqueous and other solvents is the pH effect on interface roughening and on crystal growth rate. The data of Figs. 2.9(a) and (b) show both the growth rate of the {010} and {001} faces and the crystal habit change as a function of pH for the material cytosine monohydrate. A possible or at least partial explanation for this type of behavior can be seen by considering the quartz dissolution data of Fig. 2.9(c). Here we see that the dissolution rate is

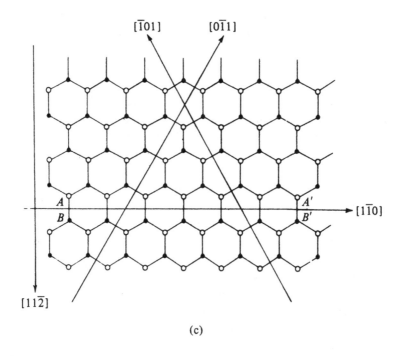

$[\bar{1}01]$ $[0\bar{1}1]$

$[1\bar{1}0]$

$[11\bar{2}]$

(c)

Fig. 2.10. (c) Top view of a (111) plane showing the two types of [110] ledges. AA' and BB'. The BB'-type ledge has two outward directed bonds while the AA'-type ledge has only one. F. Spaepan and D. Turnbull in *Laser Annealing of Semiconductors*, Eds. J. M. Poate and J. W. Mayer (Academic Press, New York, 1982) p. 30.

near zero when the pH of the solution is the same as the isoelectric point (IEP) of the crystal and that it increases strongly with ΔpH from the IEP (IEP ≈ 2.2 for quartz). We see also that the dissolution rate increases as the electrolyte content increases. This appears to be a general phenomenon for a wide range of minerals and appears to relate to the effect of H^+ adsorption and electrolyte absorption to produce interface weakening of the bridge bonds between molecules so that they have a smaller barrier to break in order to go into solution. It is not unlike the negative heat of mixing, ΔH_{mix}, effect discussed earlier, where the adsorption of solvent at the interface occurs and ε_1 is reduced thereby. Thus, although Fig. 2.9(c) for quartz refers to the dissolution condition, if the driving force is altered to be consistent with growth then, because the surface and ledge roughening has increased due to the ΔpH and electrolyte condition, the growth rate will also be faster.

As we can see from the foregoing, the crystal face and ledge roughening processes are complex and a simple α-parameter approach is an insufficient description. Once again detailed surface computations are needed to provide a clearer picture.

2.1.2   *Vicinal faces and the γ-plot*

For most systems, interface orientations slightly removed from the close-packed plane are composed of steps and, for departures of the average surface orientation from the (111) plane of Si, in Fig. 2.10 we see the formation of vicinal surfaces at angle θ from these (111) layers. In Fig. 2.10(a), the vicinal surface has an orientation deviating from the [111] in the direction of the [110] and the ledges are aligned with the [1$\bar{1}$0]. In Fig. 2.10(b), the vicinal surface has an orientation deviating from the [111] in the direction of the [100]; here, the BB' ledge aligned along the [1$\bar{1}$0] can break up into $AA' - AA'$ segments aligned along the [$\bar{1}$10] and [0$\bar{1}$1]. In Fig. 2.9(c) a top view of the (111) plane is given to show the two types of [1$\bar{1}$0] ledges, AA' and BB'; the BB'-type ledge has two broken half-bonds while the AA'-type ledge has only one.

The excess energy of this vicinal face is given by

$$\gamma(\theta) = \gamma_f \cos\theta + \gamma_\ell(\theta, h)\sin\theta \qquad (2.7a)$$

where the face energy, γ_f, is assumed to be constant but the ledge energy, γ_ℓ, will be a function of ledge–ledge separation distance (θ) and of ledge height (h). Although in Chapter 7 we see some justification for this based on the quasi chemical energy factor, it can also be seen from a stress interaction viewpoint. Recent calculations have shown that a perfect Si(111) surface without ledges has a strong compressive stress tensor ($\sigma_c \approx -50,000$ atm) but the addition of a ledge introduces a dilatational stress relaxing some of the surface compressive stress and lowering the energy of the system. The value of γ_ℓ is thus reduced below γ_ℓ^0 by this reconstruction-type contribution due to stress relaxation. The magnitude of the effect is found to depend upon both θ and h. By including the quasi chemical and stress energy factors, one can begin to see why γ_ℓ is much smaller than γ_f for a face of the same orientation. For a material like GaAs, which has a strong piezoelectric coefficient, the interface stress tensor generates an electric field tensor leading to long-range electric field effects between ledges as well as long-range stress interaction effects.

At small values of θ relative to the facet plane, Eq. (2.7a) becomes

$$\frac{\gamma(\theta)}{\gamma_f} = 1 + \frac{\gamma_\ell}{\gamma_f}|\theta| + \cdots \qquad (2.7b)$$

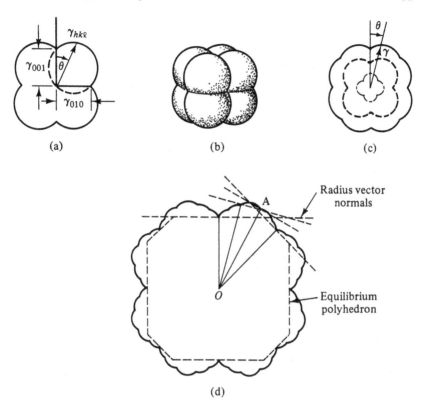

Fig. 2.11. Anisotropy plots of the surface tension, γ for a cubic crystal at $T = 0$ K: (a) in polar coordinates for a two-dimensional section; (b) spatial polar diagram in the first NN-only approximation; (c) two-dimensional section in the first plus second NN approximation (the *dashed* line denotes the 1st NN contribution, and the *dot-dashed* line denotes the second NN contribution); and (d) a general two-dimensional section with the equilibrium crystal habit (dashed polygon) generated by the Wulff construction method.

Thus, at 0 K, a plot of γ/γ_f with θ would yield a sharp cusp with a jump in slope of $2\gamma_\ell/\gamma_f$ at $\theta = 0$. If we also considered the perpendicular direction, a conical cusp would be present at $\theta \to \phi \to 0$. At temperatures $T > 0$ K, a rounding would appear at the bottom of the cusp for configurational entropy reasons. Polar plots of $\gamma(\theta, \phi)$ versus θ and ϕ are called "γ-plots" or Wulff plots and give a pictorial representation of the variation of interfacial excess free energy with respect to crystallographic orientation of the surface. Figs 2.11(a) and (b) illustrate a γ-plot for the SC system for a NN-only interaction (neglecting stress relaxation and reconstruction effects). Figs 2.11(c) and (d) illustrate the modification to be expected from adding only second NN and many NN interactions, respectively. We see from this that the addition of higher-order neighbor interactions introduces successively smaller cusps into the surface of the γ-plot (at 0 K). Thus, at 0 K, the general γ-plot will be a raspberry-shaped figure with symmetry features related to the crystallographic features of the crystal. At higher temperatures, entropy effects will wash out some of these cusps to produce a simpler figure. As mentioned earlier, stress relaxation effects due to ledges also influence γ_ℓ and thus may lead to an alteration in the degree of cusping present in the γ-plot. Thus, as T decreases, the degree of energy anisotropy increases. It needs to be pointed out here that $\gamma(\theta)$ is expected to be a strong function of h for some systems, even though $\theta = $ constant. Thus, the equilibrium γ-plot is expected to involve some optimum $h = h^*(\theta)$ and this will be a function of the environmental conditions.

From Eq. (2.7b), we see that the importance of the cusp in the γ-plot depends upon the magnitude of γ_ℓ/γ_f. For "tight-binding" systems like covalent solids, the force interaction range is short so γ_ℓ/γ_f may be large in the vicinity of the densest packed plane (provided no significant ledge relaxation occurs) and the cusping will be large. For metallic systems, the force interaction range is large so γ_ℓ/γ_f will be small and the γ-plot will contain very little cusping. In this latter case, the γ_ℓ contribution to γ will also tend to increase the isotropy of the γ-plot. Considering Eq. (2.7a), one can ask the question "Using the terrace/ledge/kink (TLK) model, what is required to provide a completely circular γ-plot in two dimensions?". Since $\partial\gamma/\partial\theta = 0$ is required, this constraint leads to

$$\tan\theta = \frac{\gamma_\ell}{\gamma_f - \dfrac{\partial\gamma_\ell}{\partial\theta}} = \frac{h}{\lambda_\ell} \qquad (2.7c)$$

for constant step height, h, and ledge spacing, λ_ℓ. Since $\partial/\partial\theta = (\partial/\partial\lambda_\ell)(\partial\lambda_\ell/\partial\theta)$, Eq. (2.7c) becomes

$$\frac{\partial(\gamma_\ell/\gamma_f)}{\partial(\lambda_\ell/h)} = \frac{(\lambda_\ell/h)(\gamma_\ell/\gamma_f) - 1}{1 + (\lambda_\ell/h)^2} \qquad (2.7d)$$

Thus, if the ledge–ledge interaction is of such a form as to satisfy this first-order differential equation, the γ-plot will be completely isotropic. Extension to the three-dimensional case is straightforward. In the limit that $\gamma_\ell/\gamma_f \rightarrow$ very small values such as one might find for a metallic

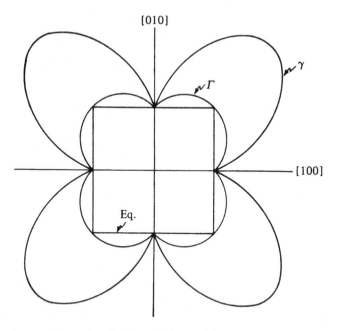

Fig. 2.12. γ-plot, Γ-plot and the equilibrium shape for the section of a cubic crystal normal to the [001].

system, Eq. (2.7d) leads to

$$\frac{\gamma_\ell}{\gamma_f} = \left(\frac{\gamma_\ell}{\gamma_f}\right)_{\theta=\frac{\pi}{4}} - \left[\left(\frac{\lambda_\ell}{h}\right) + \frac{1}{3}\left(\frac{\lambda_\ell}{h}\right)^3 - \frac{4}{3}\right] \qquad (2.7e)$$

since $\lambda_\ell/h = 1$ at $\theta = \pi/4$.

2.1.3 Equilibrium form

If we take a pure spherical crystal in a nutrient phase (vapor, melt, solution) at temperature T^* and allow it to come into equilibrium with the nutrient phase only by changing its shape with no addition or loss of atoms, what will be the crystal shape that develops? This crystal shape is the equilibrium form and is sometimes called the crystal habit. To obtain this minimum free energy shape, we construct Wulff planes at every point of the γ-plot. A Wulff plane is a plane through any point A on the γ-plot that is perpendicular to the radius vector OA as illustrated in Fig. 2.11(d). The inner envelope of the Wulff planes for all possible points on the γ-plot is the equilibrium shape. We see from Fig. 2.11(d) that this is an eight-sided figure for this planar cut through the γ-plot with the larger face corresponding to the deeper cusp. It can

Table 2.4. *Dependence of crystal habit on $\gamma_{max}/\gamma_{min}$*

$\gamma_{max}/\gamma_{min}$	Equilibrium form characteristics
1.01–1.02	Sphere-like with no flat surfaces
1.04–1.1	Flat areas surrounded by curved regions
1.15–1.25	Polyhedral form with rounded corners
$\gtrsim 1.3$	Polyhedral

be shown that $\gamma_1/\gamma_2 = A_2/A_1$ where A refers to the area of the face in the polyhedral equilibrium form. If the 0° and 45° cusps in the γ-plot of Fig. 2.11(d) had not been so steep, then the flats of the equilibrium form would intersect outside the γ-plot rather than inside the γ-plot which means that the corners would be rounded and would contain those orientations of the γ-plot falling between the large face intersection lines. In Fig. 2.12, we introduce the Γ-plot which is the minimum cusped γ-plot giving the identical equilibrium form.[8] Such a shape is called the pedal of the equilibrium γ-plot. Obviously, although one can go from the γ-plot to derive a unique equilibrium form, one cannot go from the equilibrium form to derive a unique γ-plot. However, $\gamma_{max}/\gamma_{min}$ in the γ-plot can be connected to crystal shape in a semiquantitative way with the correlation of Table 2.4 being found. We thus see that observation of equilibrium crystal habit does provide some useful information concerning the three-dimensional γ-plot. We can also see that a similar situation occurs for the equilibrium shape of a crystal island on a facet plane and that observation of this shape provides information concerning the magnitude of $\gamma_\ell(max)/\gamma_\ell(min)$ for that plane.

From the foregoing, we could define γ_ℓ/γ_f as an "anisotropy factor" which qualitatively indicates the type of three-dimensional equilibrium form to be found for a particular system. In a like fashion, γ_k/γ_ℓ should be considered as the anisotropy factor for an equilibrium island shape on a crystal plane where γ_k refers to the energy of the kink position in the type of ledge under consideration. As γ_ℓ/γ_f decreases, the equilibrium form becomes more rounded approaching sphere-like character at very small values of γ_ℓ/γ_f such as one finds for metallic systems (long-range forces). As γ_ℓ/γ_f increases the equilibrium form becomes more polyhedral with sharp corners. For the two-dimensional islands on a particular crystal face, as γ_k/γ_ℓ increases the circumference of the islands becomes composed of straight segments. This is most likely to occur with a bonding situation involving only short-range forces. In the Si crystal/melt system, the Si(100) exhibits too small a cusp at high temperatures to be present on the equilibrium form as a facet plane. However, in the GaAs crystal/melt system, the GaAs(100) does exhibit a large enough cusp to appear on the equilibrium form.

In many crystal growth situations, minor chemical constituents are present that adsorb at the crystal–nutrient interface. If the average adsorption energy for these species is ΔG_{ads}, the equilibrium interface concentration, C_{ads}^*, is just related to the adlayer thickness times the bulk nutrient concentration, C_∞, by a Boltzmann factor, $\exp(|\Delta G_{ads}| /\kappa T)$. We thus see that the degree of adsorption increases strongly as the crystal growth temperature is lowered. For this species, ΔG_{ads} will generally be different for face, ledge, or kink adsorption so that the respective values of γ_f, γ_ℓ or γ_k will be decreased accordingly. For our present purpose, if only existing face adsorption occurs, then the value of γ_ℓ/γ_f will increase and the equilibrium form will become more polyhedral with the size of that face increasing accordingly. The face size increase is directly correlated with the magnitudes of C_∞ and $|\Delta G_{ads}|$. If only existing ledge adsorption occurs, γ_ℓ/γ_f decreases and a more rounded equilibrium form will be observed. Of course, to obtain the ledge concentration in number per centimeter of length one must also multiply by ledge adsorption layer thickness. If both face and ledge adsorption occur, it is the percentage change of each energy that governs the crystal shape change. As expected, similar conclusions apply to the shape of islands on particular crystal planes when one considers ledge versus kink adsorption. In some cases, adsorption occurs only at specific kink and ledge sites with the net consequence that a new cusp develops in the γ-plot (or an old shallow cusp becomes significantly deeper) so that a new face appears in the equilibrium form. In fact, the primary facet plane for a system can be changed by appropriate adsorption conditions.

An alternate description of surface energetics is via the use of a parameter $\tilde{\beta}$ defined as the surface energy per unit area of projected low index face.[8] In terms of the slope, p, of the surface, we have

$$\tilde{\beta}(p) = \tilde{\beta}_0 + \tilde{\beta}_1 p + \tilde{\beta}_2 p^2 + \cdots \qquad (2.8)$$

Comparing Eqs. (2.8) and (2.7a) and the requirement that

$$\gamma(p) = \tilde{\beta}(p)\cos\theta \qquad (2.9)$$

we identify $\tilde{\beta} \sim \tilde{\beta}_0$ with γ_f, $\tilde{\beta}_1$ with $\gamma_\ell(\theta=0)/h$ where p/h is the ledge frequency and $\tilde{\beta}_2$ with ledge–ledge interaction. One of the principal advantages of the $\tilde{\beta}(p)$ description is its use in determining the stability

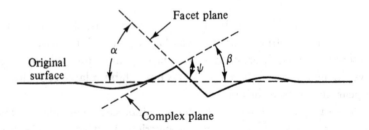

Fig. 2.13. Decomposition of an interface into a pair of energetically more favorable interfaces.

of a particular crystal face with respect to decomposition (faceting) into a pair of more energetically favorable faces. This process is illustrated in Fig. 2.13 and is known to occur when

$$\gamma + \frac{\partial^2 \gamma}{\partial \theta^2} < 0 \tag{2.10a}$$

or, alternatively in the $\tilde{\beta}$ description, when

$$\frac{\partial^2 \tilde{\beta}}{\partial p^2} < 0 \tag{2.10b}$$

The opposite sign of the inequality leads to stability of the particular face under consideration. Considering Fig. 2.12, it can be shown that any orientation for which the γ-plot lies further from the origin than the Γ-plot is unstable with respect to faceting into a hill and valley structure and the reduction in surface energy accompanying such faceting is given by the difference between the γ and Γ values for that orientation. From Eqs. (2.8) and (2.10b), one can predict that $\tilde{\beta}_2 < 0$ will lead to facet formation (and to macroscopic step formation during crystal growth). Initial calculations indicate that $\tilde{\beta}_2[\text{Si}(111)] > 0$ while $\tilde{\beta}_2[\text{GaAs}(00\bar{1})] < 0$. This direct relationship of $\tilde{\beta}$ from Eq. (2.8) to the equilibrium structure is the principal advantage of this alternate description.

The issue of whether an interface, which has orientation (θ, ϕ) relative to the closest packed plane of the system, is atomically smooth or layered with the face parallel to this closest packed plane is an important one for crystallization. If it is not layered, then the interface moves forward via a uniform attachment mechanism with negligible crystallographic features being evidenced. If it is layered, then the interface shape may be rounded or angular on a microscopic scale depending upon the magnitude of γ_ℓ / γ_f and definite crystallographic features are apparent. It is also possible for the interface to be atomically smooth at $V = 0$ but to become layered at $V > V_1$ where V_1 is some specific value. This will be discussed more fully later.

2.1.4   Gibbs–Thomson effects

In contrast to phase transformations via the unidirectional translation of an infinitely flat interface, phase transformation via the growth of small particles involves the creation of substantial amounts of new surface area. In such cases, a portion of the total driving force for phase change must be expended in the creation of this new surface. If the excess surface free energy is isotropic, for certain shapes of particles, the driving force for particle growth can be calculated by averaging over the entire particle. Using the familiar example of a cylindrical particle of radius R and unit length, we recall that, as the particle grows from R to $R + dR$, the area of surface created is $2\pi dR$ and the volume transformed is $\Delta v = 2\pi R dR$. If the volume free energy change per unit volume of transformed material at the new equilibrium transformation temperature is ΔG_V, the total free energy change, ΔG, is given by

$$\Delta G = \Delta G_V + \Delta G_E = 2\pi R dR \delta G_V + 2\pi \gamma dR \tag{2.11}$$

The free energy change per unit volume of transformed material is

$$\frac{dG}{dv} = \frac{\Delta G}{\Delta v} = \delta G_V + \frac{\gamma}{R} \tag{2.12}$$

and thus the new equilibrium temperature, which derives from taking $dG/dv = 0$, is given by

$$T_E = T^* - (\gamma / \Delta S_F) \mathcal{K} \tag{2.13}$$

where $\mathcal{K} = 1/R$ in this case and T^* is the equilibrium transformation temperature for the bulk phase. This is the well-known Gibbs–Thomson equation.

If we next consider the growth of a cylinder of square cross section instead of circular cross section and compute the average free energy change per unit volume of transformed material, we again obtain Eqs. (2.12) and (2.13) provided γ is isotropic and we replace $\mathcal{K}$ with $1/\bar{R}$ in Eq. (2.13). Thus, the average free energy change for a square particle is the same as for a circular particle. However, unlike the circular cylinder which exhibits a constant curvature at any point of its surface, the square cylinder exhibits a zero curvature along its faces and infinite curvature at its corners. The use of Eq. (2.13) on a point-to-point basis would obviously lead to a very different picture than that given by the averaging procedure.

In 1952, Herring[9] showed that the excess free energy of a molecule on a generally curved surface over that on a plane surface was given by

$$\Delta G_E = \gamma \left(\frac{1}{\tilde{R}_1} + \frac{1}{\tilde{R}_2} \right) + \frac{\partial^2 \gamma}{\partial \hat{n}_x^2} \frac{1}{\tilde{R}_1} + \frac{\partial^2 \gamma}{\partial \hat{n}_y^2} \frac{1}{\tilde{R}_2} \tag{2.14}$$

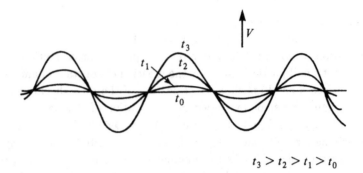

$$t_3 > t_2 > t_1 > t_0$$

Fig. 2.14. Illustration of time-dependent growth in amplitude of a harmonic shape distortion for a macroscopic front drifting at velocity V.

where $\tilde{R}_1$ and $\tilde{R}_2$ are the two principal radii of curvature; $(1/\tilde{R}_1 + 1/\tilde{R}_2) = \mathcal{K}$; and $\hat{n}_x, \hat{n}_y$ are the projections of the variable unit vector $\hat{n}$, on which γ depends, onto a plane tangent to the surface at the point in question, the x-axis being chosen in the direction of the principal curvature $1/\tilde{R}_1$ and the y-axis in that of $1/\tilde{R}_2$. This result has been extensively used to evaluate the driving force for the migration of molecules from a surface region of one value of curvature to that of another. Such a use of Eq. (2.14) is quite correct. However, since that time, it has also become general practice to use Eq. (2.14) to account for the excess free energy associated with surface curvature changes during a phase transformation; i.e., during motion of the interface. To illustrate the general error of this procedure, one need only consider the time-dependent growth in amplitude of an interface having a harmonic shape as illustrated in Fig. 2.14 and moving at an average velocity V. Clearly, in this case, the average curvature is zero so the average excess energy via Eq. (2.14) is zero and the average interface undercooling via the Gibbs–Thomson equation would be zero. However, since the interfacial area is increasing with time, some average interface undercooling must develop to provide the potential for its formation. If the shape change of Fig. 2.14 occurs slowly, but V is large, then the average interface undercooling can be small. For the reverse situation, the average interface undercooling must be large. If the amplitude of the interface undulation is δ, the average interface undercooling must increase as $\dot{\delta}/V$ increases. In addition, instead of an average undercooling, we need a point-to-point undercooling which accounts for this effect. Such a procedure will be developed in the next chapter.

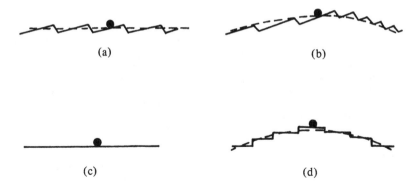

Fig. 2.15. Illustration of different adatom environments (a) terrace of a flat vicinal surface; (b) terrace of a curved vicinal surface; (c) flat facet surface; and (d) curved facet plane surface (on average).

In addition to expressing surface creation effects in other than curvature terms, it is necessary to distinguish between the continuum approximation for the surface and the terrace-ledge-kink (TLK) approximation for the surface. The chemical potential of the surface atom dependence on surface curvature is well defined for the continuum picture of a surface but not so well defined for the TLK picture of a surface. For example, considering Figs. 2.15(a) and (b), the adatom under consideration sits on the same terrace in both cases, but the average surface curvature is decidedly different. If the atomic forces are very short range and adjacent ledges sufficiently far apart that they are not interacting, then the adatom is unable to sense any difference in its environment for the two cases. The same situation applies for the adatom in Figs. 2.15(c) and (d), where a facet surface of the same macroscopic curvature change as (b) has been formed. We note also that the creation of the excess surface area in Fig. 2.15(b) relative to Fig. 2.15(a) requires much less energy than for the case of Fig. 2.15(d) relative to Fig. 2.15(c). In the former case, the total amount of terrace and ledge is the same and only the ledge spacing is different so the energy cost is small. In the latter case, the amount of terrace is the same but the amount of ledge has been increased so the energy cost may be large. However, on a continuum surface picture, ΔG_E for the case of Fig. 2.15(b) will be larger than that for the case of Fig. 2.15(d) when creation of the extra surface is considered. These features reveal the importance of crystallography and the TLK picture to interface morphology considerations. We will deal more fully with this in Chapter 3.

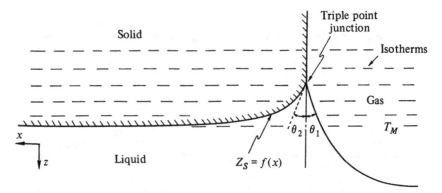

Fig. 2.16. Schematic illustration of interface shape, $Z_S = f(x)$, in the vicinity of the triple junction (gas, liquid, solid) relative to idealized planar isotherms.

To close this section, it is perhaps useful to define the curvature, $\mathcal{K}$, of a surface in the TLK approximation. If we adopt an origin at the center of a terrace of width λ_0 with the adjacent right and left hand terraces of width λ_{+1} and λ_{-1}, respectively, the curvature at the origin is given, for small curvature, by

$$\mathcal{K} = \frac{-\partial^2 z}{\partial x^2} = -\frac{\left[(\partial z/\partial x)_+ - (\partial z/\partial x)_-\right]}{\Delta x} = \frac{[h/\lambda_{+1} - h/\lambda_{-1}]}{\lambda_0 + \frac{1}{2}(\lambda_{+1} + \lambda_{-1})} \quad (2.15)$$

$$\approx \frac{h(\lambda_{+1} - \lambda_{-1})}{2\lambda_0^3}$$

where h is the step height for all terraces and $\Delta\lambda/\lambda_0 \ll 1$.

2.1.5 *Static interface shapes*

We can utilize Eq. (2.14) to obtain the undercooling, ΔT_E, of a pure melt, needed for it to be in equilibrium with a curved region of crystal–melt interface based upon the continuum approximation; i.e., $\Delta T_E = \Delta G_E/\Delta S_F$. In particular, we can utilize this value of ΔT_E to determine the contour of the equilibrium interface ($V = 0$) in the CZ crystal pulling arrangement under idealized thermal conditions. Later, we will extend this understanding to the growth interface ($V > 0$).

In Eq. (2.14), $\tilde{R}_2 \sim$ crystal radius $>> \tilde{R}_1$ and thus can be neglected so only the two-dimensional interface cut by a plane passing along the crystal axis need be considered; i.e., $Z_S = f(x)$ in Fig. 2.16. For simplicity, assume that the isotherms are flat and parallel to the macroscopic interface so that the T_M isotherm touches the interface at $z = 0, x = 0$ (the crystal center). These isotherms will actually bend near the periphery so that our deduced results will only be qualitatively correct. Further, let the temperature gradient normal to the T_M isotherm in the solid and in the meniscus region of the liquid be constant as x is varied and be given by $\mathcal{G}_n$. Then, any actual temperature variation along the interface must be balanced by the curvature generated contribution, ΔT_E.

It can be readily seen that the interface cannot be perfectly flat in the meniscus region because of the force balance requirement at the triple point where the three surfaces meet. At equilibrium, the angles θ_1 and θ_2 are given from the solution of

$$\gamma_{LV}\cos\theta_1 + \gamma_{SL}\cos\theta_2 = \gamma_{SV} \tag{2.16a}$$

$$\gamma_{LV}\sin\theta_1 = \gamma_{SL}\sin\theta_2 \tag{2.16b}$$

So long as $\theta_2 < \pi/2$, some region of Z_S must be at a temperature below T_M and, therefore, must be curved to be in equilibrium with the liquid. The temperature balance in this case is given by

$$T(0,0) - T[x, f(x)] = -\mathcal{G}_n Z_S = -\frac{\gamma}{\Delta S_F}\mathcal{K} = -\frac{\gamma}{\Delta S_F}\left[\frac{Z_S''}{(1 + Z_S'^2)^{3/2}}\right] \tag{2.17}$$

The solution to this equation for $Z_S = f(x)$ is in the form of elliptic integrals with some general results being plotted in Fig. 2.17 for several values of the parameter $K^2 = \gamma/\mathcal{G}_n\Delta S_F$.[10] The outer surface of the crystal (vertical line in Fig. 2.16) must meet the appropriate curve at the angle θ_2 and will determine the crystal radius, $R = R(\theta_2) < R(0)$. From Fig. 2.17, the size scale of the crystal edge surface contour is $\sim 10\ \mu$m and increases as $\mathcal{G}_n$ decreases and as γ increases. We note also that a rim of supercooled liquid exists near the meniscus wherein stray crystals have a small but finite probability of nucleating. In Fig. 2.18, we see that the volume of this supercooled liquid increases as the freezing point isotherm shape changes from convex to planar to concave to the liquid with the specimen axis being parallel to the external surface. Here, A and C correspond to the concave and convex interfaces, respectively. Similar interface contouring with its attendant zone of supercooling exists for crystal growth in a container both during the steady state growth regime and during the seeding regime when the crystal dimension is increasing. Further, the same type of interface contouring occurs at an internal boundary such as a grain boundary. Fig. 2.19 illustrates an asymmetric grain boundary groove profile between crystal I having anisotropic $\gamma = \gamma_1$ and crystal II having $\gamma = \gamma_1$ for all orientations except that for the flat interface where $\gamma = \gamma_0$. Two values of anisotropy

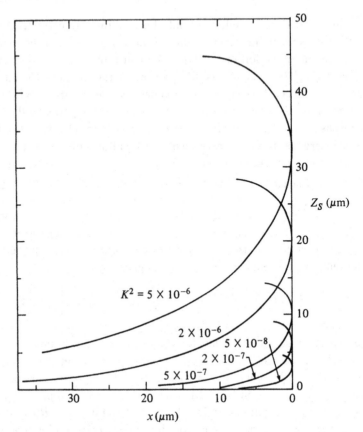

Fig. 2.17. Calculated interface shape for isotropic γ and several values of $K^2 = \gamma/\mathcal{G}_n \Delta S_F$.

ratio γ_0/γ_1 are illustrated and the corresponding equilibrium directions of grain boundary alignment shown. This illustrates one mechanism of texture development for castings of very pure materials wherein the steeper groove wall encroaches laterally on the shallow groove wall.

From Fig. 2.17 with $K^2 = 5 \times 10^{-7}$ ($\gamma/\Delta S_F \approx 1.2 \times 10^{-5}$, $\mathcal{G}_n \approx 25$ for Si), $Z_S(max) \approx 10$ μm and the maximum supercooling for the adjacent volume of liquid is only 0.025 °C. This is a typical value for different $\mathcal{G}_n$ even though the volume of supercooled liquid changes appreciably as $\mathcal{G}_n$ changes. If all crystal growth could occur under such conditions, it is unlikely that one would need to worry about the nucleation of stray crystals in the meniscus region. However, as we will see later, two new factors enter the picture to change the results. First, for dislocation-free

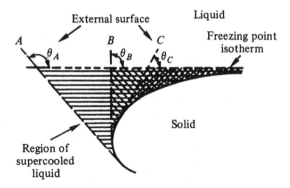

Fig. 2.18. For specimen axis parallel to the external surface A, B or C, the volume of supercooled liquid for different wetting angle, θ, is illustrated.

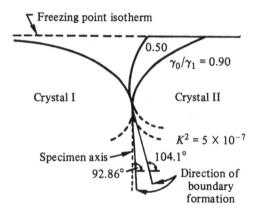

Fig. 2.19. Illustrating the direction of grain boundary formation when adjacent grains produce an anisotropic groove shape for surface energy reasons. The angle of encroachment increases as γ_0/γ_1 decreases.

Si, growing in the $\langle 100 \rangle$ or $\langle 111 \rangle$, layer sources at the additional $\{111\}$ poles located in the rim region operate to provide the layers that feed the main interface. The supercooling, ΔT_K, needed to drive these layer sources can be appreciably larger than ΔT_E for large V. This consideration is general and applies to most materials, not just Si. Second, any furnace power fluctuations are felt most strongly in the meniscus region which causes a large $\delta T(t)$ wave to impinge on the layer sources in the rim region causing alternately melting and freezing on a microscopic scale. Due to this cyclic temperature perturbation, one might anticipate that ΔT_K could reach $\gtrsim 1\,°\mathrm{C}$ at the rim layer sources. For a

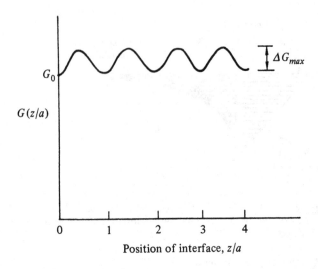

Fig. 2.20. Variation of system free energy, G, at $T = T_M (\Delta G_v = 0)$ as a flat interface between solid and liquid is moved to the right through successive equilibrium positions $0, 1, 2, \ldots$ of the interface centroid.

1 °C temperature fluctuation at the crucible walls, the thermal diffusion decay of the wave as it travels towards the interface via the bulk liquid is large so the main interface "sees" only a small temperature perturbation. However, radiant heat transfer from the crucible wall to the meniscus region is rapid and unattenuated so the 1 °C wall fluctuation is essentially transferred in total to the meniscus region.

2.2   Molecular attachment kinetics

From Section 2.1 and later in Chapter 7 of this book, we learn that the equilibrium interface roughness depends on the ratio $\varepsilon_1 / \kappa T^*$, where ε_1 is the NN bond strength between the two phases and T^* is the equilibrium interface temperature. For large values of $\varepsilon_1 / \kappa T^*$, neither the interface terrace nor the interface ledges are rough; however, as $\varepsilon_1 / \kappa T^*$ decreases, the terrace and ledges become increasingly more rough until they are maximally rough. For this extreme state at T^*, where $\varepsilon_1 / \kappa T^* = (\varepsilon_1 / \kappa T^*)_c$, interface adatoms or advacancies can be added anywhere with no change of free energy for the system so that the macroscopic interface can advance (crystal growth) via a uniform attachment (UA) mechanism. For $\varepsilon_1 / \kappa T^* < (\varepsilon_1 / \kappa T^*)_c$, if we now wish to consider growth of the crystal at zero bulk driving force ($\Delta G_\infty = 0$) by random attachment of atoms on the crystal face, we must add +1 states beyond the fraction X_1^* to move the centroid of the macroscopic interface a distance δz. This requires an increase of the interfacial energy, γ. Thus, as one moves the centroid of the interface an integral number of atom distances, a, the free energy of the system oscillates periodically as illustrated in Fig. 2.20. The magnitude of ΔG_{max} depends on ε_1 in that, if ε_1 is small, X_1^* is large and only a small change in X_1 is needed to move the interface centroid through the barrier between equilibrium states so ΔG_{max} will also be small. On the other hand, if ε_1 is large, X_1^* is small and a large change in X_1 is needed to move the interface centroid $(\Delta X_1 \sim X_1(max) - X_1^*)$ so ΔG_{max} will be large.

In order to see under what conditions the random attachment of atoms to the crystal face is an allowed process, let us follow the approach of Cahn [11] and consider such attachment at some interface undercooling ΔT_K. The free energy change, δG, due to interface centroid movement of δz is given by

$$\delta G = \left(-\Delta G_i + \frac{d\gamma_f}{dz}\right)\delta z \qquad (2.18a)$$

where $d\gamma_f/dz \equiv dG/dz$ from Fig. 2.20. Only if $\delta G < 0$ can the process occur spontaneously, which requires, for uniform face attachment (UFA) that

$$\Delta G_i > \left(\frac{d\gamma_f}{dz}\right)_{max} \qquad (2.18b)$$

As usual, $\Delta G_i = \Delta G_K + \Delta G_E$ where ΔG_E is due to trapped defects or strain in the crystal layers being formed by this process. The conditions given by Eq. (2.18b) occur only at very large driving forces for most materials and it is thought by this author that they are almost never encountered in single crystal growth except perhaps for some metals under conditions of very rapid solidification or during film growth by molecular beam epitaxy (MBE) at high flux, J, relative to the equilibrium flux, J^*. In the more common, small driving force regime,

$$\Delta G_i < \left(\frac{d\gamma_f}{dz}\right)_{max} \qquad (2.18c)$$

and the UFA process cannot occur spontaneously. In this regime, atoms can only attach onto the crystal at layer edges so that the macroscopic interface can only move forward by the lateral passage of layers across the interface. This is called the layer edge attachment (LEA) regime.

Of course, even for the layer edge, we must ask the question, "What are the conditions favoring uniform ledge attachment (ULA)?" Since the layer edge is always rougher than the face $(X^{'*} \sim X^{*1/2})$, there is more

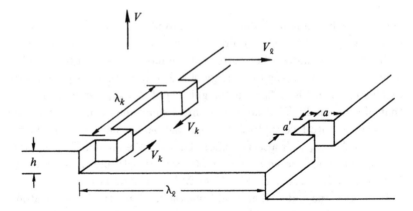

Fig. 2.21. Schematic representation of a vicinal surface illustrating average ledge spacing, λ_ℓ, average kink spacing, λ_k, and the key velocities V, V_ℓ, and V_k.

likelihood that it will be maximally rough; however, we can set up an equation analogous to Eq. (2.18a) for the layer edge and find that ULA occurs only when

$$\Delta G_i > \left(\frac{\mathrm{d}\gamma_\ell}{\mathrm{d}x}\right)_{max} \tag{2.19a}$$

where x is the ledge normal coordinate in the face. When ΔG_i is too small to satisfy Eq. (2.19a), then molecules can join the crystal only at kink sites to give kink only attachment (KOA); i.e., when

$$\Delta G_i < \left(\frac{\mathrm{d}\gamma_\ell}{\mathrm{d}x}\right)_{max} \tag{2.19b}$$

For attachment at a kink site, no increase in free energy of the system occurs, provided that no long-range kink–kink interaction is present, so this type of process is allowed at all values of $\Delta G_i > 0$. From the foregoing, we now see that the macroscopic interface velocity, V, depends upon both (a) the rate of movement of its two important structural elements, (i) the kink velocity, V_k, and (ii) the layer ledge velocity, V_ℓ, illustrated in Fig. 2.21 and (b) the probability of these structural elements being present on the interface. We must eventually consider the nucleation frequency of new ledges on a flat face and the formation rate of kinks on a perfectly smooth ledge to develop its state of equilibrium roughness.

2.2.1   Layer motion-limited kinetics

Considering Fig. 2.21, the macroscopic growth velocity, V, is

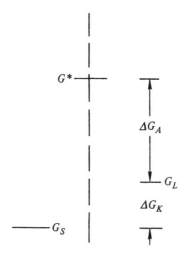

Fig. 2.22. Key free energies of a molecule G_L, G_S and G^* involved in the transfer from one phase to the other.

given in terms of the ledge density, $\rho_\ell = \lambda_\ell^{-1}$, by

$$V = h\rho_\ell V_\ell \qquad (2.20a)$$

where h may be one or more atoms high. The layer velocity, V_ℓ, is given by the molecular distance, a, normal to the ledge divided by the average time τ_ℓ for kinks to growth together when the kinks are separated on the average by a distance $\lambda_k (\lambda_k = a'/X_1'^*)$. Thus, we have

$$V = h\rho_\ell\, a/\tau_\ell = 2ha\, V_k/\lambda_\ell\lambda_k \qquad (2.20b)$$

To determine V_k, we need to consider transition state theory wherein, for a liquid Si atom to become a solid Si atom, the liquid atoms adjacent to the kink sites must (i) give up their communal entropy and make a small translational motion into the crystal site adjacent to the kink, (ii) make a significant electronic state change from metallic-type bonding to covalent-type bonding and (iii) give up their latent heat to the vibrational modes of the lattice. It is also possible that unreconstruction of the ledge adjacent to the kink and dissociation of an interface adsorbed layer Si-complex are additional requirements.

The key free energies involved in the process are represented in Fig. 2.22 where G^* is the free energy of the activated state. Thus, for the n_L atoms at free energy G_L that are situated to make the transition to the crystal site adjacent to each kink, we have the rate of transition given by[12] (see Eq. (8.12))

$$\bar{k}_- = n_L a_L \hat{\nu}_L \exp(-\Delta G_A/\mathcal{R}T) \qquad (2.21a)$$

where $\hat{\nu}_L$ is the average vibrational frequency of the liquid atoms and a_L is a geometrical factor that indicates the fraction of vibrations when these n_L atoms are going in the proper direction to land in the site adjacent to the kink site. Similarly, the atoms in the new kink sites have a melting reaction rate, $\bar{k}_+$, given by

$$\bar{k}_+ = a_S \hat{\nu}_S \exp[-(\Delta G_A + \Delta G_K)/\mathcal{R}T] \qquad (2.21b)$$

The net rate of transfer to the crystal state, $\bar{k} = \bar{k}_- - \bar{k}_+$, is given by

$$\bar{k} = n_L a_L \hat{\nu}_L \exp(-\Delta G_A/\mathcal{R}T) - a_S \hat{\nu}_S \exp[-(\Delta G_A + \Delta G_K)/\mathcal{R}T]$$
$$= B \exp(-\Delta G_A/\mathcal{R}T)[1 - \exp(-\Delta G_K/\mathcal{R}T)]$$
$$(2.21c)$$

since $\bar{k} = 0$ when $\Delta G_K = 0$ leading to $n_L a_L \hat{\nu}_L = a_S \hat{\nu}_S = B$.

In the small driving force regime, $\Delta G_K/\mathcal{R}T \ll 1$ and Eq. (2.21c) yields

$$V_k = a'\bar{k} = \frac{a'B}{\mathcal{R}T} \exp(-\Delta G_A/\mathcal{R}T)\Delta G_K = \beta_0 \Delta G_K \qquad (2.22a)$$

so that

$$V = \frac{2ha\beta_0}{\lambda_\ell \lambda_k} \Delta G_K = \beta_2 \Delta G_K \qquad (2.22b)$$

Thus, layer edge attachment, which is layer motion-limited rather than layer source-limited, varies linearly with ΔG_K at small driving forces. At large driving forces we have the calculated Cu results shown in Figs. 2.23(a) and (b) for $a'\bar{k}_-, a'\bar{k}_+$ and $a'\bar{k}$. We note that $\bar{k}/\bar{k}_- < 10^{-3}$ for $a'\bar{k} \lesssim 1$ cm s^{-1} so that many back and forth state changes occur before the liquid atom is frozen into its kink site position. From Eq. (2.21c) we have, in general,

$$V = \left(\frac{2haa'B}{\lambda_\ell \lambda_k}\right) \exp(-\Delta G_A/\mathcal{R}T)[1 - \exp(-\Delta G_K/\mathcal{R}T)] \qquad (2.22c)$$

From Eqs. (2.22b) and (2.22c), we note that V varies inversely with λ_k which, in turn, depends upon ε_1 and T^*. Since, for growth from the vapor, ε_1 is an order of magnitude larger than for growth from the melt, λ_k will be several orders of magnitude larger for vapor growth than melt growth and V will be correspondingly several orders of magnitude smaller for the same ΔG_K. Likewise, as T^* for the growth process decreases, V will also strongly decrease for a pure system. If there are surface active minor constituents present like O_2, Te, Se, etc., they will tend to adsorb preferentially at kink sites and deactivate them as transformation sites. This factor will also increase as T^* decreases so that λ_k will further increase and V further decrease due to such adsorption effects at fixed ΔG_K. Due to continued growth, these adsorption states

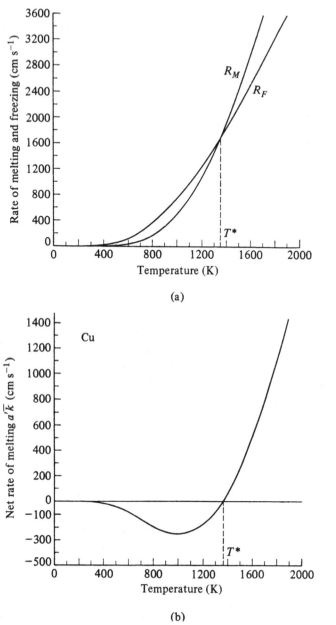

Fig. 2.23. Variation of key rates for Cu with temperature: (a) melting, $R_M = a'\bar{k}_+$, and freezing, $R_F = a'\bar{k}_-$; (b) net rate of melting (or freezing), $a'\bar{k} = R_M - R_F$.

will be overgrown and imbedded into the crystal with an accompanying ΔG_E requirement on ΔG_i.

Turnbull[13] uses an alternate approach to the TLK model for describing interface attachment kinetics and it is useful to describe it briefly here since some use of it will be made later. It is an extension of the UFA model wherein V is given by

$$V = \lambda \hat{f} k_u [1 - \exp(-\Delta G_K/\mathcal{R}T)] \qquad (2.23a)$$

k_u is the frequency of configurational rearrangements, $\hat{f}$ is the fraction of interface sites at which rearrangement can occur and λ is the distance moved by the interface in each rearrangement. Thus, $\hat{f}$ relates to the fraction of kink sites for KOA in the TLK model and allows one to extend the approach to transformations occurring at non-crystalline interfaces. One finds that k_u often scales with the frequency of diffusive transport, $k_u = k_D = D_C/<\lambda_D^2>$, or of viscous flow in the melt, $k_u = k_\eta = a_{\bar{\eta}}\mathcal{R}T/\bar{\eta}\Omega$, where λ_D is the jump distance in the melt, $\bar{\eta}$ is the viscosity, Ω is the molecular volume and $a_{\bar{\eta}}$ is the molecular diameter. The Stokes–Einstein equation ($\bar{\eta} = \kappa T/3\pi\lambda D_C$) connects these two frequencies. For metal melts, k_u seems to be largely controlled by the frequency of impingement of atoms from the fluid onto the interface ($\hat{f}k_u > 2\times10^{13}$ s^{-1} experimentally) and crystal growth is largely limited by heat transport. For covalent systems, k_u is generally controlled by the rate of interfacial rearrangements

$$k_u = n_r \hat{\nu}_r \exp(\Delta S_A/\mathcal{R}) \exp(-\Delta H_A/\mathcal{R}T) \qquad (2.23b)$$

where $\hat{\nu}_r$ is the normal frequency of the reaction model, n_r is the number of rearrangements resulting from a single activation, ΔS_A is the entropy of activation and ΔH_A is the activation enthalpy. The thermodynamic factor of Eq. (2.23a) yields $\Delta G_K = n'\Delta G_{\hat{c}}$ where $\Delta G_{\hat{c}}/\mathcal{N}_A$ is the free energy of crystallization per atom ($\mathcal{N}_A$ is Avogadro's number) and n' is the number of atoms crystallized per rearrangement. In most chain reactions of interest to us, $n' >> 1$, and the bracket in Eq. (2.23a) may approach unity at rather small undercoolings.

2.2.2   Layer source-limited kinetics

At the other end of the scale to layer motion limitations, we can have the crystal growth limited by the generation of layers from several possible sources. In practice, the three chief layer sources are (1) two-dimensional nucleation, (2) screw dislocations, and (3) twin-plane reentrant corners. These three mechanisms are illustrated in Fig. 2.24. From Chapter 8, for the nucleation of pill-box-shaped layers, we find

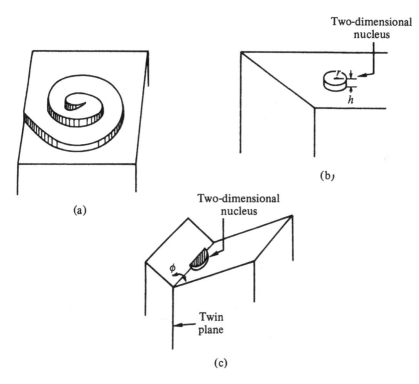

Fig. 2.24. Schematic illustration of the primary layer source mechanism involved in crystal growth: (a) screw dislocation; (b) two-dimensional pill-box nucleation and (c) twin-plane reentrant corners.

that

$$V = a'' A I \qquad (2.24a)$$

where

$$I = I_0 \exp(-\pi a'' \gamma_\ell^2 / \kappa T \, \Delta G_i) \qquad (2.24b)$$

is the nucleation frequency per unit area and A is the area available for nucleation. In Eq. (2.24a), it is assumed that layer motion is infinitely rapid compared to the rate of formation of new pill-boxes so that each nucleation event results in the crystallization of an entire layer. If this does not hold, then the RHS of Eq. (2.24a) must be multiplied by a factor < 1 to yield an effective value of A wherein the above condition is assumed to hold. Obviously this factor will be a function of ΔG_i and will decrease as ΔG_i increases.

To evaluate this factor, we can ask the question, "What is the radius of disk, R, which has unit probability that a new nucleus will form on top of the disk in the time τ_{R_1} that it takes for the original disk to grow to $R = R_1$?" With $\bar{V}_\ell$ the average disk edge velocity, we have the relationships

$$I(\pi R_1^2)\tau_{R_1} = 1 \tag{2.25a}$$

and

$$\tau_{R_1} \approx \frac{R_1}{\bar{V}_\ell} = \frac{1}{\bar{V}_\ell} \int_0^{\tau_{R_1}} V_\ell dt \tag{2.25b}$$

which leads to

$$R_1 = (\bar{V}_\ell/\pi I)^{1/3} \tag{2.25c}$$

and

$$V = a'' \bar{V}_\ell^{2/3} (\pi I)^{1/3} \tag{2.25d}$$

Thus, our effective area is just πR_1^2 and V becomes

$$
\begin{aligned}
V \approx \pi^{1/3} a'' I_0^{1/3} & \left(\frac{2aa'B}{\lambda_k} \right)^{2/3} \exp\left(-\frac{2\Delta G_A}{3\mathcal{R}T} \right) \\
& \times \exp\left(\frac{-\pi a'' \gamma_\ell^2}{3\kappa T \, \overline{\Delta G_i}} \right) \left[1 - \exp\left(\frac{-\overline{\Delta G_K}}{\mathcal{R}T} \right) \right]^{2/3}
\end{aligned}
\tag{2.25e}
$$

We have used $\bar{V}_\ell$ in Eq. (2.25) instead of V_ℓ because, during the growth of the disk, all of the ΔG_i is not available for molecular attachment. A portion of it must be consumed in creating the new surface associated with disk enlargement and this decreases as the radius of the disk increases. Thus, the lateral velocity of the disk, V_ℓ, is zero at $R = r_c$, the critical radius for pill-box nucleation and increases as R increases (since $\Delta G_K = \Delta G_i - \gamma/R$) so that more and more of the ΔG_i is shifted to the ΔG_K portion of the driving force picture as R increases.

For the screw dislocation source (see Fig. 2.24), the nose of the dislocation has a radius of curvature, r_c, equal to the radius of the critical sized pill-box for two-dimensional nucleation ($r_c = \gamma_\ell/\Delta G_K$). The spacing between the layers far from the nose is given by

$$\lambda_\ell = 4\pi r_c = 4\pi \gamma_\ell/\Delta G_K \tag{2.26a}$$

Thus, for this case at small driving forces

$$V = \frac{ha\beta_0}{2\pi\gamma_\ell\lambda_k} |\Delta G_K| \, \Delta G_K \tag{2.26b}$$

This square law dependence of V on ΔG_K is written this way to illustrate that $V < 0$ when $\Delta G_K < 0$. For large driving forces, $\Delta G_i/\mathcal{R}T \gtrless 1$,

$$V = \frac{haa'B}{2\pi\gamma_\ell\lambda_k} \Delta G_K \exp\left(\frac{-\Delta G_A}{\mathcal{R}T} \right) \left[1 - \exp\left(\frac{-|\Delta G_K|}{\mathcal{R}T} \right) \right] \tag{2.26c}$$

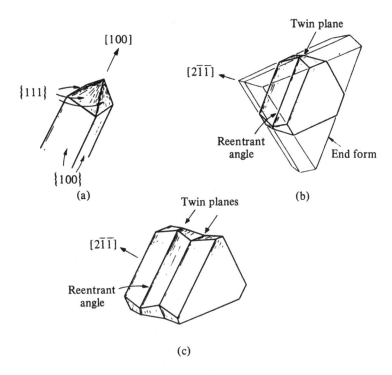

Fig. 2.25. Schematic illustration of some diamond cubic growth forms as a function of rotation twin content: (a) no twins, (b) one twin and (c) two twins which leads to growth in $\langle 2\bar{1}\bar{1} \rangle$ directions.

For the twin-plane reentrant corner mechanism (TPRE), layers are again produced by nucleation but now at a much higher frequency because the reentrancy lowers the energy of formation for the partial pill-box at the reentrant corner in Fig. 2.24(c). For this case, the active surface region is reduced to a width $2r_c$ centered on the twin plane and the free energy of formation of the half pill-box is given by

$$\Delta G = -\left(\frac{\pi r^2 a}{2}\right)\Delta G_K + \pi r a \gamma_\ell + 2ra[\gamma_{tw} - \Delta\gamma(\phi)] \quad (2.27a)$$

$$= \frac{1}{2}[-\pi r^2 a \,\Delta G_K + 2\pi ra\gamma^*] \quad (2.27b)$$

and

$$\gamma^* = \gamma_\ell + \frac{2}{\pi}[\gamma_{tw} - \Delta\gamma(\phi)] \quad (2.27c)$$

Here, γ_{tw} is the twin boundary energy and $\Delta\gamma(\phi)$ is the change in the surface energy of the flat edge which is in contact with the adjacent tilted face relative to the $\phi = 180°$ case. Obviously, when ϕ approaches $90°$, $\Delta\gamma(\phi) \to \gamma_\ell$ and, when ϕ approaches $180°$, $\Delta\gamma(\phi) \to -\gamma_\ell$. Thus, for the case of infinitely rapid layer growth compared to layer nucleation, we have

$$V = a'' \, \ell_{tw} \left(\frac{2\gamma^*}{\Delta G_K} \right) \exp \left[-\pi a \left(\frac{\gamma^{*2}}{2\kappa T |\Delta G_i|} \right) \right] \qquad (2.27d)$$

where ℓ_{tw} is the length of the twin plane per unit area of crystal.

A non-twinned Ge or Si filament crystal would grow in the $< 100 >$ at normal velocities under relatively large supercooling via the two-dimensional pill-box nucleation mechanism (see Fig. 2.25(a)) on the {111} facet planes. Since the rotation twin energy in Si and Ge is very low, the formation of a faulted two-dimensional embryo with a rotation twin between the substrate and the pill-box is an event of reasonable probability. Thus, as ΔG_i increases, it is just a matter of time before such an embryo forms and becomes a critical nucleus. The formation of one such twin leads to very rapid initial growth and a change of the crystal habit from Fig. 2.25(a) to Fig. 2.25(b) which is also a slowly growing form because the reentrancy has grown out (faster growth than adjacent faces) and the bounding faces of the form have produced a roof structure with $\phi > \pi$. When a second parallel rotation twin forms, then a body having a $\phi < \pi$ and a $\phi > \pi$ roof structure forms on each major face as illustrated in Fig. 2.25(c). Thus, the fast growing face segment ($\phi < \pi$) cannot be grown out because, to decrease its size on one face, just increases the segment size on the two adjacent faces and vice versa. The structure of Fig. 2.25(c) is thus stable and grows very rapidly as long ribbons in the $< 2\bar{1}\bar{1} >$ directions.

A great many crystal systems develop a twinned character when the molecular attachment aspects of particle growth begin to become significant. For the growth of long dendrite ribbons such as one finds in Si, it is necessary that the twin plane be parallel to a plane of the equilibrium form. For crystal systems exhibiting equilibrium forms completely closed by one set of facet planes, at least two parallel twin planes are required for continuous growth. For crystal systems where stable equilibrium facets do not form a closed body, only one twin plane may be needed for continuous growth. The minimum number of twin planes needed for continuous or ribbon growth in various crystallographic systems is given in Table 2.5.

2.2.3   Growth regimes for increasing ΔG_K

From Eqs. (2.24)–(2.27), and neglecting any ΔG_E effect, we note

Table 2.5. *Conditions for continuous ribbon growth*

Crystal structure	Equilibrium face	Twin plane	System	Minimum no. of twins
Diamond cubic and	{110}	{111}	Closed	2
zinc blende	{111}, {113}	{111}	Open	1
Face-centered cubic	{111}	{111}	Closed	2
Body-centered cubic	{101}	{112}	Closed	
Hexagonal close-packed	{0001}	{10$\bar{1}$2}	Open	
Hexagonal close-packed	{0001}	{0001}	Open	1
Orthorhombic	{110}	{110}	Open	1
Rhombohedral	{110}	{110}	Open	1

that, as a function of ΔT_K (or ΔG_K), there are four different regimes of growth to be considered: (a) for $0 < \Delta T_K < \Delta T_K^{***}$, the classical layer growth mechanism should be observed and the details of the V versus ΔT_K shape in this region will depend upon which mechanism is rate limiting. In this region, the growth is characterized by a step energy γ_ℓ and a kink density X'^* that are relatively independent of ΔT_K. The layer motion-limited, the screw dislocation and the two-dimensional nucleation source-limited cases are illustrated in Fig. 2.26(a). The term $\bar{\eta}/\bar{\eta}_0$ corrects for any temperature dependence of the viscosity ($\exp[\Delta G_A/\mathcal{R}T]$ term). (b) For $\Delta T_K^{***} < \Delta T_K < \Delta T_K^{**}$, the same situation applies as in regime (a) but now the kink density has increased and the ledge energy has decreased since the transition to a uniform ledge attachment condition has occurred. A portion of this regime is also shown in Fig. 2.26(a). (c) For $\Delta T_K^{**} < \Delta T_K < \Delta T_K^*$, the lateral growth mechanism must be modified to allow for a decrease in γ_ℓ due to the overlap of the diffuse regions at the step edges. In this region a gradual transition is made from the classical lateral growth to the UFA regime. (d) For $\Delta T_K^* < \Delta T_K$, the interface can advance normal to itself without need for the lateral motion of steps. These four regimes are illustrated in Fig. 2.26(b). The rising growth rate in the UFA region comes from the $\exp(\Delta G_K/\mathcal{R}T)$ term in Eq. (2.22c). Of course, for some crystal systems $\Delta G_K/\mathcal{R}T >> 1$ even in the vicinity of ΔT_K^{***} so that there will be no flat plateaus in Fig. 2.26(b) to the right of that value of ΔT_K. In Fig. 2.26(c), the change in topology of a vicinal surface with increasing ΔT_K is shown to illustrate that the overall transition to UFA occurs via gradual roughening of the ledges.

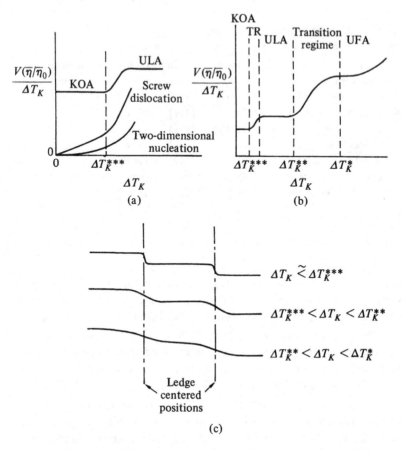

Fig. 2.26. Crystallization velocity, scaled for viscosity changes, and interface shape (side view) for a vicinal surface as a function of ΔT_K: (a) very low driving force domain; (b) total driving force domain; (c) ledge roughness change as ΔT_K increases. (TR in (b) refers to the transition regime where the kink spacing on the ledge λ_ℓ decreases due to uniform ledge attachment.)

Many investigators neglect the KOA region of Fig. 2.26 and the transition to the ULA region and use the two alternate plots of Fig. 2.27 to represent kinetic data for the screw dislocation mechanism. It should be noted that these plots are also limited to the driving force domain were $\Delta G_K / \mathcal{R}T \ll 1$. In reality, this condition may obtain at $\Delta T_K < \Delta T_K^*$ so that no regions of constant slope will be found on this type of plot. Of course, for crystal growth from solution where ΔT_K is replaced by the relative supersaturation, ΔC_K, regions of constant slope will be predicted at high driving force ($\Delta G_i = \mathcal{R}T\ell nC/C^*$). In any

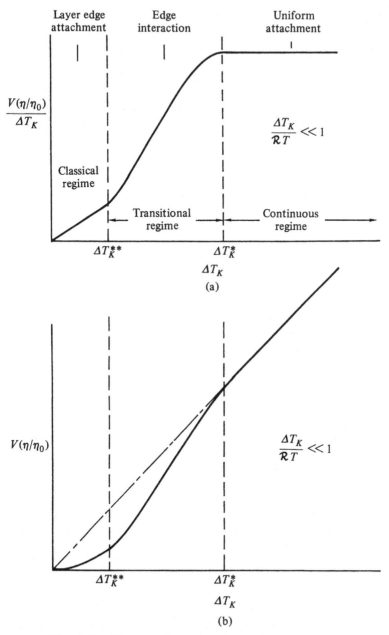

Fig. 2.27. Theoretically predicted growth rate curve as a function of undercooling, ΔT_K, for interfaces with emergent dislocations used by many investigators. The growth rate in the ordinate is adjusted for the temperature dependence of melt viscosity.

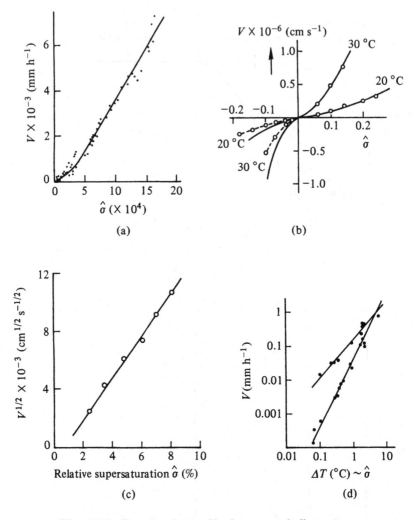

Fig. 2.28. Growth velocity, V, plots versus bulk nutrient supersaturation, $\hat{\sigma}$, for various systems: (a) sodium chlorate crystals; (b) (111) face of the benzophenol crystal (solid lines) compared to theoretical calculations from the BCF theory (0); (c) barium strontium niobate crystals; and (d) urtropine crystals (arsenolith and iodine are similar).

event, for a purely harmonic variation of $G(z/a)$ in Fig. 2.20, Cahn *et al* [14] show that one expects to find $\Delta T_K^{**} = \Delta G_{max}/\Delta S_F$ and $\Delta T_K^* = \pi \Delta G_{max}/\Delta S_F$. Growth kinetic data for a variety of crystals grown from solution are given as a function of supersaturation, $\hat{\sigma} = \Delta C_K/C^*$, in Fig. 2.28.

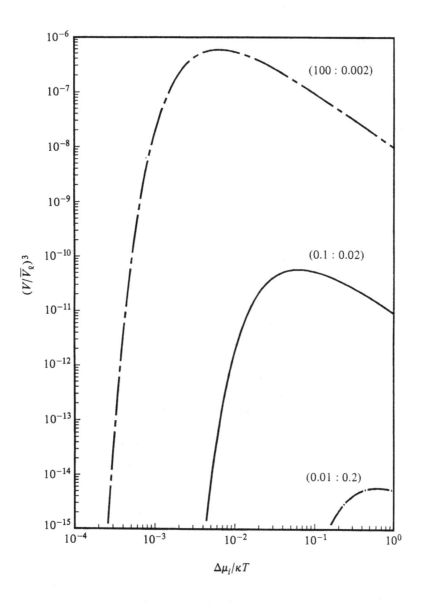

Fig. 2.29. Plot of the effective hillock slope, $V/\bar{V}_\ell$, as a function of interface driving force, $\Delta\mu_i/\kappa T$, for variations of the two key material parameters, $\alpha_2 = a\gamma_\ell^2/\kappa T$ (two-dimensional nucleation) and $\alpha_1 = \lambda_k/\beta_2\kappa T$ (lateral layer spreading), listed on the curves as $(\alpha_1 : \alpha_2)$.

2.2.4   Micro-hillocks

From Eqs. (2.25), one can see the interplay that occurs between the two-dimensional nucleation process and the lateral layer flow process. This interaction gives rise to micro-contouring of a facet surface centered about the two-dimensional nucleation sites and falling off in a conical fashion away from these centers. Of interest is the average angle $\bar{\theta} = \tan^{-1}(V/\bar{V}_\ell) \approx V/\bar{V}_\ell$ of the conical hillocks. If $\bar{\theta}$ is large, then the two-dimensional nucleation process is dominant; if $\bar{\theta}$ is small, then the layer spreading process is dominant. From Eq. (2.25d), we have

$$\tan\bar{\theta} = \frac{V}{\bar{V}_\ell} = \left(\frac{\pi a''^3 I}{\bar{V}_\ell}\right)^{1/3} \qquad (2.28a)$$

so that

$$\ell n\left(\frac{V}{\bar{V}_\ell}\right) = \frac{1}{3}\left[\ell n\left(\frac{a''^2 I_0 \lambda_k}{\beta_2 \Delta G_i}\right) - \frac{\pi a'' \gamma_\ell^2}{\kappa T \Delta G_i}\right] \qquad (2.28b)$$

Here, ΔG_K ranges from ~ 0, at a radius close to the critical radius for two-dimensional nucleation, to $\sim \Delta G_i$ at very large radii so that an average value was taken yielding $\bar{V}_\ell = \beta_2 \overline{\Delta G}_K \approx \beta_2 \Delta G_i/2$. In Fig. 2.29, a plot of $\ell n(V/\bar{V}_\ell)$ is given as a function of $\Delta\mu_i/\kappa T$ as the two key material parameters, $a\gamma_\ell^2/\kappa T$ and λ_k/β_2, are varied ($a \approx a'' \approx 2 \times 10^{-8}$ cm and $I_0 \sim 10^{28}$). As expected, (1) when λ_k/β_2 is small, there are many kinks and easy ledge activation so that $\bar{\theta}$ is very small and (2) when, $a\gamma_\ell^2/\kappa T$ is small, two-dimensional nucleation becomes very easy and $\bar{\theta}$ can grow to large values. For the screw dislocation case, $\bar{\theta}$ is already given from the geometry of the screw in that the radius of the second turn of the spiral is essentially R_1 and, from Eq. (2.25c), $V/\bar{V}_\ell \approx a/R_1$.

Problems

1. Consider that Fig. 2.5 refers to a pure face-centered cubic (FCC) crystal in contact with its vapor and we are looking at the (100) face, (010) ledges and (001) sidewalls. The heat of varporization for this material is $\Delta H_V = 100$ kcal mole^{-1}. Calculate the excess energy, ΔE_{ex}, for each delineated atomic state in Fig. 2.5, relative to an atom deeply buried in the crystal, using a first NN-only bond model. For a temperature of 1000 °C, what is the Boltzmann probability ($\exp(-\Delta E/kT)$) for finding states $(-1, 1)$ relative to state 0 and states $(5, 6)$ relative to state 4?

2. Consider the crystal–melt interface roughness for an FCC material with $\Delta H_F = 12.1$ kcal mole^{-1} and $T_M = 1412\ °C$. Based on a 3-state interface model, calculate the number of first, second and third NN bonds in the plane, above the plane and below the plane for the $\{111\}, \{100\}$ and $\{110\}$ planes. Use this data to calculate X_1^* for each of the planes at $T = T_M$ using a $(1,2,3)$NN bond-only model with $\varepsilon_2/\varepsilon_1 = 1/4$ and $\varepsilon_3/\varepsilon_1 = 1/27$. For each of these planes, calculate X'^*, for the two most densely packed ledges at $T = T_M$.

3. For the problem of question 2, calculate $\Delta F_R(X^*)$ for the $\{111\}$ plane and $\Delta F'_R(X^*)$ for the two densest ledges on the $\{111\}$ plane. Also determine the value of ΔF_R^M and $\Delta F'^M_R$ for the state of maximum roughening on this face and these ledges ($X^{*M} \sim X'^{*M} \sim 1/3$).

4. Referring to the results of question 3, assume that $-\Delta G_{fmax} = \Delta F_R^M - \Delta F_R(X^*)$ and $-\Delta G_{\ell max} = \Delta F_r'^M - \Delta F'_R(X'^*)$ and that each varies harmonically with distance with the period of a lattice spacing, either perpendicular to the face or ledge, respectively. Calculate the value of ΔT_i at which uniform ledge attachment and uniform face attachment occur during crystallization of this material from the melt.

5. Consider the example of problem 4 for the case of crystallization from the vapor at $T \sim T_M$ where $\Delta H_V = 100$ kcal mole^{-1} and where the volume supersaturation is increased by increasing the species vapor pressure (recall that $\Delta G_V = RT\ell n(P/P^*)$). Calculate the values of P/P^* at which uniform ledge attachment and uniform face attachment occur during crystallization from this vapor.

6. In questions 2–5, the material constants used refer to Si but an FCC structure was chosen for simplicity. Now consider the diamond cubic structure and see what differences you find. Note the important differences associated with this structure change.

7. For GaAs the crystal structure is zinc blende with $\Delta H_V = 150$ kcal mole^{-1}, $\Delta H_F = 16$ kcal mole^{-1} and $T_M = 1238\ °C$. What new energy states need to be listed in Fig. 2.5 for the system? What new information is needed to evaluate X_1^* and $X_1'^*$ for the important crystal faces? After specifying values for this information, solve question 2 for this system.

8. For Si in the FCC form ($\Delta H_V = 100$ kcal mole^{-1}, $\Delta H_F = 12.1$ kcal mole^{-1}, $T_M = 1412\ °C$), from a first NN-only bond picture at $T \sim T_M$, what vibrational entropy difference, ΔS_V, must exist between the Si(111) and Si(100) planes in order for the $\{100\}$-type plane to be

a facet plane in the equilibrium form? Do this for both a Si crystal in contact with its melt and with its vapor. Is this a reasonable value?

9. Redo question 8 but with the Si in the diamond cubic form.

10. Redo question 9 but for GaAs in the zinc blende form. ($\Delta H_V = 150$ kcal mole^{-1}, $\Delta H_F = 16$ kcal mole^{-1}, $T_M = 1238\,°C$).

3

Dynamic interface shape effects in bulk crystals

3.1   Kinetic versus equilibrium viewpoint

Often when one views a particular crystal shape, one wonders whether it has been determined by equilibrium considerations of minimum free energy or by kinetic considerations of terrace planes and layer flows or whether these considerations even lead to different results. To assess this matter, let us consider a diamond cubic (DC) crystal. From an equilibrium viewpoint, if γ_f refers to $\gamma_{\{111\}}$ and γ_ℓ is very small, the γ-plot will have no cusps so the equilibrium form will tend to be sphere-like. As γ_ℓ increases, the sphere will first change to sphere-like form with small flats at the $\{111\}$ poles and, as $\gamma_\ell \to \gamma_f$, the shape will change to a cube-like form with well-developed $\{111\}$ faces and [110]-type ridges. This picture will emerge independent of whether we adopt a continuum or a TLK model of the interface.

From the kinetic viewpoint, if we adopt a continuum view of interfaces, a crystal sphere with isotropic γ, growing under driving force conditions that fall into the UFA regime ($\Delta G_i > \Delta G_K^*$), will continue to grow as a sphere because the crystal surface is an isotherm under such conditions. Using the TLK picture in the small driving force regime with γ_ℓ small, the growing sphere becomes sphere-like with small flats at the $\{111\}$ poles for the generation of new layers, faint ridges form along the [110]-type lines where layer edges from two sources annihilate and faint nipples form at the $\{100\}$ poles where layer edges from four sources annihilate. As γ_ℓ increases, the interface roughening decreases and the

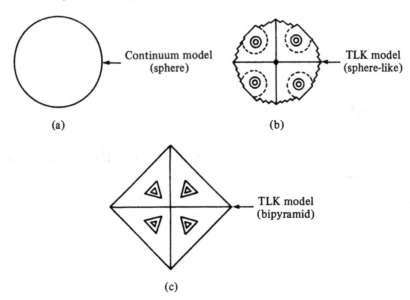

Fig. 3.1. Sphere-like growth forms: (a) continuum model or TLK model with extremely small γ_ℓ; (b) development of flats, ridges and nipples on a sphere based on the TLK model with small γ_ℓ; and (c) bipyramid growth form based on TLK model with intermediate to high γ_ℓ.

sphere-like form changes to cube-like with large {111} flats and strong [110]-type ridges because now a large driving force is needed to generate the new layers at the {111} poles. Thus, we see great similarities between the growth form and the equilibrium form for this system and note the transition from sphere-like to cube-like (bipyramid) character as γ_ℓ/γ_f increases (see Fig. 3.1). This great similarity develops because the layer source generation mechanism was considered to be two-dimensional nucleation. However, if the layer source mechanism is a physical defect like a twin as in Si, Ge, etc., then triangular or ribbon-shaped growth forms develop which are quite different from the equilibrium form. If the defect is a screw dislocation, then whiskers often develop and there is no doubt that the anisotropic kinetic factors largely determine the shape.

If we start with the spherical shape of Fig. 3.1(a), then, at high γ_ℓ but very small supersaturation, $\hat{\sigma}$, insufficient driving force exists for layer source creation on the {111} poles, so existing kinks and ledges move by atomic attachment. Flats begin to develop on the sphere by movement of the existing ledges and this process eventually leads to the equilibrium form which we shall assume to be the bipyramid form of

Fig. 3.1(c), with the original sphere being inscribed by the form, and then all the growth stops. For increasing $\hat{\sigma}$, a value of $\hat{\sigma} = \hat{\sigma}_c$ will be crossed at which the nucleation frequency on the flats becomes non-zero and the body becomes a slowly growing bipyramid. For large $\hat{\sigma}$, the face nucleation frequency becomes so high and the heat evolution so great that dendritic growth begins to develop at the corners (points) which are the regions of greatest ΔG_i. If γ_ℓ had been very small in this example, then, at small $\hat{\sigma}$, the nucleation frequency on the flats would have been moderate while the lateral edge growth rate would be high so the facet size would stay small and the body would remain somewhat spherical with [110]-type ridge lines.

To illustrate the problem associated with using Eq. (2.14), designed for a continuum surface viewpoint, to describe the TLK situation, consider Fig. 3.1(c) the growing bipyramid form. At the center of the faces, $\Delta\mu_i = \Delta\mu_K$ is needed to provide the initial triangular-shaped nuclei via a two-dimensional nucleation process so that the initial ledge energy is paid for by this value of $\Delta\mu_K$. As the triangular nuclei expand, a portion of the local $\Delta\mu_i$ goes to pay for the expanding ledge area while the remainder becomes $\Delta\mu_K$ (for no defect generation). As the ledges on adjacent {111} faces approach the ridge lines, they meet and extend both the area of the faces and the length of the ridge line; thus, the local $\Delta\mu_i$ must pay for this change in energy. From this description, we can see that, although the radius of curvature in the plane of the {111} face clearly enters the analysis, the radius of curvature in the perpendicular direction does not explicitly enter the analysis. In this perpendicular direction, no additional energy features beyond those already included as ledge–ledge and ledge–ridge interactions seem to be needed. This is perhaps not so surprising because layer flow is essentially a planar phenomenon and all the excess energy of the non-planar surface can be accounted for by ledge effects within each plane.

One of the interesting and important features that is apparent from the foregoing description of layer generation and layer flow on a sphere is the fact that a nipple develops at those orientations well-suited for dendritic growth. The growing sphere has pre-made distortions pointing in these important directions (as a natural consequence of the TLK picture) with the nipple amplitude growing as the overall sphere growth rate increases. It is through such symmetrical layer-fed shape distortions that the unique crystallographic features of dendrite growth directions are manifest. The break-away condition of unique dendrite development

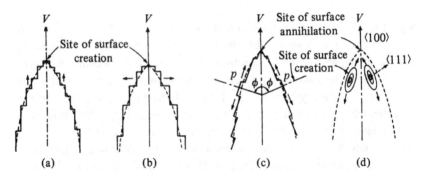

Fig. 3.2. Schematic illustration of constant curvature dendrite tips with different layer configurations and layer growth directions: (a), (b), (c) are plane curves while (d) is a body of revolution for the FCC system.

from the sphere awaits the critical value of ΔG_i for the local layer generation rate at that location to exceed that which feeds the main sphere.

The conflict between use of the curvature factor and the local instantaneous surface creation factor to determine the excess free energy involved in the development of various growth forms, is perhaps best illustrated by considering the growth of three cylindrical filaments of the same macroscopic curvature but different layer flow configurations as shown in Figs. 3.2(a)–(c). In Fig. 3.2(a), the tip region moves forward by layer flow in the axial direction providing elements of new surface that are being annihilated by adjacent steps moving in the axial direction. Thus, the new surface is being created at the tip and tends to be displaced laterally to the outer surfaces of the filament by the progressive and controlled flow of the adjacent layers to maintain the parabolic cylinder shape. In Fig. 3.2(b), the new steps are also being created in the tip region but they flow in a direction normal to the filament axis and eventually form the lateral surface of the macroscopic body. In Fig. 3.2(c), the surface creation occurs at the crystallographic poles p making angles ϕ with the filament axis. Surface annihilation of the ledges moving along the $(\pi/2 - \phi)$ directions takes place at the tip region of the filament. Flow of these ledges in the reverse direction forms the lateral surface of the macroscopic filament.

For those three cases, the local curvature is the same on a macroscopic scale; however, on a microscopic scale we have very different patterns of local surface creation for the same volume transformed. Thus, we must expect the dynamic interface temperature distribution to be very different. For Figs. 3.2(a) and (b), the lowest temperature occurs at the tip and increases with distance away from the tip. For Fig. 3.2(c), the temperature must be lowest at the ϕ-poles and must increase with distance away from these zones. Because of the ledge annihilation at the filament axis, free energy is being released in a fluctuation-type process at that location so the temperature will be highest there.

In Fig. 3.2(d), if we consider an FCC or a DC lattice and a paraboloid of revolution filament growing in the $\langle 100 \rangle$ direction, there are four-point sinks of free energy (sources of layer edges) at the $\langle 111 \rangle$ poles, one point-source of comparable strength (sink for layer edges) at the axial $\langle 100 \rangle$ pole and four line-sources of half strength (sinks for layer edges) along the [110]-type lines formed by the {111} intersections.

If we had adopted a continuum picture of the interface for the growing body of revolution filament, we would have noted that: (i) the excess energy of the crystal atoms is highest closest to the tip but that new surface creation is needed at the lateral regions of the filament to form the extending sidewall; (ii) the dynamic interface temperature is highest at the sidewall and lowest at the tip; and (iii) no reason is apparent for the sharply preferred crystallographic orientation observed for dendrites in nature. In Fig. 3.3(a), a schematic illustration is given of heat flow from a paraboloidal filament tip that is an isotherm (assuming $\Delta T_E = \Delta T_K = 0$). In Fig. 3.3(b), the heat flow portion for the continuum interface, non-isothermal paraboloidal filament is illustrated; i.e., additional heat is channeled along the filament axis by conduction from the hotter side walls to the tip of the filament. In Fig. 3.3(d), we see that the filament velocity, V, is greatly decreased (factor of ~ 5–10) by this very small temperature difference between the tip of the filament and the sidewalls ($\Delta T_i \sim 10^{-2}$ °C). Because the distance is so small ($\delta \ell \sim 10^{-4} - 10^{-6}$ cm), this gives local temperature gradients comparable to or larger than the macroscopic gradients and reduces the ability of the tip to remove latent heat of fusion at the exact tip center.

When we turn to the TLK picture of the filament interface (Fig. 3.2(d)), we note that (i) the dynamic temperature is highest at the tip and decreases towards the {111} poles, (ii) a sharply defined dendrite direction is developed as the one symmetrically located between the maximum number of operating layer source planes in the system that make the steepest angle to the filament axis other than 90° and (iii) sidewall surface creation develops by lateral movement of the ledges from the {111} sources. From Fig. 3.3(c) we note that the local heat flow at the filament tip is reversed to that in Fig. 3.3(b) so that the

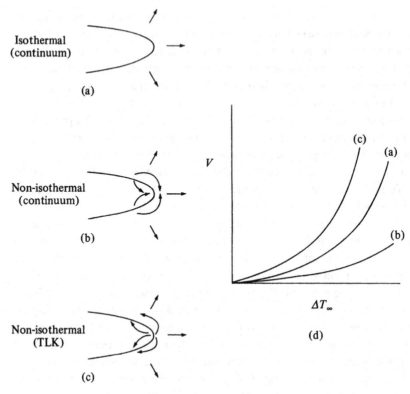

Fig. 3.3. Comparison of equal shape paraboloid of revolution filament growth forms: (a) isothermal filament based on the continuum model; (b) non-isothermal filament based on the continuum model; (c) non-isothermal filament based on the TLK model for the FCC system; and (d) dendrite growth velocity, V for these three filaments as a function of bath supercooling.

macroscopic temperature gradient in the liquid at the filament tip now has an increased capacity to conduct away latent heat of fusion and the V versus ΔT_∞ curve of Fig. 3.3(d) shows that such a non-isothermal filament will grow faster than the isothermal filament (Fig. 3.3(a)).

Once again we see that the TLK picture with the surface creation factor rather than the simple curvature factor is more in accord with experimental observations than the continuum picture and, here, we see that it restores important crystallographic features to macroscopic crystal forms. Because of the foregoing, it is useful to consider that perhaps the metallic systems and transparent organic systems that freeze with smooth surface contours at the optical microscopy level are really

TLK systems with atomic scale ledge heights and layers that have very little tendency to bunch. Certainly, for the TLK mechanism, as γ_ℓ goes to very small values one expects smoothly-rounded γ-plots and smooth interface morphologies at an optical resolution size scale. Although one may never be able to obtain an atomic scale resolution of a crystal–melt interface for metallic systems, there is certainly pedagogical merit in considering *all* crystals to grow via the TLK model with metallic systems occupying the domain where the ledges are very diffuse and where γ_ℓ is going asymptotically towards $\gamma_\ell = 0$. At least with such a picture, some crystallographic features would be restored to the crystallization process for all systems.

3.2 Adding surface creation to the equation

At this point, we must extend this surface creation picture to treat the general interface and develop a mathematical expression for the dynamic interface temperature, T_i. For dynamic shape changes during crystal evolution, it is necessary to account for local changes in surface energy associated with correlated local changes in the volume of nutrient phase transformed. It is this transformed volume that must be under-cooled or supersaturated sufficiently to provide the volume free energy change needed to compensate for the change in surface energy storage. This is completely analogous to the Gibbs–Thomson equation treatment of Chapter 2 (see Eqs. (2.11)–(2.13)).

For a moving interface, on a point-to-point basis, the key quantity of interest is the ratio of rates $\Delta \dot{G}_E / \Delta \dot{v}$ where $\Delta \dot{v}$ is the local rate of volume transformation per unit area of interface.[1] The local rate of surface free energy change per unit area of interface, $\Delta \dot{G}_E$, involves (i) creation of new layer edge, (ii) changes in the interaction of layer edges as the spacing changes and (iii) annihilation of ledges; i.e.,

$$\Delta \dot{G}_E = \delta \dot{G}_{C^*} + \delta \dot{G}_{I^*} + \delta \dot{G}_{A^*} \tag{3.1}$$

where the subscripts C^*, I^* and A^* refer to creation, interaction and annihilation, respectively. For a general surface, some regions are totally dominated by ledge creation, others by ledge interaction and still others by ledge annihilation. Defining s as a surface coordinate on a general cylindrical surface relative to some origin and θ as the angle of layer flow relative to a surface plane, the local dynamic interface temperature, T_i, for a pure system is given by

$$T_i(s,\theta) = T_0^*(s,\theta) - \Delta T_K(s,\theta) - \frac{\Delta \dot{G}_E(s,\theta)}{\Delta S_F \, \Delta \dot{v}(s,\theta)} \qquad (3.2a)$$

where ΔS_F is the entropy of fusion and T_0^* is the equilibrium temperature for that curvature. At those values of $s = s_{C^*}$ where new layers are being generated by two-dimensional nucleation, $\delta G_K = 0$ because the term is already accounted for by the $\delta \dot{G}_{C^*}$ contribution in $\Delta \dot{G}_E$. For those values of $s \neq s_{C^*}, s_{A^*}$ where layers are neither being created nor annihilated, δG_K is given as ΔG_K in terms of the local velocity via Eqs. (2.22). For those values of $s = s_{A^*}$ where layer edges are being annihilated, the released free energy may be utilized by the system to drive the attachment process and δG_K will be given by $\delta G_K < \Delta G_K$ utilized in Eqs. (2.22) (see below).

If we define our $\Delta \dot{G}_E$ and $\Delta \dot{v}$ to be on the same unit interface area basis, then Eq. (3.2a) becomes

$$T_i(s,\theta) = T_0^* - \Delta T_K - \frac{\Delta \dot{G}_E(s,\theta)}{\Delta S_F \, V(s,\theta)}. \qquad (3.2b)$$

For parts of the surface other than where new layers are being created (see Fig. 3.4), it is best to consider unit projected area normal to the local layer flow direction so that $V(s,\theta) = V_\theta(s)$; i.e., the layer edge velocity. However, for the regions of layer creation, one should use $V(s_{C^*},\theta) = V_{\theta+(\pi/2)}(s_{C^*})$; i.e., the layer nucleation velocity. When specific cases are dealt with later, the full power of this approach will be seen.

As a simple illustration for calculation purposes, let us consider the two-dimensional example of a small amplitude harmonic undulation from flatness of an interface between a crystal and its melt (see Fig. 3.4) for the three cases (a) the continuum approximation, (b) the TLK model where the flat surface is a facet plane and (c) the TLK model where the flat surface is a vicinal surface making an angle ϕ with the facet plane. The harmonic undulation will be defined by $z = z_0 + \delta \sin \omega x$. In all three cases, one sees periodic regions of positive and negative curvature; thus, the average curvature of the interface is zero and, if the Herring expression of Eq. (2.14) for a cylinder is used to account for the formation of surface, the average interface undercooling will be zero regardless of the amplitude δ of the wave. This fact reveals the inadequacy of using Eq. (2.14) to account for surface area changes during crystallization. Such undulations may develop during the unconstrained growth of a planar interface.

(a) To account properly for surface creation in the continuum case we

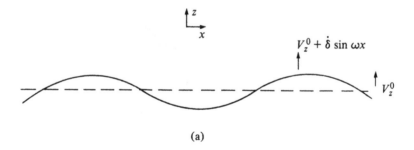

(a)

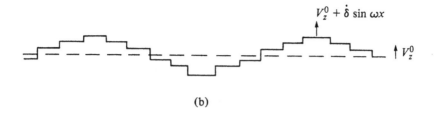

(b)

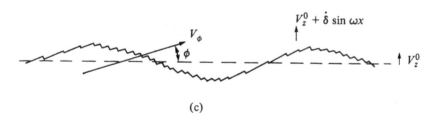

(c)

Fig. 3.4. Interface undulation development relative to a planar interface moving at velocity, V_z^0, based on several models: (a) continuum model; (b) TLK model with the terrace plane parallel to the planar interface; and (c) TLK model with the terrace plane making an angle ϕ with the planar interface.

recognize that, for a time increment dt and a surface element dx at x,

$$\frac{\Delta \dot{G}_E}{\Delta \dot{v}} = (\gamma_0 + \gamma_0'')\frac{\Delta s}{\Delta v} \tag{3.3a}$$

$$\approx (\gamma_0 + \gamma_0'')\frac{(\{1 + [z'(t+dt)]^2\}^{1/2} - \{1 + [z'(t)]^2\}^{1/2})dx}{[z(t+dt) - z(t)]dx} \tag{3.3b}$$

Since $z'(t) = \omega\delta\cos \omega x$ and $z'(t+dt) = \omega(\delta + \dot{\delta}\, dt)\cos \omega x$, insertion into

Eq. (3.3b) leads to

$$\frac{\Delta \dot{G}_E}{\Delta \dot{v}} = (\gamma_0 + \gamma_0'') \frac{(\omega^2 \delta \dot{\delta} \, dt \cos {}^2 \omega x + \cdots) dx}{(V_z^0 + \dot{\delta} \sin \omega x) dt dx} \tag{3.3c}$$

$$= (\gamma_0 + \gamma_0'') \frac{\omega^2 \delta \dot{\delta}}{V_z^0} \cos {}^2 \omega x \left(1 - \frac{\dot{\delta}}{V_z^0} \sin \omega x + \cdots \right) \tag{3.3d}$$

Inserting Eq. (3.3d) into Eq. (3.2a) yields

$$T_i = T_M - \Delta T_K(x,t)$$
$$- \frac{(\gamma_0 + \gamma_0'')}{\Delta S_F} \omega^2 \delta \left[\sin \omega x + \frac{\dot{\delta}}{V_z^0} \cos {}^2 \omega x \left(1 - \frac{\dot{\delta}}{V_z^0} \sin \omega x \right) + \cdots \right] \tag{3.3e}$$

The use of the approximation sign in Eq. (3.3b) relates to the fact that differential slices along x, rather than along the surface normal direction, are used. For small amplitude undulations these are essentially the same; however, for large amplitude undulations they are different and Eq. (3.3b) is not a good approximation for such a case.

If we now take the spatial average of T_i, we obtain the average surface creation contribution given by

$$\left(\frac{\Delta \dot{G}_E}{\Delta S_F \, \Delta \dot{v}} \right)_{Avg} = \frac{(\gamma_0 + \gamma_0'') \omega^2 \delta}{2} \left(\frac{\dot{\delta}}{V_z^0} \right) + \cdots \tag{3.3f}$$

We note the important dependence on $\dot{\delta}/V_z^0$ so that rapid local shape changes on a slowly drifting main interface have larger consequences than the same shape changes on a rapidly drifting interface.

(b) In Fig. 3.4(b), a great and abrupt change in undercooling is needed to allow the nucleation frequency to increase sufficiently to overcome the large area decrease associated with each undulation, relative to the flat surface, even for $\dot{\delta} = 0$. At the peak of the undulations we have, from Eqs. (2.24),

$$T_i = T_0^* - \frac{\pi a'' \gamma_\ell^2}{RT_0^* \, \Delta S_F \, \ell n[(V_z^0 + \dot{\delta})/2a'' X_c]} \tag{3.3g}$$

where $X_c \approx (a''/\omega^2 \delta)^{1/2}$ and V_z^0 is the velocity of the flat surface. The area A available for nucleation at the tip of each undulation is $2X_c$. Away from the peak of the undulations, in the positive half-cycle, the surface energy increases as x increases ($\gamma = \gamma_f + \gamma_\ell \mid \theta \mid + \cdots$ and $\theta \approx \omega \delta \cos \omega x$) and then decreases again with further increase of x in the negative half-cycle. If δ is sufficiently small that γ_ℓ is independent of ledge spacing at all θ, then $\Delta \dot{G}_E \approx 0$ and T_i is given by

$$T_i(x,t) \approx T_0^* - \Delta T_K(x,t) \tag{3.3h}$$

Of course, V_ℓ for the assumed shape may lead to a value of $\Delta T_K(x,t)$ via Eqs. (2.22) such that $T_i(x,t) \sim T_i(\lambda/4,t)$. For a fixed temperature relationship along the surface, the interface shape is likely to adjust away from the simple harmonic form. In the troughs, $\Delta \dot{G}_E$ is negative due to ledge annihilation so that V_ℓ can be expected to accelerate in this region even with ΔT_K small or zero.

For this case, if we evaluate the total rate of excess energy increase due to surface creation per unit area of interface, we find that

$$\frac{4}{\lambda} \int_0^{\pi/2} \Delta \dot{G}_E \, dx = \frac{2}{\pi} \left(\dot{\delta} \, \omega^2 \gamma_{\ell_0} + \frac{\pi}{2} \delta \dot{\delta} \, \omega^3 \gamma_\ell' \right) \tag{3.3i}$$

where

$$\gamma_\ell = \gamma_{\ell_0} + \gamma_\ell' \theta + \gamma_\ell'' \theta^2 + \cdots \tag{3.3j}$$

Thus, in this case, we are clearly accounting for the surface creation factor.

(c) For the case where the median plane of the undulations makes an angle ϕ with the layer plane, the layer edges are close enough to interact strongly and, using Eq. (3.3j), we find that

$$\Delta \dot{G}_E = \gamma_\ell \dot{n}_\ell + n_\ell \dot{\gamma}_\ell \approx n_\ell \dot{\gamma}_\ell \tag{3.3k}$$

$$\approx \left(\frac{\partial \gamma_\ell}{\partial |\theta|} \right)_\phi \omega \dot{\delta} \cos \omega x \tag{3.3l}$$

where n_ℓ is the number of ledges per unit distance perpendicular to the layer flow direction ($n_\ell = h^{-1}$ for layer height h) and this does not change as δ changes. So long as ϕ is not near a cusp in the γ-plot, Eq. (3.3j) should hold at any point on the interface and the total rate of excess energy increase due to surface creation per unit area of interface will be zero. However, on a point-to-point basis we have

$$T_i(x,t) = T_0^* - \Delta T_K(x) - \left(\frac{\partial \gamma_\ell}{\partial \theta} \right)_\phi \frac{\omega \dot{\delta} \cos \omega x}{\Delta S_F V_\phi} \tag{3.3m}$$

where $V + \dot{\delta} \sin \omega x = V_\phi \sin \phi$ and ΔT_K comes from Eqs. (2.22). In this case, the undulation drifts in the ϕ direction as the amplitude grows or shrinks, the sign of the effect depends upon the portion of the wave under consideration while the magnitude of the effect depends upon the velocity factor, $\dot{\delta}/V_\phi$, and the anisotropy factor, $(\partial \gamma_\ell/\partial \theta)_\phi$, which could be close to zero for metals. We note also that the ω dependence is linear here in contrast to the square dependence for Eq. (3.3e).

(d) For the opposite extreme, where the interface undulations have a large amplitude, δ relative to their wavelength, λ, with an amplitude growth rate of $\dot{\delta}$, surface creation is best accounted for by an average undercooling effect over the interface. This may be evaluated for unit cross-sectional area of crystal by equating the needed volume free energy change per unit time of the interface fluid with the energy stored in new interfacial area created during that period of time. For such a situation, this leads to

$$T_i = T_0^* - \Delta T_K - \frac{2\gamma\dot{\delta}}{\lambda\Delta S_F V} \tag{3.3n}$$

where λ is the average wavelength of the undulations in the x direction. This is essentially $(\dot{\delta}/V)$ times twice the static capillarity contribution present in T_0^*. Thus, $\dot{\delta}/V = 0.5$ essentially doubles the excess energy contribution present in T_i. If the interface is also undulating in the y direction, then an additional term must be added to account for the wavelength effects in that direction. When one neglects to include this $\dot{\delta}/V$ contribution, the value of $\gamma = \gamma_{eff}$ needed to match experiment and theory can vary within a factor of 2 for $\dot{\delta}/V \lesssim 0.5$ with one-dimensional undulations.

3.3   Si⟨100⟩ crystal pulling: moving interface case

We can now apply the foregoing considerations to the dislocation-free Si⟨100⟩ crystal pulling example discussed earlier for the static interface case and ask "Where are the layer sources for the interface and where is the growth ledge-attachment limited versus ledge-creation limited?" The layers needed for the growth of the interface are mostly generated in the small curved rim of interface at the outer edge of the crystal near the meniscus contact line and are located at the crystallographic positions for the basal planes (see Fig. 3.5(d)). As discussed earlier with the sphere example (Fig. 3.1), the layer flow from these sources can produce flat regions and perhaps peaked regions at the crystal rim. The larger is γ_ℓ, the more sluggish will be the layer generation kinetics and the larger will be this rim region on the interface (generally $\sim 1 - 50 \mu m$ in size). These layer sources generate the external cross-sectional shapes of the crystal (square, hexagonal, triangular, etc.) leading to a round crystal with flats and ridges along the length of the crystal. The {111} planes at the four poles make an angle $\phi = 54.74°$ with the (100) interface so the following kinetic conditions

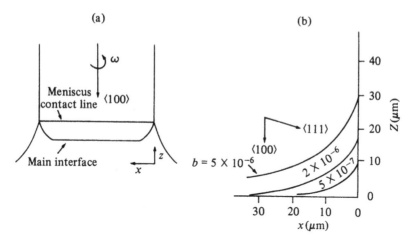

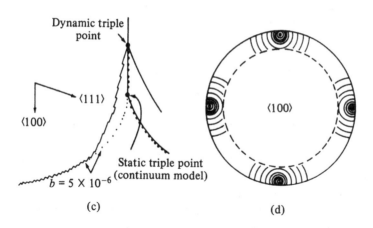

Fig. 3.5. The TLK model approach to the prediction of crystal shape from layer creation and flow in the meniscus region: (a) macroscopic view of interface; (b) continuum shape of static interface in the rim region (Fig. 2.17); (c) TLK side view of a facet region in the rim during growth; and (d) interface cross section revealing the location of the four {111} layer sources on Si(100).

must obtain at small driving forces

$$V_{(100)} = V_\ell \cos\left(\frac{\pi}{2} - \phi\right) = \beta_2 \sin\phi\ \Delta G_K(100) \tag{3.4a}$$

$$V_{(111)} = V_{(100)}\cos\phi = V_{2D} = a \int I_0 \exp\left[-\frac{\pi a'' \gamma_\ell^2}{RT\ \Delta G_i(x, Z_S)}\right] dA \tag{3.4b}$$

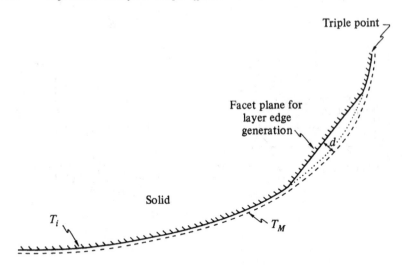

Fig. 3.6. Supercooled region, d, needed at the facet planes in the rim region for a sufficient layer generation rate to satisfy V.

where, on the facets, ΔG_i will be a function of position and dA is the incremental area. Beyond the edge of the facet, ΔT_K will be $\Delta T_K(100)$ and, in the center of the facet, $\Delta T_K \approx \mathcal{G}_n d$ (see Fig. 3.6) where d is the distance from the center of the facet to the freezing point isotherm and $\mathcal{G}_n$ is the local temperature gradient in the liquid normal to the facet. From this we can see that, as $\mathcal{G}_n$ decreases for the same V, the facet size must increase because d must increase. Likewise, as V increases for the same $\mathcal{G}_n$, the facet size must also increase (but not as strongly because of the exponential dependence in Eq. (3.4b)).

At the facet planes, the probability, p_f, of forming a stacking fault, rotation twin, etc., of some generic excess free energy, ΔG_F, will be given by

$$p_f \propto \exp[-(\Delta G_F - \Delta G_i)/\mathcal{R}T] \qquad (3.5)$$

This expression applies to fault formation anywhere on the interface but it is on the facet plane where ΔG_i is largest. Thus, the number of faults formed per unit time will increase where ΔG_i is largest and will increase as ΔG_i increases and ΔG_F decreases so that large V and small $\mathcal{G}$ lead to increased fault density. For the same reason, higher fault density is expected for growth from the vapor than for growth from the melt.

During crystal seeding and outward growth to the desired crystal diameter, the contour of the interface is often such that more growth facets will be displayed in the meniscus region than just those associated with

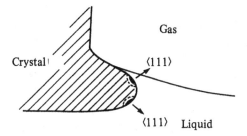

Fig. 3.7. Operating facet planes in the rim region during crystal diameter expansion after seeding.

one-half of the crystallographic sphere (see Fig. 3.7). For seeding of (111) crystal growth, this would lead to the presence of six ridges and flats on the initial crystal surface rather than the three one would expect for growth of a constant diameter crystal. We can also anticipate from Fig. 3.7 that p_f will be much higher during this seeding stage than for the constant diameter growth stage.

Experimentally, one finds that (i) the longitudinal flats or spines occur on the external surface at the same crystallographic orientations as these layer sources so they must arise by some variant of the mechanism illustrated in Figs. 3.5(c) and (d), (ii) the cross section of a Si(100) crystal becomes more circular as the crystal rotation rate, ω, is increased and (iii) the cross section of large diameter Si(111) crystals is triangle-like but with the presence of large flats or oval development on the crystal surface depending upon the amount and type of doping. We shall see that these are all products of the layer generation and flow process as influenced by temperature gradient changes and doping changes. In particular, these features are determined by the difference in freezing temperature of the triple-point junction between the facet location and the main rim region at off-facet locations.

If the isotherms are flat and T_i is exactly the same in the layer source region of the triple junction as on the rest of the rim region, a completely round crystal with no distinctive surface features should be observed. The formation of ridges and flats involves a difference in T_i between these two locations; i.e.,

$$\Delta T_F = T_i^{\ell s} - \bar{T}_i^R = -\left\{\frac{\gamma^{\ell s}\mathcal{K}^{\ell s} - \bar{\gamma}^R\bar{\mathcal{K}}^R}{\Delta S_F}\right\} - (\Delta T_K^{\ell s} - \Delta \bar{T}_K^R) - m_L(C_i^{\ell s} - \bar{C}_i^R) \tag{3.6}$$

Near the layer source (ℓs) locations, the liquid meniscus periodically attaches at a ledge corner and wets the terrace before breaking free and sliding down the terrace to attach at the next major ledge corner. The vertical velocity of these ledges is expected to be appreciable so that $\Delta T_K^{\ell s}$ may be significant. As one moves laterally from the layer source region around the rim (R), the meniscus makes contact first along a serrated face comprised of terrace and ledge with a gradually reduced fraction of terrace as one moves away from the layer source location until the contact is at the high-index orientation associated with the ledge. For such orientations the meniscus is not bound by a *corner* equilibrium but rather can slide easily down the serrated interface. Of course, one might expect $\gamma^{\ell s} < \bar{\gamma}^R$ since the former relates largely to the low-index orientation while the latter relates to the high-index orientation. Because the angle between the high-index orientation at the rim and the main interface is expected to be large, one might assume that $\Delta \bar{T}_K^R << \Delta T_K^{\ell s}$. Thus, the more sluggish the layer motion kinetics, the larger should be this contribution to the ridge formation on the crystal surface. When the combination of the three contributions in Eq. (3.6) leads to $\Delta T_F < 0$, an extra lifting of the meniscus occurs at the layer source regions leading to an outward bulge or ridge. When $\Delta T_F > 0$, an inward bulge or flat is the consequence. We are now in a position to explain the three experimental observations: (a) For pure dislocation-free Si, the presence of a ridge or a flat depends primarily on the size of the facet required to produce the needed nucleation frequency of two-dimensional pill-boxes relative to the radius of curvature of the ledges for the average rim material at the three-phase contact line. For a steep temperature gradient in the meniscus region, it is very likely that $\gamma^{\ell s} \bar{\mathcal{K}}^{\ell s} > \bar{\gamma}^R \bar{\mathcal{K}}^R$ even though $\gamma^{\ell s} < \gamma^R$ so the first term in Eq. (3.6) is negative and $\Delta T_F < 0$. This is because of the smaller radius of curvature at the facet. Thus, the inclined ledge flow from the source will lift the meniscus above that for the main rim material before it freezes so that a raised ridge of Si will be formed at these locations. It is interesting to note that, if screw dislocation generation occurs in the Si interface region by some mechanism, a more efficient layer source is present so the two-dimensional nucleation mechanism ceases to operate at some of the {111} poles. This causes the disappearance of a local ridge or ridges on the crystal surface and this ridge disappearance has been taken by the crystal growth operator to mean the loss of the dislocation-free state.

For a shallow temperature gradient in the meniscus region, the facet plane will become greatly enlarged in order to have sufficient under-

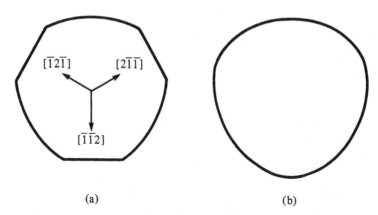

Fig. 3.8. Observed cross sections of large diameter Si(111) crystals:
(a) intrinsic and heavily B-doped; and (b) heavily Sb-doped.

cooling for the needed two-dimensional nucleation frequency. This leads to a reduction in $\gamma^{\ell s}\mathcal{K}^{\ell s}$ relative to $\bar{\gamma}^{R}\bar{\mathcal{K}}^{R}$ and allows the first term to become positive. Thus, the inclined ledge flow–meniscus contact region has a higher freezing temperature than the main rim because of its lower curvature due to the large facet. This leads to the development of a flat at the layer source location.

When the crystal diameter is being decreased rapidly, these layer source planes become exposed as inward-inclined facets. When the crystal diameter is being increased rapidly (as in seeding, etc.), the unused facet planes on the other half of the crystallographic sphere come into play and become exposed as outward-directed facets. For Si(100), these inward- and outward-directed flats lie along the same rib lines; however, for Si(111), the inward and outward-directed flats lie along symmetrical sets of rib lines rotated from each other by 60°.

(b) When the crystal rotation rate, ω, is increased, the convected heat flux from the liquid to the interface increases and this causes the main interface to recede to a position higher in the meniscus. At this location, convection is diminished but a larger interface $\mathcal{G}_n$ is needed to satisfy the local heat balance. This effect causes the layer source areas to decrease in size for the same V so the crystal becomes more rounded.

(c) During the growth of larger diameter Si crystals Lin and Hill[2] observed the crystal cross section to change with doping as illustrated in Fig. 3.8(b). For lightly doped or heavily B-doped crystals, the periphery of the crystal was observed to be somewhat triangular with three large flats in the ⟨2$\bar{1}\bar{1}$⟩-type directions. However, for heavily Sb-doped crystals, the outer crystal morphology was significantly changed and large peaks had developed to replace the large flats.

To grow large diameter crystals, one generally reduces ω and $\mathcal{G}$ for mechanical stability and thermal stress reasons, respectively. Thus, the layer source facet areas should be quite large and produce large surface flats for the reasons already indicated if the Si is pure. Although this is the expectation for lightly doped materials, the picture may be different for heavily doped alloys because the solute partitioning will be more restricted in the entire rim region than on the main interface. For the same reason, one expects $C_i^{\ell s} > \bar{C}_i^R$ for solute with $k_0 < 1$.

Since B has a k_0-value close to 1 in Si and should only affect $\gamma^{\ell s}$ or γ^R in a small way, a negligible solute redistribution effect is expected for B and the heavily doped Si should behave in essentially the same way as lightly doped Si ($C_i^{\ell s} \sim \bar{C}_i^R$). For Sb doping, the picture is quite different because $k_0 \sim 0.02$ for this case. Considerable solute enrichment will occur in the layer source region because of the upward movement of the layers, compared to the rim region where the layers are moving laterally, so that ΔT_F becomes dominated by the solute term. Thus, for Sb doping, $|\Delta T_F|$ is so large that the outward-inclined ledges keep growing and lift the meniscus far beyond that for the average rim material to generate the cross-sectional shape shown in Fig. 3.8(b).

Because the thermal fluctuations that reach the interface will be largest in the rim region, periodic melting and refreezing is likely to be largest in the rim region with the facet size alternately shrinking and growing to accommodate the need for layer sources. In this region, because of the fluctuations, one might find the local layer nucleation velocity to be an order of magnitude larger than the average and with a correspondingly larger facet size and interface supercooling. It is under this condition that one might expect twin formation to have a high probability.

In closing this topic, if one extends these considerations to a material like GaAs, which is thought to have a much larger value of γ_ℓ than Si, one would expect all the shape anisotropies such as surface ridge and flat formation to be much more pronounced than for Si.

3.4   Conditions for macroscopic facet development during constrained crystallization

An important question, that has not been given sufficient attention by the crystal growth community, is whether or not a moving

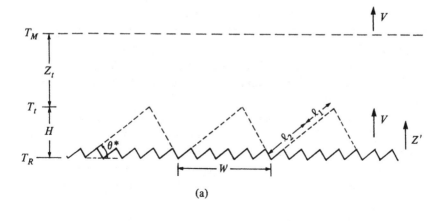

(a)

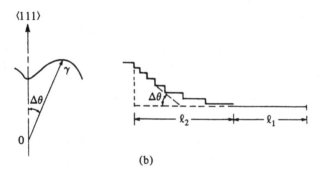

(b)

Fig. 3.9. (a) Macroscopic facet development (width W and height H) for an interface growing at V; and (b) γ-plot and surface ledge spacing change from the root to the tip of a macroscopic facet.

interface may facet in a different fashion than a static interface. If one includes the $\Delta G_K(V, \theta)$ contribution associated with the movement at velocity V of an interface of orientation θ relative to a facet plane of the γ-plot, and one also includes the dynamics of terrace and ledge reconstruction as a function of V_ℓ to yield $\gamma = \gamma(\theta, h, V_\ell)$ it would seem that one could make the definition

$$\gamma_{eff}(V, \theta) = \gamma(\theta, h, V_\ell) + \Delta G_K(V, \theta) \qquad (3.7a)$$

and one could draw a $\gamma_{eff}(V, \theta)$-plot. Thus, one should now expect facet formation to occur following an expansion of the rules described earlier in Eqs. (2.10); i.e., faceting will occur in the dynamic situation when

$$\gamma_{eff} + \frac{\partial^2 \gamma_{eff}}{\partial \theta^2} < 0 \qquad (3.7b)$$

Thus, the anisotropy of the interface attachment mechanism may combine with the anisotropy of the interfacial energy either to strengthen the static facet system or to generate a new facet system. This is likely to be especially true for the cases exhibiting very little anisotropy in $\gamma(\theta)$. These considerations should apply equally well to unconstrained or constrained crystallization.

A second long-standing question has concerned whether or not a vicinal or high-angle interface will grow with monomolecular high ledges or whether macroscopic size facets might develop under certain conditions. This latter situation is illustrated in Fig. 3.9(a) for the two-dimensional case where facets of width W and height H are presumed to develop during constrained growth of a Si(100) interface at a velocity V. Such macroscopic $\{111\}$ facet development should occur if the system free energy can be lowered while still satisfying the attachment kinetic requirements. For simplicity we will neglect distortions of the isotherm shape due to the macro-facet formation (not correct at large V). Thermodynamically this process is expected to proceed provided that

$$W\gamma_{100} > \frac{W}{\cos\theta^*}\gamma_{111} + \Delta\gamma_{110}^{\hat{R}} - \Delta S_F \int_0^H (\Delta T_K^{\hat{R}} - \mathcal{G}_L z') \left(\frac{H-z'}{2\tan\theta^*}\right) dz' \tag{3.8a}$$

where $\Delta\gamma_{\{110\}}^{\hat{R}}$ is the energy change involved in forming the macroscopic $\{110\}$ ridge, and $\Delta T_K^{\hat{R}}$ is the value at the root of the facet. It is the elimination of the supercooling between the macroscopic and microscopic facets that largely drives the process. Evaluating the integral, we have the condition

$$\gamma_{100} - \frac{\gamma_{111}}{\cos\theta^*} - \frac{\Delta\gamma_{100}^{\hat{R}}}{W} \gtrsim -\frac{\Delta S_F}{16}W\tan\theta^*\left(\Delta T_K^{\hat{R}} - \frac{\mathcal{G}_L W \tan\theta^*}{6}\right) \tag{3.8b}$$

Since $\theta^* = 70°53'$ for Si, $\gamma_{100}/\gamma_{111} \sim 1.4(\gamma_{100} \sim 1450$ erg cm^{-2} and $\gamma_{111} \sim 1000$ erg cm^{-2} for the solid–vapor case) and $\gamma/\Delta S_F \sim 1.2\times10^{-5}$, the RHS of Eq. (3.8b) must be negative so that $\Delta T_K^{\hat{R}} > \mathcal{G}_L W \tan\theta^*/6$ is required. This will be true for sufficiently small W. Thus, the minimum value of W is found by neglecting $\mathcal{G}_L W \tan\theta^*/6$ with respect to $\Delta T_K^{\hat{R}}$ and this leads to $W_{min} \sim 10^{-4}/\Delta T_K^{\hat{R}}$ cm. We thus see that, from a purely thermodynamic viewpoint, values of W in the micron range are allowed and these will occur when V is large and $\mathcal{G}_L$ is small.

Now we must assess the kinetic stability of such a macro-facet configuration. There are two prime kinetic mechanisms for converting excess energy effects into available driving force for attachment: (i) layer edge annihilation at the tip which provides the extra driving force needed over the length ℓ_1 (see Fig. 3.9(a)) for the ledges to move at the average velocity $\bar{V} = V/\sin\theta^*$; and (ii) surface contour change as in Fig. 3.9(b) so that γ_ℓ decreases as the edge moves along the facet plane over the length ℓ_2 and this change $\Delta\gamma_\ell$ is converted to driving force for attachment so that a steady state layer flow at the average velocity $\bar{V} = V/\sin\theta^*$ is possible.

The local undercooling for molecular attachment at position s along the facet is thus given by

$$\Delta T_K(s) \approx \Delta T_K^{\hat{R}}(0) - \mathcal{G}_L s \sin\theta^* + \frac{1}{\ell_2 \Delta S_F} \int_0^{\ell_2} \left(\frac{\partial\gamma}{\partial\theta}\right)\left(\frac{\partial\theta}{\partial s}\right) dx$$
$$+ \frac{\gamma_\ell}{\ell_1 \Delta S_F} + \Delta T_{NP} \qquad (3.8c)$$

In Eq. (3.8c), ΔT_{NP} is the temperature change due to non-planar isotherms while the third and fourth terms are averages. To produce a sharp, well-developed facet, the average ledge velocity must be greater than or equal to its value at the root, $s = 0$. Thus, neglecting ΔT_{NP}, we require that

$$-\mathcal{G}_L \sin\theta^* \frac{(\ell_1 + \ell_2)}{2} + \frac{1}{\ell_2 \Delta S_F} \int_0^{\ell_2} \left(\frac{\partial\gamma}{\partial\theta}\right)\left(\frac{\partial\theta}{\partial s}\right) dx + \frac{\gamma_\ell}{\ell_1 \Delta S_F} \gtrsim 0 \quad (3.8d)$$

The value of the integral will be slightly less than γ_ℓ, but, for simplicity, we can approximate it as γ_ℓ which leads to $\ell_2 \approx \ell_1$ so that Eq. (3.8d) yields

$$\ell_1 \approx \ell_2 \approx \left(\frac{\gamma_\ell}{\mathcal{G}_L \sin\theta^* \Delta S_F}\right)^{1/2} \qquad (3.8e)$$

Since $\gamma/\Delta S_F \sim 1.2 \times 10^{-5}$ for Si and $\gamma_\ell/\gamma \sim 0.1$ say, we see that $\ell_1 + \ell_2$ will be in the 1–10 μm range for $\mathcal{G}_L$ in the 100–1 °C/cm^{-1} range. Thus, W again increases as $\mathcal{G}_L$ decreases.

Since both the thermodynamics and the kinetic stability conditions lead to predictions for W in the same range, we can conclude that these effects are capable of stabilizing facets in the size range equal to or greater than 1 μm. Such considerations will become important in discussions of micro-segregation traces, interface stability and rapid laser melting and regrowth of thin Si layers.

3.5   Kinematic theory of ledge motion for growth or dissolution

3.5.1   *Ledge spacing dependence only*

Although the Frank[3] kinematic theory holds in three dimen-

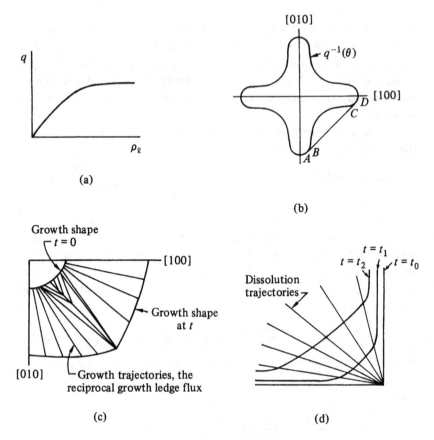

Fig. 3.10. Ledge kinematics: (a) ledge flux, q, versus ledge density, ρ_ℓ; (b) reciprocal growth flux, q^{-1}, as a function of orientation for a cubic crystal; (c) crystal interface profile as a function of time illustrating that orientations B and C in (b) have almost completely disappeared; and (d) a crystal interface dissolution profile.

sions, for simplicity we shall consider only a two-dimensional case corresponding to the motion of straight ledges on a vicinal surface (see Figs. 2.10, 2.21). With λ_ℓ the step spacing and h the step height, we call $\rho_\ell = \lambda_\ell^{-1}$ the step density on the surface leading to a surface slope of $\partial z / \partial y = h \rho_\ell$ and a step flux q of V/h (see Fig. 2.21). If $q = q(\rho_\ell)$ only, as illustrated in Fig. 3.10(a), the growth or dissolution of the surface can be simply analyzed based on two important velocities, $v(\rho_\ell) = q/\rho_\ell$, the ledge velocity and $c(\rho_\ell) = dq/d\rho_\ell$, the "kinematic wave velocity" of the ledges.

The conservation equation for steps (continuity equation)

$$\frac{\partial q}{\partial y} + \frac{\partial \rho_\ell}{\partial t} = 0 \tag{3.9}$$

yields a step pattern where, in the (y, t) plane projection, ρ_ℓ is constant and q is constant along a line of slope c which is called a "characteristic"; i.e., the trajectory in space of a particular orientation of a crystal surface is a straight line for either growth or dissolution. Thus, if one plots the polar diagram of V^{-1}, the trajectory of a point of given orientation will be parallel to a normal to this polar diagram at the point of corresponding orientation. Therefore, as long as the assumption of $q = q(\rho_\ell)$ only is valid, knowing the plot of $q(\rho_\ell)$ one can also plot the dissolution or growth profile of an arbitrary crystal shape as a function of time and vice versa. Typical plots of q^{-1} and the corresponding growth shape are shown in Figs. 3.10(b) and (c) for a cubic crystal. The equivalent dissolution shape is shown in Fig. 3.10(d).

From this approach it follows that a corner, or discontinuity in slope from $(h\rho_\ell)_1$ to $(h\rho_\ell)_2$, follows a trajectory in the (y, t) plane of a shock wave with velocity

$$\frac{dy}{dt} = \frac{q_2 - q_1}{\rho_{\ell 2} - \rho_{\ell 1}} \tag{3.10}$$

Thus, if a fluctuation occurs to produce a "bunching" of the ledges into a zone of higher density than the remainder of the surface, the ledges at the trailing edge of the moving bunch will converge while those at the leading edge will diverge so that the bunch will die out after a long time provided $d^2q/d\rho_\ell^2 < 0$. In this case, the individual ledges continuously move through the bunch or shockwave as growth or dissolution proceeds. In the case where $d^2q/d\rho_\ell^2 > 0$, a sharp corner will develop at the leading edge while the trailing edge will be rounded as illustrated in Fig. 3.11. Thus a macroscopic ledge or etch pit with a clearly defined leading edge can only be stable if $d^2q/d\rho_\ell^2 > 0$. For an evaporating surface with small values of $d\lambda_\ell/dy$, the calculated flux q due to the surface adsorbate population near the ledges (Fig. 3.10(a)) is such that $d^2q/d\rho_\ell^2 < 0$ for all ρ_ℓ so that macroscopic ledges should not form during evaporation.

3.5.2 *Other factors influencing ledge motion*

The foregoing kinematic theory requires that $q = q(\rho_\ell)$. However, when $d\lambda_\ell/dy$ is large, q will depend upon both ρ_ℓ and $d\rho_\ell/dy$ so that, near an initially sharp corner, $q \neq q(\rho_\ell)$ solely and the above form of the kinematic theory does not hold. If q is time-dependent as it would be when capillarity is involved (the ΔG_E term is important), again the

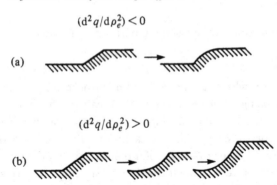

Fig. 3.11. Dissolution profiles at a discontinuity in ledge spacing for two different functions $q(\rho_\ell)$.

theory is not fully applicable. In the general case where q will depend upon $\Delta G_E, \Delta G_K$ and ΔG_{sv}, we cannot expect this simple approach to be fully applicable. However, it does provide insight and is thus useful for our overall perspective. In closing this section it is important to note that time-dependent impurity adsorption will lead to a decrease in ledge velocity and to piled-up bunches of ledges. The presence of stacking faults or other imperfections in the growing surface is also expected to alter the q–ρ_ℓ relationship and lead to macroscopic step formation.

3.6   Adsorption and adsorbate trapping effects

When a solute crystallizes from its supersaturated solution, the presence of a third component, j, can often have a dramatic effect on both the *crystal growth kinetics* and the crystal habit even at extremely low concentrations ($10^{-9} < X^j < 10^{-3}$) suggesting that an adsorption mechanism is operating. Following Davey,[4] the reaction path for the crystallizing unit can be considered to be as illustrated in Fig. 3.12(a). A solvated unit (shown as a six-coordinate species) arrives at the crystal surface and is adsorbed onto it (position 1). Surface migration then occurs followed by desorption back into the solution or, as a step is reached, attachment at the step (position 2) and incorporation into the crystal lattice at a kink (position 3). The activation free energy barriers associated with each step in the process are shown in Fig. 3.12(b). As illustrated in Fig. 3.12(c), impurity adsorption can occur at kink, ledge or face sites. Adsorption isotherm results for $j = Cd^{2+}$ ions onto the (100) and (111) faces of NaCl crystals revealed only three types of adsorption sites: (i) one of high energy which is occupied almost instantly

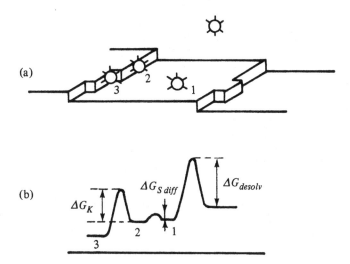

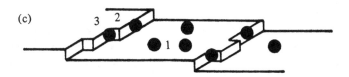

Fig. 3.12. Adsorption effects: (a) key surface structure elements; (b) reaction path for a growth unit; and (c) adsorption sites for an impurity species.

for mole fractions $X^j \sim 10^{-9}$, (ii) one of intermediate energy and one of low energy, both in the range $10^{-9} < X^j < 10^{-4}$. Using a first NN-only bond model with an adsorption energy of ε_a per bond, on the (100) faces the fractions of the face, ledge and kink sites occupied by adsorbate are

$$X_f^j = (C_\infty^j h_{af}/N_0) \exp[(\epsilon_a - T\Delta S_{af})/\kappa T] \qquad (3.11a)$$

$$X_\ell^j = (C_\infty^j h_{af} h_{a\ell}/N_\ell) \exp[(2\epsilon_a - T\Delta S_{a\ell})/\kappa T] \qquad (3.11b)$$

and

$$X_k^j = (C_\infty^j h_{af} h_{a\ell} h_{ak}/N_k) \exp[(3\epsilon_a - T\Delta S_{ak})/\kappa T] \qquad (3.11c)$$

where N_0 is the number of adsorption sites per square centimeter on the (100), N_ℓ is the number of adsorption sites per centimeter on an (010) ledge, N_k is the number of kink sites along a (010) ledge per centimeter of length, h_{aq} is the width of the adsorption layer under consideration ($q = f, \ell$ or k) while ΔS_{aq} is the entropy change as-

(a)

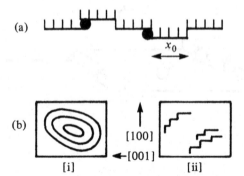

(b) $\uparrow$
                      [100]
                   $\leftarrow$[001]

[i]   [ii]

Fig. 3.13. (a) Impurity adsorption at kink sites and (b) step bunches on the {100} faces of $(NH_4)_2SO_4$ for (i) pure solution and (ii) solution with 5 ppm Cr^{3+}.

sociated with the adsorption for the three states under consideration. There is often a tendency to neglect the ΔS_{aq} contribution; however, in many cases, it can be a very significant factor. ΔS_a involves an entropy of mixing change plus a vibrational entropy change. Let us just consider the entropy of mixing change for the case where the solution concentration is $j \sim 1$ ppb $\sim 10^{13}$ atoms cm^{-3}, the solution temperature is 500 °C, $h_{af} \sim 2 \times 10^{-8}$ cm, $N_0 \sim 2 \times 10^{15}$ cm^{-2} and $\epsilon_a^j = 1.5$ eV. If we neglect the ΔS_{af} contribution, then Eq. (3.11a) leads to $X_f^j \sim 0.5$. Thus, we wish to calculate the ΔS_{af} mixing change on a per atom basis associated with a surface coverage change from $X_f^j \sim 10^{13} \times 2 \times 10^{-8}/2 \times 10^{15} = 10^{-10}$ to $X_f^j \sim 0.5$. From the simple entropy of mixing formula $\Delta S_{af} = 23\kappa - 1.38\kappa \sim 21.6\kappa$ so that $\Delta S_{af}T \approx 1.44$ eV. This is comparable to the ϵ_a contribution and thus significantly affects the numerical values of X_f^j. The point to be made here is that at least the entropy of mixing contribution in ΔS_{aq} should be taken into account when considering the quantitative aspects of interface adsorption.

In addition to the entropy effect, let us further suppose that the adsorbate is $b\%$ larger in radius than the crystal molecule so stored strain energy is developed in the crystal as the j species are progressively walled into the crystal. We shall denote the excess energy as $\Delta E_b/3$ for each coordinate direction that is walled-in. This occurs first by completion of a row, next by coverage with the adjacent row and finally by coverage with the next layer. Now, let us consider the specific adsorption sites in a little more detail.

3.6.1 Kink site adsorption

Kinks are spaced at distance λ_k in pure solutions and kink adsorption illustrated in Fig. 3.13(a) has two effects: (1) solute entry into the kink requires desorption of the adsorbate so that the activation energy for such kink site entry is increased to $\Delta G_K + \Delta G^j$ where the ΔG^j is for impurity desorption which may depend on the fraction of decorated sites; and (2) new pseudo-kink sites are created at which a growth event would require walling in a j species. In this case, the free energy for kink site entry is $\Delta G_K + \Delta G_p$. The effect of (1) alone is to increase λ_k while the importance of (2) depends upon the parameter $p =$ lifetime of adsorbed species/relaxation time for solute entry into a pseudo-kink site. If $p > 1$, such kinks will play a role in the growth process and λ_k will be decreased from its pure solution value. If $p < 1$, the pseudo-kinks can be neglected and only (1) need be considered. This latter case is the most common result.

Generally, kink site adsorption causes λ_k to increase so that the ledge velocity, V_ℓ, decreases. This increase in λ_k causes polygonization (facet development) of previously non-polygonized growth steps since, for the ledge velocity, V_ℓ to be independent of crystallographic orientation, every growth unit arriving at a ledge must be able to enter a kink; i.e., the mean surface diffusion distance of a solute species, $\bar{X}_s$, must be much greater than λ_k. However, impurity adsorption that increases λ_k can create a situation where $\bar{X}_s < \lambda_k$ so that V_ℓ will be limited by kink site density resulting in polygonization (see Fig. 3.13(b)).

When pseudo-kink sites play a role, ΔG_K will still have its pure solution value and λ_k will decrease. This causes V_ℓ to increase and polygonized steps to become unpolygonized. Fig. 3.14 shows such an effect for the growth of (100) sucrose faces in the presence of raffinose. In pure solution, the step bunches were polygonized perpendicular to the [010] direction while 0.5% raffinose added to the solution produced rounding of the bunches in this direction. Here, raffinose can be considered as a substituted sucrose molecule (raffinose = sucrose + D-galactose residue) so the mechanism of pseudo-kink formation could be used to describe the observation. However, raffinose causes a simultaneous decrease of V_ℓ which would indicate that there are also other factors involved. Of course, this might just be a ΔG_E effect which would reduce the effective driving force available for kink site attachment so V_ℓ will decrease if the ΔG_E effect outweighs the $\Delta \lambda_k$ effect. We might expect that the growth of Si from Sn or Ge solutions or the growth of 3–5 compounds from alloy solutions leads to pseudo-kinks.

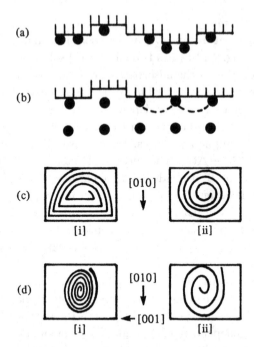

Fig. 3.14. Step site impurity adsorption: (a), (b) proposed models; (c) step bunches in sucrose {100} faces for (i) pure solution and (ii) solution with 0.5% raffinose; and (d) step bunches on $NH_4H_2PO_4${100} faces for (i) pure solution and (ii) 10 ppm added $CrCl_3 \cdot 6H_2O$.

Defining V_∞ as the limiting ledge velocity in an impure solution and V_0 as the ledge velocity in a pure solution, the actual ledge velocity is expected to depend upon adsorbate coverage θ in the following way

$$V_\ell = V_0 - (V_0 - V_\infty)\theta \qquad (3.12a)$$

When θ is given by a Langmuir isotherm, Eq. (3.12a) becomes

$$\frac{V_0 - V_\infty}{V_0 - V_\ell} = 1 + \frac{1}{KX_\infty^j} \qquad (3.12b)$$

where X_∞^j is the mole fraction of impurity in the solution and $K = \exp(-\Delta G_a/\mathcal{R}T)$ where ΔG_a is the free energy of adsorption of the impurity on the ledge. Good fit to Eq. (3.12b) was found for the growth of the {111} faces of NaCl and KCl in the presence of Cd^{2+} and Pb^{2+} which led to $\Delta G_a = -11.2$ and -9.9 kcal mole^{-1}, respectively. Since the {111} faces of these salts are denoted as K-faces, these values must be related to kink site adsorption which is consistent with the kink site ad-

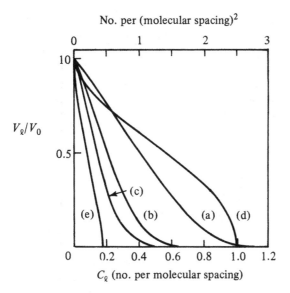

Fig. 3.15. Calculated adsorption at a step: (a), (b), (c) from Albon and Dunning[5] model with $2r^* = 2, 6, 10$ molecular spacings while (d)(c) are from Cabrera and Vermilyea[6] model with $2r^* = 2$ and 5.

sorption of Cd^{2+} on $NaCl(100)$. In this latter case, the adsorption states were found to be -11.2 kcal mole^{-1} for the kink sites, -8.7 kcal mole^{-1} for the intermediate level and -6.2 kcal mole^{-1} for the low energy sites.

3.6.2 Ledge site adsorption

For a ledge to advance between adsorbed impurity species, each separate length of step must "bow out" and increase its curvature so as to reduce the step velocity according to the relationship

$$V_\ell = V_0 \left(1 - \frac{r^*}{r}\right) \tag{3.13a}$$

where r^* is the radius of the critical nucleus and r is the radius of the curved step. Impurity adsorption onto the ledge reduces the effective length of ledge along which kinks can form and may provide steric barriers for the entry of growth units. When the separation of impurity atoms along the ledge is less than $2r^*$, V_ℓ must go to zero via Eq.(3.13a). For the general case where C_ℓ is the concentration of adsorbed species in number per molecular spacing along the ledge, we have

$$V_\ell = V_0[2r^* - (1 - C_\ell)2r^* + (1 - C_\ell)](1 - C_\ell)^{2r^*} \tag{3.13b}$$

Fig. 3.15 shows this V_ℓ/V_0 versus C_ℓ relationship for four different values

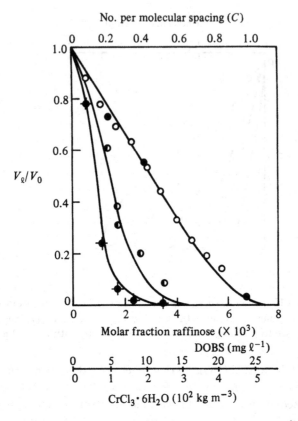

No. per molecular spacing (C)

Molar fraction raffinose ($\times 10^3$)

Fig. 3.16. Data match to the Albon and Dunning[15] model: (0) sucrose in the presence of raffinose at $\hat{\sigma} = 0.02$, (◆) same at $\hat{\sigma} = 0.0065$, (●) $Na_5P_3O_{10} \cdot 6H_2O$ in the presence of DOBS and (◐) same in the presence of $CrCl_3 \cdot 6H_2O$.

of r^*. By assuming that C_ℓ is related to X_∞^j via a Freundlich isotherm ($C_\ell = K(X_\infty^j)^{1/n}$ with $n \approx 1$), this model gives a good fit to the experimental data for the [010] step bunch velocity on the (100) faces of sucrose in the presence of raffinose. In Fig. 3.16, we see that good correlation also occurs for the [001] ledge bunch velocity on the (100) faces of ammonium dihydrogen phosphate (ADP) in the presence of $CrCl_3 \cdot 6H_2O$ and for the data on the $\{0\bar{1}0\}$ face of $Na_5P_3O_{10}$ in the presence of sodium dodecylbenzene sulfate (DOBS).

It was observed by Davey that 15 ppm of $CrCl_3 \cdot 6H_2O$ increased the step bunch spacing on the (100) faces of ADP from 33 μm to 100μm while 50 ppm $CrCl_3 \cdot 6H_2O$ inhibits entirely the production of step bunches

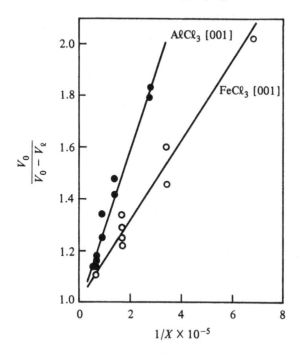

Fig. 3.17. Data for ATP in the presence of FeCℓ_3 and AℓCℓ_3.

(Fig. 3.14(d)). Assuming that the creation of step bunches depends on cooperation with dislocation groups, it follows that a necessary condition for step bunch creation is that the separation, d, of dislocations within the group must be less than the diameter of the critical nucleus; i.e., $d < 2r^*$. For impurity adsorption at a step, γ_ℓ decreases so r^* decreases and, although $d < 2r^*$ for pure solutions may hold, we may have $d > 2r^*$ for impure solutions so that cooperation within the dislocation group ceases and visible step bunches are no longer created.

3.6.3   Face adsorption

Impurity adsorption to the face reduces the flux of crystallizing units to the ledge so that V_ℓ decreases. If the adsorbed j species concentration is high and its lifetime is much greater than the lifetime of the adsorbed crystallizing units on the surface, the mean surface diffusion distance, $\bar{X}_s$, may be decreased. Thus, ledges which are polygonized in pure solution $(\bar{X}_s < \lambda_k)$ will remain polygonized while steps which were non-polygonized $(\bar{X}_s > \lambda_k)$ in the pure solution may become polygonized in the impure solution.

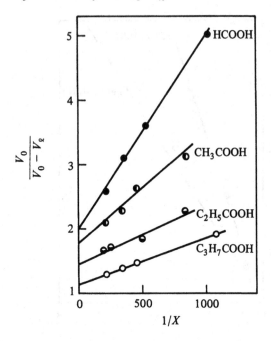

Fig. 3.18. Growth of KBr in the presence of HCOOH, CH_3COOH, C_2H_5COOH and C_3H_7COOH.

If the number, n_s, of adsorbed growth units is directly proportional to the fraction of free surface, the step velocity is

$$V_0 = 2f_0(a/t_k)(n_s - n_0) \tag{3.14a}$$

where n_0 is the number of growth units adsorbed at equilibrium, t_k is the relaxation time to enter a kink, a is the distance between neighboring growth units in the crystal and f_0 is the area occupied by a growth unit in the crystal. The values of n_s and n_0 are then reduced by the factor $1-\theta$ in the presence of an adsorbed impurity and, if the surface coverage is given by a Langmuir isotherm, we have

$$\frac{V_0}{V_0 - V_\ell} = 1 + \frac{1}{KX_\infty^j} \tag{3.14b}$$

This result is similar to Eq. (3.12b) except that it predicts a zero ledge velocity at $KX_\infty^j \to \infty$ while Eq. (3.12b) allows for cases where a limiting value (V_∞) is attained. Fig. 3.17 shows data for ledge advance in the [001] on the (100) faces of ADP conforming to Eq. (3.14b) while Fig. 3.18 shows data conforming to Eq. (3.12b).

The slopes of the lines in Fig. 3.17 give free energies of adsorption

Table 3.1. *Estimated ledge free energies from growth kinetics*

Material	Relative $\hat{\sigma}$	γ_ℓ(erg cm^{-1})
Sucrose {100} faces	0.0065	4.5×10^{-8}
	0.020	1.3×10^{-8}
ADP {100} faces	0.052	9.3×10^{-8}
$Na_5P_3O_{10} \cdot 6H_2O$ {01$\bar{0}$} faces	0.8	4.5×10^{-7}

Table 3.2. *Estimated values of adsorption free energies*

System (aq. solution)	Impurity	T (°C)	$\hat{\sigma}$ ($\Delta C/C$)	ΔG^a (kcal mole^{-1})
ADP	FeCℓ_3	23.96	0.052	-3.8
	AℓCℓ_3	23.96		-3.7
KBr	Phenol	26	0.046	-2.6
	HCOOH	19	0.046	-2.8
	CH$_3$COOH	19		-3.4
	C$_2$H$_5$COOH	19		-3.8
	C$_3$H$_7$COOH	19		-4.1
KCℓO$_3$	Ponceau 3R	20	–	-3.6
NaCℓO$_3$	Na$_2$SO$_4$	17	0.035	-3.2

of -7.6 and -7.4 kcal mole^{-1} in the presence of FeCℓ_3 and AℓCℓ_3, respectively. Since both negative and positive ions are involved for charge neutrality, if we assume equal adsorption energies, then for a single positive or negative ion the adsorption energy is -3.8 and -3.7 kcal mole^{-1}, respectively. For the (100) face growth of KBr in the presence of phenol, the adsorption energy is found to be -2.6 kcal mole^{-1} while, for the case of KBr in the presence of the carboxylic acids shown in Fig. 3.18, the lines do not go through the origin which suggests that $V_\infty \neq 0$. Tables 3.1 and 3.2 provide some available data for these systems. It is interesting to note that face adsorption energies are generally ~ -3 kcal mole^{-1}.

3.6.4 Adsorbate trapping

When we turn to consider the trapping of the adsorbate in the crystal, we realize that the adsorbate molecule does not acquire any

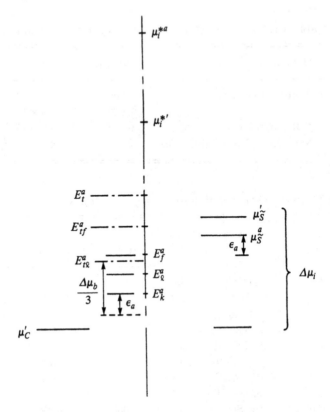

Fig. 3.19. General energy level diagram for adsorption from a solution onto a crystal at face, ledge and kink sites: E_ℓ^a refers to adsorbate at a ledge while $E_{t\ell}^a$ refers to the adsorbate trapped or just walled-in to the ledge; μ'_C and $\mu'_{\tilde{S}}$ refer to host molecule while μ_i^* refers to the activated species.

stress-raised energy state until it is first walled-in along a row. At that point, its energy relative to the solution level of $\mu_{\tilde{S}}^a$ is $E_{t\ell}^a = \mu_{\tilde{S}}^a - 4\varepsilon_a + \Delta\mu_b/3$ and this may be greater or less than $\mu_{\tilde{S}}^a$ depending upon the ratio $\Delta\mu_b/\varepsilon_a$ (see Fig. 3.19). Let us suppose that the time of jumping of the adsorbate out of the filled ledge is τ_{1a}, which will depend upon the energy difference $\mu_i^* - E_{t\ell}^a$ and the time for over-growth of a new row is $\tau* = \lambda_k/2V_k$. If $\tau_{1a} \lesssim \tau^*/2$, then essentially all the adsorbate can jump onto the top of the new row and it is not incorporated in the old row. If $\tau_{1a} \gtrsim 2\tau^*$, then essentially all the adsorbate is trapped in the plane and its new energy level is now $E_{tf}^a = \mu_{\tilde{S}}^a - 5\varepsilon_a + 2\Delta\mu_b/3 \sim \mu_{\tilde{S}}^a$. Let us further suppose that the time of jumping out of the filled face is τ_{2a}

but the time of over-growth of the face is $\tau^{**} = \lambda_\ell / V_\ell$. If $\tau_{2a} \lesssim \tau^{**}/2$, then essentially all the adsorbate can jump onto the top of the new face and it is not incorporated into the crystal. If $\tau_{2a} \gtrsim 2\tau^{**}$, then essentially all the adsorbate is trapped in the crystal and its energy level is now $E_t^a = -6\varepsilon_a + \Delta\mu_b$. The average excess stored energy due to adsorbate trapping at a trapped fraction of X_{at} is just

$$\Delta\mu_E = X_{at}[(E_{ta} - \mu_C') - \kappa T \, \ell n \, X_{at}] \tag{3.15a}$$

so that the average kinetic attachment driving force for the crystallizing species (denoted by the primed quantities in Fig. 3.19) is

$$\Delta\mu_K' = \mu_{\tilde{S}}' - \mu_C' + \Delta\mu_E \tag{3.15b}$$

Fig. 3.19 is the energy level diagram illustrating this overall process. The chemical potential for the six adsorbed states with energies E_f^a, E_ℓ^a, E_k^a, $E_{t\ell}^a, E_{tf}^a$ and E_t^a depends upon the different X_a^j for these various states. The first three are in equilibrium with the solution so they are all given by $\mu_{\tilde{S}}^a$. The other three are kinetically determined and require knowledge of $\tau_{1a}, \tau_{2a}, \tau^*$ and τ^{**}.

3.6.5   *Adsorption model of periodic impurity entrapment*

Consider a pure crystal growing from a solution which also contains an impurity that is a strong adsorber to the surface orientation under consideration. The bulk supersaturation is being continuously increased by a constant cooling rate $\dot{T}$. For the correct set of conditions, the impurity adsorbs to the interface and saturates kink sites on the layer edges so that X'^* decreases. As X'^* decreases, V decreases while ΔG_K increases. Because V is lowered, more absorbing species reach the interface which further decreases X'^* and further lowers V (for fixed ΔG_K). Eventually, $X'^* \to 0$ and $V \to 0$ for the proper set of environmental conditions. However, the total bulk driving force, $\Delta G_\infty(t)$, continues to increase with time and, at some point in time, ΔG_i becomes sufficiently large that two-dimensional nucleation of a new layer can occur to over-lie the adsorbed layer. Once nucleated, this new layer sweeps over the old layer and is able to grow at this driving force because now X'^* is again significant in magnitude. As this excess driving force becomes consumed, the interface begins to slow down so that adsorption of the kink-poisoning impurity begins to increase again and the process repeats itself *ad infinitum*. This leads to a crystal containing a periodic set of impurity sheets with the spacing between the sheets being determined by the rate of increase of supersaturation, the adsorbate energetics and the two-dimensional nucleation characteristics. As a function of time,

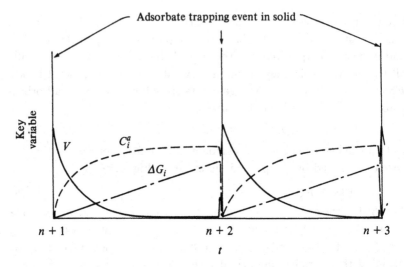

Fig. 3.20. Schematic illustration of periodic adsorbate trapping during crystal growth; $C_i^a(t)$ causes the decaying $V(t)$, while the increasing $\Delta G_i(t)$ leads to the periodic bursts of overgrowth.

the total sequence of events for $V(t), \Delta G_i(t) = \Delta G_K(t) + \Delta G_E(t)$ and the surface concentration of adsorbate, C_i^a, are illustrated in Fig. 3.20. Trapping events occur at $t = 1, 2, 3$, etc.

Problems

1. You are growing an elemental FCC crystal from a transparent solution at a temperature such that $\Delta H_{\tilde{s}}/\kappa T_i \sim 2.5$ and you know that the γ-plot for the system has small cusps at both the $\langle 111 \rangle$ and $\langle 100 \rangle$-type poles with the remainder of the orientations being of fairly constant γ. You place a dislocation-free sphere of 1 cm diameter into the transparent solution as a seed and observe the crystal's growth at various supersaturations via a high-resolution microscope. First, you look at the macroscopic level and see flat regions, round regions and ridged regions, developing on the crystal. Why is this? Next, you look at the microscopic level and, at large supersaturations, the crystal layers show rounded edges as they travel across the surface. However, at small supersaturations, they are comprised of straight and angular segments. Why is this?

What would you expect to see on both a microscopic and a macroscopic level if (a) the crystal had been grown from the vapor phase, (b) the crystal had been grown from solution at a lower temperature but the same ΔG_i, (c) the solution contained a strongly adsorbing impurity that either (i) absorbs only at kink sites or (ii) adsorbs only at face sites and (d) the crystal had a single screw-dislocation aligned across the diameter and exiting on the $[111]$ and $[\bar{1}\bar{1}\bar{1}]$ faces?

2. You are pulling a small diameter Si(100) dislocation-free fiber ($2R \sim 10 - 100$ μm) from a pedestal melt at $V = 10^{-1}$ cm s^{-1} and $\mathcal{G} \sim 10^3$ °C cm^{-1}. What is the size of the rim region, $\overline{\delta R}$, where new layers form when $\gamma/\Delta S_F \approx 10^{-5}$ cm °C? What is the approximate size of the facet plane needed at the $\langle 111 \rangle$ poles of the rim when $\gamma_\ell = 10^2$ or 10 erg cm^{-2} and $\beta_2 \Delta S_F = 1$ cm s^{-1} °C^{-1}? (Assume $d/\delta R \ll 1, I_0 \sim 10^{28}, a \sim$ 3A and $\Delta S_F = 6.5\kappa \times 5 \times 10^{22}$.) Does this distort the rim shape very much?

3. Using Eq. (3.3n) to compare the surface creation undercooling effect with the curvature undercooling effect, what value of $\overset{\bullet}{\delta}/V$ is needed to have the former equal twice the latter? For the data of problem 2, compare these two contributions to ΔT_E with ΔT_K for $\lambda = 10$ μm and $\overset{\bullet}{\delta}/V = 1$.

4. You are growing a pure Si{111} crystal of 1 cm^2 cross section from the vapor phase at temperature T and supersaturation $\hat{\sigma}$. The environment contains a contaminant that strongly absorbs to the {111}. Although the thermodynamic equilibrium fraction of this contaminant in the Si crystal should be 10^{-9}, one finds a periodic sheet-like trapping pattern for the contaminant. Within the trapping sheet, the contaminant site fraction is 10^{-1} and the spacing of the sheets is 10^3 times the sheet thickness. Between the sheets, the equilibrium fraction is present. You want to grow 1 cm^3 of crystal and you don't care how long it takes. What is the minimum value of $\hat{\sigma}$ that will allow you to do this if you assume that you have a small number of screw dislocations threading the {111} surface?

4

The solute partitioning process

In the general crystallization literature, there are two solute partition coefficients between the nutrient and crystal phases that have received a great deal of attention. These are the phase diagram solute distribution coefficient, k_0, and the effective solute distribution coefficient, k. For a more complete understanding of the crystallization process, it is necessary to define and utilize two other solute partition coefficients. These are k_i, the interface distribution coefficient and $\hat{k}_i$, the net interface distribution coefficient. In this chapter, we shall discuss each of these quantities in turn and will show the connections between them as their individual roles in the overall solute partitioning process are revealed.

4.1   Phase diagram distribution coefficient (k_0)

The prime characteristic of a solute species, j, that allows its manipulation during freezing is that, under equilibrium conditions, the atom fractions of solute in the bulk solid, X_S^j, and the bulk liquid, X_L^j, are different. The parameter used to represent this difference is the equilibrium phase diagram distribution coefficient, $\tilde{k}_0^j$, defined by

$$\tilde{k}_0^j = X_S^j / X_L^j \tag{4.1a}$$

By convention, the ratio of the concentration (weight % or volume %) in the solid phase, C_S, to the concentration in the liquid phase, C_L, is more

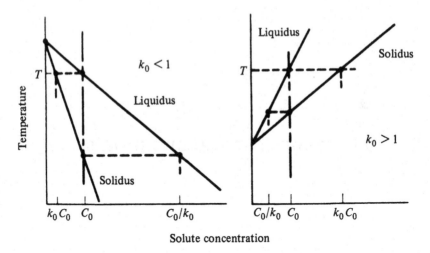

Fig. 4.1. Portions of phase diagrams for the case where the solute either lowers ($k_0 < 1$) or raises ($k_0 > 1$) the melting point of the alloy relative to the pure solvent.

commonly used and is called the equilibrium distribution coefficient.

$$k_0^j = C_S^j / C_L^j \tag{4.1b}$$

The different solute species can be separated into two distinct classes depending upon whether k_0 is greater or less than unity. Solutes with $k_0 < 1$ lower and solutes with $k_0 > 1$ raise the melting point of the solvent. Examples of k_0 are shown in Fig. 4.1 for the two cases. In both cases, for liquids of solute concentration C_∞, equilibrium freezing begins when the temperature of the liquid has been decreased to $T_L(C_\infty)$. At this point, solid of concentration $C_S = k_0 C_\infty$ begins to form (provided all interface fields are zero). For cases where the solute and solvent species are very similar, k_0 is close to unity. For cases where they are very different k_0 may be very small ($k_0 \stackrel{<}{\sim} 10^{-4}$). Table 4.1 shows a range of k_0 values for a variety of solute species in Si. For a polycomponent system, the values of C_S, C_L and $\bar{K}_0$, the gross equilibrium distribution coefficient, should be considered as multidimensional with $\bar{K}_0$ consisting of $n - 1$ components, $k_0^1, k_0^2, \ldots, k_0^j, \ldots, k_0^{n-1}$, arising from the $n - 1$ alloying constituents; i.e.,

$$C_S[1, 2, \ldots, n - 1] = \bar{K}_0 \bullet C_L[1, 2, \ldots, n - 1]$$
$$= \underline{i} k_0^1 C_\infty^1 + \underline{j} k_0^2 C_\infty^2 + \cdots + \underline{s} k_0^{n-1} C_\infty^{n-1} \tag{4.2a}$$

where, in general, we expect to find

$$k_0^j = f[C_\infty^1, C_\infty^2, \ldots, C_\infty^{n-1}] \tag{4.2b}$$

Table 4.1.k_0-*values at the melting point of Si*

Element	k_0	Element	k_0
Li	0.01		
Cu	4×10^{-4}	N	$< 10^{-7}(?)$
Ag		P	0.35
Au	2.5×10^{-5}	As	0.3
		Sn	0.023
		Bi	7×10^{-4}
Zn	$\sim 1 \times 10^{-5}$	O	0.1 to 5.0
Cd		S	10^{-5}
		Te	
B	0.80	V	
Aℓ	0.0020	Mn	$\sim 10^{-5}$
Ga	0.0080	Fe	8×10^{-6}
In	4×10^{-4}	Co	8×10^{-6}
Tℓ		Ni	
Si	1	Ta	10^{-7}
Ge	0.33	Pt	
Sn	0.016		
Pb			

Here, $\bar{K}_0$ is the tie-line connecting a specific C_S to a specific C_L at the liquidus temperature T_L which can be the same for a wide range of compositions in the multicomponent field. For very dilute alloys, the k_0^j are generally independent of the different C_∞^q.

As we see from Chapter 6, equilibrium between any two phases requires equating the extended electrochemical potentials for each species for these two phases. Provided all interface fields are zero and the species are uncharged, this becomes

$$\mu_S^j = \mu_L^j, \quad j = 0, 1, 2, \ldots, n - 1 \tag{4.3a}$$

where element 0 is chosen as the solvent. Results for ideal, dilute and general solutions are given in Chapter 6 and we will restrict our attention to the following approach here.

The chemical potential of the jth species, μ_β^j, in the β phase is

$$\mu_\beta^j = \mu_{0\beta}^j + \kappa T \ell n \hat{\gamma}_\beta^j X_\beta^j; \quad \beta = S, L \tag{4.3b}$$

and, inserting Eq. (4.3b) into Eq. (4.3a) yields

$$\tilde{k}_0^j = \frac{\hat{\gamma}_L^j}{\hat{\gamma}_S^j} \exp \left(\frac{\Delta \mu_0^j}{\kappa T_L} \right); \quad j = 1, 2, 3, \ldots, n-1 \tag{4.4a}$$

and

$$\tilde{k}_0^0 = \frac{X_S^0}{X_L^0} = \frac{1 - \sum_{j=1}^{n-1} k_0^j X_L^j}{1 - \sum_{j=1}^{n-1} X_L^j} \tag{4.4b}$$

$$= \frac{\hat{\gamma}_L^0}{\hat{\gamma}_S^0} \exp \left(\frac{\Delta \mu_0^0}{\kappa T_L} \right)$$

where

$$\Delta \mu_0^j = \mu_{0L}^j - \mu_{0S}^j; \quad j = 0, 1, 2, \ldots, n-1$$

$$= \Delta S_{F0}^j (T_M^j - T_L) + \int_{T_M^j}^{T_L} \Delta c_p^j dT - T \int_{T_M^j}^{T_L} \frac{\Delta c_p^j}{T} dT \tag{4.4c}$$

Here, $\Delta \mu_0^j = 0$ at T_M^j where T_M^j is the melting temperature of pure j and the $\hat{\gamma}$s are activity coefficients. We note from Eqs. (4.4) that $\tilde{k}_0^j \to 1$ as $\Delta \mu_0^j \to 0$ (since this would also mean that $\hat{\gamma}_S^j \to \hat{\gamma}_L^j$). This is likely when solute and solvent have the same molecular size and the same valence. Since it is shown in Chapter 6 that $\Delta \mu_0^j$ is proportional to ΔH_{s^*}, the heat of sublimation, we expect $\tilde{k}_0^j$ to increase with the heat of sublimation of j. This arises because we have used pure j as our standard state point of reference. If we had used an infinitesimal amount of j in the solvent as our reference point then a different conclusion would be made.

All the non-idealities associated with placing the j solute in the solvent environment (size and valence differences, etc.) are incorporated into the $\hat{\gamma}_\beta^j$ terms. In order to match the Trumbore plot (Fig. 6.30) in a quantitative way, the solid must exhibit a large Henry's law constant for j ($\hat{\gamma}_S^j \gg 1$) whereas $\hat{\gamma}_L^j$ may be close to ideal ($\hat{\gamma}_L^j \sim 1$).

Before leaving this section it is important to recognize that most dopant elements in elemental semiconductors (e.g., P, B, As or Sb in Si) can occupy three types of sites in the crystal – substitutional, interstitialcy and interstitial. At equilibrium, the chemical potential of the species in any of these sites is the same and the relative population of the species in any two types of sites 1 and 2 is proportional to

$\exp[(\Delta G_{f_1} - \Delta G_{f_2})/\mathcal{R}T_m]$ where ΔG_f is the free energy difference between taking the dopant from the gas phase to the particular crystal site. The important point to be noticed here is that the conversion of dopant species from one type of site to the other involves either a vacancy, V_{Si}, or an interstitial, Si_I, of the host. Thus, the dopant equilibrium also involves point defect equilibrium and, under truly equilibrium conditions, crystallization leads to dopant partitioning such that $k_0 = k_{01} + k_{02} + k_{03}$. However, if this equilibrium is upset and the formation of type 2 and type 3 states is somehow blocked, one may see that $k_0 \to k_{01}$. If all the type 1 sites were not totally available then k_0 for the species will have a different value from that found for the perfect equilibrium situation. If we extend this type of thinking to compound crystals, then a dopant species can sit on any lattice site plus interstitial and interstitialcy sites so that $k_0 = k_{01} + k_{02} + k_{03} + \cdots$ and we might anticipate some "odd" behavior when the growth conditions are out of equilibrium in a particular way.

An additional kind of partitioning effect must be anticipated when growing crystals of compounds from a nutrient phase. First, it is reasonable to expect partial dissociation of the compound (e.g. $LiNbO_3 \rightleftarrows \frac{1}{2}Li_2O + \frac{1}{2}Nb_2O_5$) and it is also reasonable to expect partial ionization of each dissociated species (e.g., $Li_2O \rightleftarrows Li^+ + OLi^-$, etc.). Thus, even with a stoichiometric melt (e.g., $LiNbO_3$) and no dopants, one may actually have 1–7 species of various concentrations and k_0 values partitioning at the crystallizing interface. This certainly makes crystal growth of such compounds "interesting"!

4.1.1 Charged species effects

For systems containing electrically charged ions which react amongst themselves to form neutral molecules, the electrical form of work must be taken into account in forming k_0^j. Since the electrochemical potential is given by

$$\eta_\beta^j = \mu_{0\beta}^j + \kappa T \ell n \hat{\gamma}_\beta^j X_\beta^j + \hat{z}_\beta^j e \phi_\beta; \quad \beta = S, L. \tag{4.5}$$

Equations (4.4a) and (4.4b) must now be replaced by

$$\tilde{k}_0^j = \frac{\hat{\gamma}_L^j}{\hat{\gamma}_S^j} \exp\left[\frac{(\Delta \mu_0^j + q^j \Delta \phi)}{\kappa T}\right]; \quad j = 0, 1, 2, \ldots, n-1 \tag{4.6}$$

where $q^j = \hat{z}^j e$ and $\Delta \phi = \phi_L - \phi_S$. Even for pure ice in contact with pure water, thermal dissociation of H_2O into H^+ and OH^- leads to the generation of a macro-potential difference, $\Delta \phi$, between the two phases (see Fig. 4.2). This arises because the dissociation constant for water,

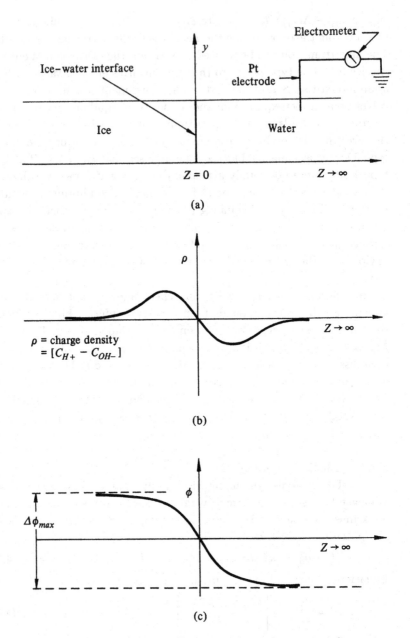

Fig. 4.2. Relevant configurations for evaluating interface potentials: (a) conceptual experimental geometry; (b) spatial distribution of charge, ρ; (c) spatial distribution of electrostatic potential, ϕ.

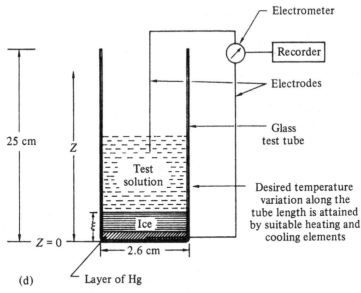

(d)   Layer of Hg

Fig. 4.2. (d) Actual experimental set up for unidirectional ice growth of thickness ξ.

K_W, is different than that for ice, K_I, and this leads to different electrochemical potentials in the two phases for H^+ and OH^-. Since H^+ is extremely mobile, it quickly redistributes to generate the macropotential difference, $\Delta\phi = \phi_W(\infty) - \phi_I(\infty)$, between the two phases. For pure ice–water, $\Delta\phi \approx -250$ mV; however, as the pH of water is changed by the addition of an acid or base, $\Delta\phi$ is found to change linearly with the logarithm of the concentration of electrolyte as shown in Fig. 4.3(a).

Even for the pure ice–water system one finds that

$$\tilde{k}_0^{H^+} = \left(\frac{K_I}{K_W}\right)^{\frac{1}{2}} \exp\left(\frac{+e\Delta\phi}{\kappa T}\right); \quad \tilde{k}_0^{OH^-} = \left(\frac{K_I}{K_W}\right)^{\frac{1}{2}} \exp\left(\frac{-e\Delta\phi}{\kappa T}\right)$$
(4.7)

Thus, if pure water is frozen into ice, OH^- will initially partition more than H^+ at the interface for negative $\Delta\phi$ so the initial solid will develop a positive charge and the interface layer of liquid will develop a negative charge. This is a type of capacitor whose capacitance, $\tilde{C}$, decreases with plate separation. The interface voltage change, $\Delta\phi_V$, due to freezing at velocity V is just $\tilde{C}^{-1}\Delta Q$ where ΔQ is the charge separation which increases almost exponentially with time up to some constant value. One can therefore expect $\Delta\phi_V$ to increase rapidly with time up to some plateau level and then decrease slowly for constant $\tilde{C}$ but decaying ΔQ.

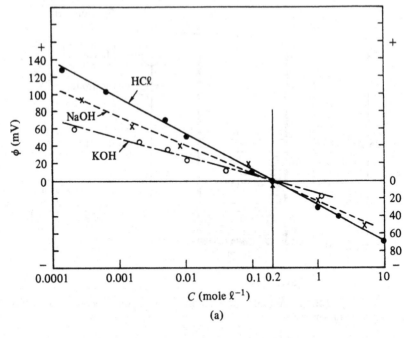

(a)

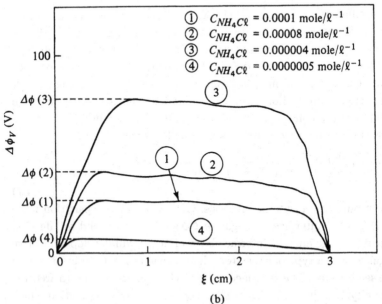

(b)

Fig. 4.3. (a) Near equilibrium ice/water potentials as a function of electrolyte content (molar), (b) typical $\Delta\phi_M$ versus ice thickness solidified, ξ, from an NH_4Cl solution onto a glass substrate.

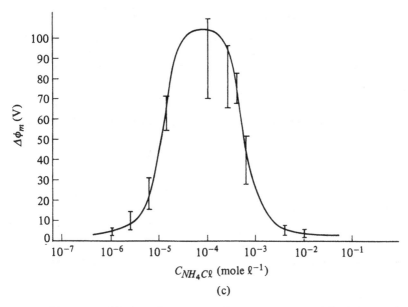

Fig. 4.3 (c) Actual variation of maximum potential, $\Delta\phi_M$, with $C_{NH_4C\ell}$ for freezing onto a mercury substrate at $V = 10^{-3}$ cm s^{-1}.

For the pure ice–water system, the voltage change associated with freezing can amount to several volts. For the $H_2O + NH_4C\ell$ system, using the apparatus of Fig. 4.2(d), $\Delta\phi_V$ changed with interface position, ξ, in the manner expected, rising to a maximum $\Delta\phi_M$ and then decaying slowly. For $V = 10^{-3}$ cm s^{-1}, $\Delta\phi_V$ rises to the peak of ~ 100 V at $C_{NH_4C\ell} \sim 10^{-5}$ mole ℓ^{-1} and then decays again (see Fig. 4.3). If the freezing is sufficiently rapid, the build-up to $\Delta\phi_M$ occurs over a very small ξ and the electric field may be sufficient to cause dielectric breakdown. Voltage build-up into the 10^4 V range has been noted for some systems so that such a process can substantially contribute to thunderstorm electricity and to the detonation of some types of crystallizing explosives.

We can expect this charged species effect to be important in the crystallization of all biological systems, especially proteins, and for insulators and/or semiconductor systems at a sufficiently low temperature that the conduction is extrinsic rather than intrinsic. The effect should increase as $\Delta E_{bg}/T$ where bg refers to the band gap of the material.

4.1.2   Dilute solution effects

For an ideal solution, no change occurs in the intermolecular potential or vibrational spectrum as solute is added to the solvent. Raoult's law is obeyed, the heat of mixing is zero and $\hat{\gamma} = 1$ ($P^j = X^j P_0^j$). For a non-ideal dilute solution where the solute species are far enough apart not to interact with each other, the solute–solvent intermolecular potential is different from that for the solute–solute interaction in its standard state. Here, the solvent follows Raoult's law while the solute follows Henry's law ($P^j = bX^j$ or $\hat{\gamma}^j = b/P_0^j$) which leads to weaker binding in the solid.

As solutions deviate from ideal, one of the results is the wide range of familiar phase diagrams shown in Chapter 6 (see Figs. 6.15–6.21). These diagrams result when we relax the restriction of a zero heat of mixing. *Negative heats of mixing* result in liquidus maxima and compound formation while *positive heats of mixing* result in liquidus minima and eutectic formation. Taking T_M^j, the melting point of pure j, as the standard state for j, for dilute solutions (ideal or real) we find from thermodynamics that

$$\tilde{k}_0^j = 1 - \frac{m_L^j \Delta H_F^j}{\mathcal{R} T_M^{02}} \tag{4.8a}$$

Here, m_L^j is the liquidus slope while ΔH_F^j is the heat of fusion of pure j. When k_0^j is not too close to zero, one can estimate k_0^j from easily determined quantities. By using Eq. (4.3a) for both solute and solvent, m_L^j can be eliminated from Eq. (4.8a) to yield

$$\tilde{k}_0^j = \exp\left[+\frac{\Delta H_F^j(T_M^j - T_M^0)}{\mathcal{R} T_M^j T_M^0} \right] \tag{4.8b}$$

Eq. (4.8b) is bound to be the more accurate expression to use for $\tilde{k}_0^j \ll 1$.

4.1.3   Curvature and pressure effects

One can readily assess the effect of surface curvature on $\tilde{k}_0$ by incorporating the excess free energy contribution due to curvature $\Delta \mu_E = \Omega^j \gamma \mathcal{K}$, into the equation for μ_S^j which leads to, for small values of $\Delta \mu_E$,

$$\frac{\tilde{k}_0^j(\mathcal{K})}{\tilde{k}_0^j(0)} = 1 - \frac{\Omega^j \gamma \mathcal{K}}{\kappa T_M^j} \tag{4.9}$$

Here, Ω^j is the molecular volume of j, γ is the interfacial energy and $\mathcal{K}$ is the curvature. Substituting numerical values into Eq. (4.9) shows that the effect begins to become significant only at radii of curvature smaller than $\sim 10^{-6}$ cm for crystal growth from the melt.

To assess the effect of a change of pressure ΔP, one adds an additional term $\Omega_\beta^j \Delta P$ to Eq. (4.3) which yields, for small values of $\Delta \mu_E$,

$$\frac{\tilde{k}_0^j(P)}{\tilde{k}_0^j(1)} = 1 - \frac{\Delta \Omega^j \Delta P}{\kappa T_M^j} \tag{4.10}$$

By inserting numerical values, one finds that the change of pressure effect begins to become significant only for $\Delta P \sim 10^2$ atm.

4.1.4 Extension to binary and ternary compound formation

An important crystal growth area is that of binary (or ternary) compound formation from a ternary (or quaternary) liquid solution. As an example, we shall consider $(GaAs)_x(GaP)_{1-x}$ layer growth from $Z-$ GaAs, GaP liquid solutions where $Z = Ga, Ge, Sn$, etc. For simplicity, we shall choose $Z = Ga$ so that only a (Ga, As, P) ternary system need be considered. The approach for a quaternary or higher degree solution is the same and only the $\hat{\gamma}_\beta^j$ change.

To obtain the thermodynamic tie-line relationships, one notes that the elements must be considered in three different states: (1) pure elements, (2) pure compounds and (3) ternary solutions. Using GaAs as our focus, from the Gibbs equilibria, we know that

$$\mu_S^{GaAs} = \mu_{0S}^{GaAs} + \kappa T \ell n \hat{\gamma}_S^{GaAs} X_S^{GaAs} = \mu_L^{Ga} + \mu_L^{As} \tag{4.11a}$$

Using the state (2) compound equilibrium at the 50/50 composition,

$$\mu_{0S}^{GaAs} = \mu_{L(2)}^{Ga} + \mu_{L(2)}^{As} - (\Delta H_F - T \Delta S_F)^{GaAs} \tag{4.11b}$$

and

$$\mu_{L(2)}^{Ga} = \mu_{0L}^{Ga} + \kappa T (\ell n \hat{\gamma}_{L(2)}^{Ga} + \ell n \frac{1}{2}), \tag{4.11c}$$

and a similar expression for the As species. Continuing in this way leads to the two tie-lines

$$\hat{\gamma}_S^{GaAs} X_S^{GaAs} = 4 \left(\frac{\hat{\gamma}_L^{Ga} \hat{\gamma}_L^{As}}{\hat{\gamma}_{L(2)}^{Ga} \hat{\gamma}_{L(2)}^{As}} \right) X_L^{Ga} X_L^{As}$$
$$\times \exp \left[\frac{(\Delta H_F - T \Delta S_F)^{GaAs}}{\mathcal{R}T} \right] \tag{4.11d}$$

and

$$\hat{\gamma}_S^{GaP} X_S^{GaP} = 4 \left(\frac{\hat{\gamma}_L^{Ga} \hat{\gamma}_L^{P}}{\hat{\gamma}_{L(2)}^{Ga} \hat{\gamma}_{L(2)}^{P}} \right) X_L^{Ga} X_L^{P}$$
$$\times \exp \left[\frac{(\Delta H_F - T \Delta S_F)^{GaP}}{\mathcal{R}T} \right] \tag{4.11e}$$

Table 4.2. *Selected entropies of fusion for III–V compounds*

III – V	ΔS_F	III – V	ΔS_F
AℓP	15.0	GaSb	15.8
AℓAs	15.6	InP	14.0
AℓSb	14.7	InAs	14.5
GaP	16.8	InSb	14.3
GaAs	16.6		

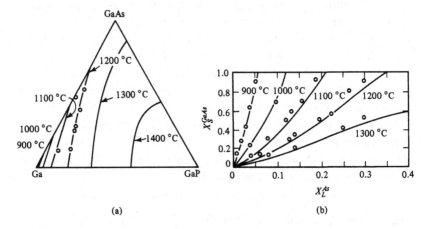

(a)   (b)

Fig. 4.4. Calculated isotherms in the Ga–As–P phase diagram (a) liquidus and (b) solidus.

The additional constraints on the system are

$$X_S^{GaAs} + X_S^{GaP} = 1 \tag{4.11f}$$

and

$$X_L^{Ga} + X_L^{As} + X_L^{P} = 1 \tag{4.11g}$$

The ΔH_F and ΔS_F data are known for all the III–V systems (see Table 4.2) so, if all the $\hat{\gamma}_\beta^j$ are known, the ratios X_S^{GaAs}/X_S^{GaP} or k_0^{GaAs} and k_0^{GaP} are known from Eqs. (4.11d) and (4.11e) and the specific equilibrium liquidus and solidus surfaces are known for a given T.

Panish and Ilegems[1] used this approach to produce Fig. 4.4, the liquidus and solidus surfaces for this system so that X_S^{GaAs} could be readily determined if X_L^{As} is known (Fig. 4.4(b)). X_L^{As} will be known by fixing

Table 4.3. *Liquidus interaction parameters for the binary systems*

$$\alpha^* = a - bT + cT^2$$

Table 4.4. *Solidus interaction parameters*

$$\alpha^*_{(III-V)/(III'-V')}$$

used for phase calculations with ternary systems

System	a	b	c	$T\,°(C)$	System	$\alpha^*_{(III-V)/(III'-V')}$
Al–P		4.5	2,800	2530	$Al_xGa_{1-x}P$	0
Al–As		12.0	600	1770	$Al_xIn_{1-x}P$	3500
Al–Sb	0.03	34.0	57,500		$Ga_xIn_{1-x}P$	3500
Ga–P		4.8	2,800	1465	$Al_xGa_{1-x}As$	0
Ga–As		9.2	5,160	1238	$Al_xIn_{1-x}As$	2500
Ga–Sb		6.0	4,700	710	$Ga_xIn_{1-x}As$	3000
In–P		4.0	4,500	1070	$Al_xGa_{1-x}Sb$	0
In–As		10.0	3,860	942	$Al_xIn_{1-x}Sb$	600
In–Sb	0.048	69.0	20,400	525	$Ga_xIn_{1-x}Sb$	1900
Al–Ga			104		$AlAs_xP_{1-x}$	400
Ga–In			1,060		$GaAs_xP_{1-x}$	400
Al–In			1,060		$InAs_xP_{1-x}$	400
P–As			1,500		$GaAs_xSb_{1-x}$	4500
As–Sb			750		$InAs_xSb_{1-x}$	2250

X_L^P at temperature T_L (Fig. 4.4(a)). Conversely, Fig. 4.4(b) provides X_L^{As} for any desired X_S^{GaAs} at a particular equilibrium temperature, T_L, while Fig. 4.4(a) tells us what X_L^P and X_L^{Ga} must be chosen with this value of X_L^{As} for equilibrium crystal growth at this T_L.

To determine these results, Panish and Ilegems[1] had to evaluate all the $\hat\gamma^j$ for the Ga solvent case. As shown in Chapter 6, an effective method for obtaining $\hat\gamma = \hat\gamma(X,T)$ is the α-parameter approach. Let us now consider this method for a ternary system. The α-parameters for each of the ij binaries associated with the ternary may be expressed as $\alpha_{ij} = e_{ij} + f_{ij}X_j$ where $e_{ij} = a_{ij} + b_{ij}/T$ and $f_{ij} = a'_{ij} + b'_{ij}/T$. If one wishes to use the regular solution approximation, then $f_{ij} = 0$ so that $\alpha_{ij} = e_{ij}$.

For a general solution, using $(\hat\gamma_1, \hat\gamma_2, \hat\gamma_3) \equiv (\hat\gamma^{Ga}, \hat\gamma^{As}, \hat\gamma^P)$, we have (note j is now a subscript rather than a superscript)

$$\ell n\hat\gamma_1 = \alpha_{12}X_2^2 + \alpha_{13}X_3^2 + (\alpha_{12} + \alpha_{13} - \alpha_{23})X_2X_3 + f_{12}X_2^2X_3 + f_{13}X_2X_3^2 - f_{23}X_2X_3^2 \tag{4.12a}$$

which simplifies for the regular solution approximation to

$$\ell n\hat{\gamma}_1 \approx \alpha_{12}X_2^2 + \alpha_{13}X_3^2 + (\alpha_{12} + \alpha_{13} - \alpha_{23})X_2X_3 \qquad (4.12b)$$

Thus, using this approximation for the Ga–As–P system, we find that

$$\left.\begin{aligned}
\ell n\hat{\gamma}_L^{Ga} &= \alpha_L^{GaP}(X_L^P)^2 + \alpha_L^{GaAs}(X_L^{As})^2 \\
&\quad + (\alpha_L^{GaP} + \alpha_L^{GaAs} - \alpha_L^{AsP})X_L^{As}X_L^P \\
\ell n\hat{\gamma}_L^{As} &= \alpha_L^{AsP}(X_L^P)^2 + \alpha_L^{GaAs}(X_L^{Ga})^2 \\
&\quad + (\alpha_L^{AsP} + \alpha_L^{GaAs} - \alpha_L^{GaP})X_L^{Ga}X_L^P \\
\ell n\hat{\gamma}_L^{P} &= \alpha_L^{GaP}(X_L^{Ga})^2 + \alpha_L^{AsP}(X_L^{As})^2 \\
&\quad + (\alpha_L^{GaP} + \alpha_L^{AsP} - \alpha_L^{GaAs})X_L^{Ga}X_L^{As}
\end{aligned}\right\} \qquad (4.13)$$

Values of $\alpha^* = \alpha/\mathcal{R}T$ were developed from experimental phase equilibrium data using the regular solution approximation and are presented in Tables 4.3 and 4.4. Most of the α values in Eqs. (4.13) can be obtained from binary diagram data. However, ternary diagram data are needed to provide α_L^{AsP} and then the $\hat{\gamma}^j$ can be obtained as a function of liquid composition and temperature. From such data and Eqs. (4.11), one can readily generate Fig. 4.4.

For a quaternary system (Z = Ge, Sn, etc.), the analogs to Eqs. (4.12) are

$$\begin{aligned}
\ell n\hat{\gamma}_1 &= \alpha_{12}X_2(1 - X_1) + \alpha_{13}X_3(1 - X_1) \\
&\quad + \alpha_{14}X_4(1 - X_1) - \alpha_{23}X_2X_3 - \alpha_{24}X_2X_4 \\
&\quad - \alpha_{34}X_3X_4 + f_{12}X_2^2(X_2 + X_4) \\
&\quad + f_{13}X_3^2(X_2 + X_4) + f_{14}X_4^2(X_2 + X_3) \\
&\quad - f_{23}X_2X_3^2 - f_{24}X_2X_4^2 - f_{34}X_3X_4^2 \qquad (4.14) \\
&\approx (\alpha_{12}X_2 + \alpha_{13}X_3 + \alpha_{14}X_4)(1 - X_1) \\
&\quad - \alpha_{23}X_2X_3 - \alpha_{24}X_2X_4 - \alpha_{34}X_3X_4
\end{aligned}$$

for a regular solution where $(1, 2, 3, 4) \equiv (\text{Z}, \text{Ga}, \text{As}, \text{P})$. Table 4.4 provides us with α_{23}, α_{24} and α_{34} while additional experimental data is needed to obtain α_{12}, α_{13} and α_{14}. Only then can the $\hat{\gamma}^j$ be obtained and used to provide analogous plots to Fig. 4.4.

4.2   Effective distribution coefficient (k)

During crystal growth at some velocity V, gross solute accumulation ($k_0 < 1$) or depletion ($k_0 > 1$) occurs in a liquid boundary layer of thickness $\delta_C \sim 10^{-4}$–10^{-1} cm at the interface as a result of both solute partitioning at the interface and convective matter trans-

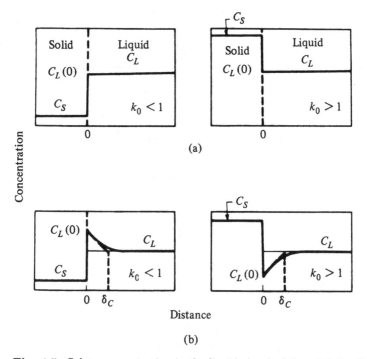

Fig. 4.5. Solute concentration in the liquid ahead of the solid–liquid interface for $k_0 \lessgtr 1$: (a) complete mixing where $k = k_0$ and (b) partial mixing where $k_0 < k < 1$.

port in the bulk liquid (see Fig. 4.5). For simplicity, we assume that mass transport occurs entirely by diffusion within this boundary layer and entirely by convection in the bulk liquid outside of this boundary layer. In reality this is not correct, we can only say that the fluid velocity is zero at the interface (for a Newtonian fluid) and increases to the bulk fluid flow velocity at δ_C. Our reason for using this boundary layer approximation is because chemical analysis of both the solid and the bulk liquid are accessible to experimental measurement and we wish to relate the concentration of the crystal at the interface to that of the bulk liquid. Because solute diffusion in the crystal is generally negligible in the growth time of a crystal, one generally assumes that the solute content measured in the crystal at a certain position is essentially the same value as was frozen in at the interface when the interface was at that position in the crystal. We thus define the effective distribution

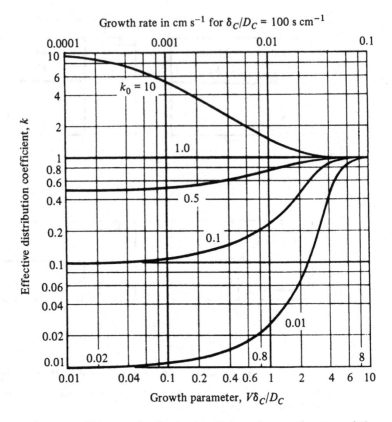

Fig. 4.6. Effective distribution coefficient, k, as a function of the dimensionless growth parameter, $V\delta_C/D_C$, for several values of the equilibrium distribution coefficient, k_0.

coefficient, k, as

$$\tilde{k}^j = \frac{X_S^j(0)}{X_L^j(\infty)}; \quad k^j = \frac{C_S^j(0)}{C_L^j(\infty)} \tag{4.15}$$

From a record of crystal pull rate with time and $X_L^j(\infty, t)$, the interface position can be correlated with time and thus k can be correlated with time.

In general, k will be a function of V, δ_C, k_0, etc., and, in the broadest sense, may also be a function of spatial coordinates. Burton, Prim and Slichter (BPS)[2] were the first to analyze this problem mathematically and found the following relationship between k and k_0

$$k = \frac{k_0}{k_0 + (1 - k_0)\exp(-V\delta_C/D_C)} \tag{4.16}$$

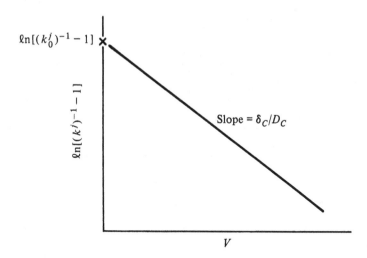

Fig. 4.7. Plot from the BPS equation of $\ln(k^{-1}-1)$ versus V for constant fluid mixing conditions $(\delta_C/D_C = \text{constant})$ used for obtaining k_0.

In the limit of complete mixing $(V\delta_C/D_C \to 0), k \to k_0$ while in the opposite limit of no mixing $(V\delta_C/D_C \to \infty), k \to 1$. In the middle range of the solute mixing parameter, $V\delta_C/D_C$, we have the normal condition of partial mixing. This formula neglects density differences between the liquid and crystal but one can readily correct for such an effect by using $V' = (\rho_S/\rho_L)V$ where ρ is the density. Fig. 4.6 provides values of k for several k_0 as a function of $V\delta_C/D_C$.

Experimentally, if one is dealing with a solute for which k_0^j is not known, Eq. (4.16) may be used in the form

$$\ln\left(\frac{1}{k^j} - 1\right) = \ln\left(\frac{1}{k_0^j} - 1\right) - \frac{V\delta_C}{D_C} \tag{4.17}$$

to obtain k_0^j. Thus, for fixed stirring conditions (fixed δ_C/D_C) one could grow portions of a solvent crystal from a melt containing a small amount of the j solute at constant V (but with several different V in each crystal) and also measure $C_L^j(\infty, t)$ so that k can be determined as a function of V for fixed δ_C/D_C. Such a result is shown in Fig. 4.7 and k_0^j can be obtained from the intercept while δ_C/D_C can be obtained from the slope. This procedure was used in the semiconductor industry to provide many of the k_0^j given in Table 4.1. An alternative procedure is to hold V fixed and vary δ_C by changing the crystal rotation rate, ω.

Hall[3] made a careful study of k for j = Sb, As and Ga in Ge and

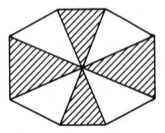

Fig. 4.8. Illustration of "sector zoning" in a crystal.

discovered two important facts. He found that k was a function of crystallographic orientation, θ, of the crystal interface for both Sb and As but not for Ga. He also found that differences in the experimental $k_0^j(\theta)$ by as much as a factor of 2–3 were needed to fit the data at the largest V and that these differences with θ tended to disappear as $V \to 0$. Thus we seem to have $k_0 = k_0^j(V, \theta)$ which should not be possible for a thermodynamic quantity like k_0.

Additional data coming from other areas of crystallization experience tended to confirm Hall's findings that something was wrong with the simple picture. Growing crystals of either Si or InSb in the $\langle 111 \rangle$ direction with Te as a dopant showed that, when the interface was slightly convex to the liquid, such that a $\{111\}$ facet formed, the distribution of Te in the crystal was much higher in the region of crystal under the facet compared to the immediately adjacent region under the non-faceted segment of interface. This ratio $k_{onF}^{Te} / k_{offF}^{Te} \sim 6$–10 for these two systems. From the technical area of liquid phase epitaxial growth (LPE), layers of different θ grown from similar melts have been found to exhibit $k(\theta)$ differing by factors ~ 2–5. Finally, in geological crystals one often finds the phenomenon called sector zoning, illustrated in Fig. 4.8, where the color of the crystal varies remarkably under the crystal growth faces of different crystallographic orientation. The Maltese cross effect of Fig. 4.8 occurs because k_0^j is very different for the two major face types exhibited by the crystal's growth habit.

In general, as the crystal's binding energy increases, this $k_0^j(V, \theta)$ effect appears more strongly, especially for crystal growth from solutions at lower temperatures. This suggests that a type of interface adsorption effect is occurring and that we must begin to recognize that the binding forces for solutes at the surface can be quite different than in either the bulk melt or the bulk crystal. This will have important consequences for

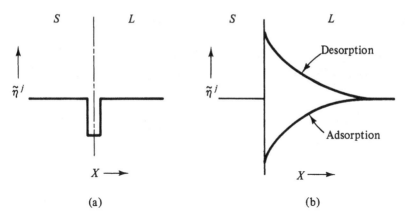

Fig. 4.9. Interface field profiles, $\tilde{\eta}^j$, for the j species: (a) square-well potential binding and (b) spatially distributed potential binding.

solute partitioning in materials of large band gap or during film growth from the vapor phase.

4.3   Interface distribution coefficient (k_i)

It is necessary to include interface adsorption effects in our description of the solute distribution coefficient. From surface studies it is well known that the chemical potential of a molecule can be quite different in an interface region compared to the bulk phases far from the interface. The change relative to the bulk can occur via both (i) the presence of an inhomogeneous field of stress, electrostatic potential, magnetic potential, etc., and (ii) the gross structural changes in the molecular environment at the interface and thus in the intermolecular potential function for an atom situated there rather than in the bulk phase. Since thermodynamic equilibrium demands a constant value for the extended electrochemical potential (defined below) of each species throughout the entire system, a redistribution of solute atoms must occur in the interface region in response to these field effects. This gives rise to the phenomenon of surface or interface segregation and involves the exchange of solute for solvent species between the bulk and the interface region. From a qualitative viewpoint, we can view the extended electrochemical potential distribution, $\tilde{\eta} \sim \tilde{\eta}(z)$, for a completely homogeneous system in the two ways illustrated in Fig. 4.9. In case (a), one considers a very local and narrow potential well for the solute at the interface. In case (b), one considers a distributed potential leading to a well-defined interface field that causes either adsorption or desorption of the species j.

Defining the total free energy of interaction between the jth species and all of the interface fields as $\widetilde{\Delta G}_0^j$, the extended electrochemical potential, $\widetilde{\eta}^j$, for species j is given by[4]

$$\widetilde{\eta}_\beta^j(z) = \mu_{0\beta}^j + \kappa T \ell n \hat{\gamma}_\beta^j X_\beta^j(z) + \widetilde{\Delta G}_{0\beta}^j(z) - \widetilde{\Delta G}_{0\beta}^0(z) \qquad (4.18)$$

Here, the μ_0s are bulk values and the interface effects are all incorporated in the $\widetilde{\Delta G}_{0\beta}^j$ term. Since the redistribution of the j species involves the concomittant redistribution of solvent because both species have a finite size, we see that it is the relative difference $\delta \widetilde{\Delta G}_{0\beta} = \widetilde{\Delta G}_{0\beta}^j - \widetilde{\Delta G}_{0\beta}^0$ between the solute and the solvent species that is involved in any redistribution of the j species.

In general, we must expect that

$$\delta \widetilde{\Delta G}_0^j = f[g^j(z), q^j(z), \phi^j(z), H^j(z), \sigma^j(z)] \qquad (4.19a)$$

$$= \widetilde{\Delta G}_0^j(z) - \widetilde{\Delta G}_0^0(z) \qquad (4.19b)$$

where $g(z)$ is due to variations in local order, local coordination and local density and extends over a distance ~ 5 molecular spacings; $q(z)$ is due to variations of local chemistry plus electron bond type and may extend over a distance ~ 100 molecular spacings; $\phi(z)$ is the electrostatic potential variation and may extend over a distance $\sim 10^{-1}$–1.0 μm; $H(z)$ is the magnetostatic potential variation and may also extend ~ 1 μm while $\sigma(z)$ is the stress potential variation and it may extend over the entire dimensions of the crystal. For a more explicit one-dimensional example, we have

$$\widetilde{\Delta G}_0^j(z) = \hat{z}^j e \phi(z) + \frac{\partial}{\partial C^j} \left[\frac{\epsilon_*}{2} \left(\frac{\partial \phi}{\partial z} \right)^2 + \frac{\mu_*}{2} H^2 + \frac{\sigma^2}{2E_*} \right] + \Delta \mu_i^j(z) \qquad (4.20)$$

In Eq. (4.20), the first term is a familiar one, the second involves the electrostatic, magnetic and strain energy storage and the third is due to bonding energy changes. In the middle term, ϵ_*, μ_* and E^* are the permittivity, permeability, and the elastic modulus of the medium respectively. The $\Delta \mu_i^j$ term will always be non-zero at the interface because of the asymmetry of the intermolecular potential fields at that location. In fact, although small in magnitude, the electrodynamic forces of the van der Waals type are still varying at $\sim 10^3$ Å from the interface. Thus, the binding potential of the crystal for a molecule in the liquid adjacent to it will always be different than the binding potential for that molecule completely surrounded by n shells of liquid molecules ($n \to \infty$). The same argument holds for a molecule on the crystal side of the interface relative to a molecule in the bulk crystal. As one might expect, the magnitudes of the $\widetilde{\Delta G}_0^j$s are related to the interfacial free energy, γ, so that, for a flat interface,

$$\gamma = \sum_{j=0}^{m} \int_{-\infty}^{\infty} [n_\beta^j(z)\widetilde{\eta}_\beta^j(z) - n_\beta^j(\infty)\widetilde{\eta}_\beta^j(\infty)]\mathrm{d}z \qquad (4.21)$$

In terms of specific examples, in Si during free surface reconstruction, the outermost (111) plane relaxes strongly inwards giving an extremely strong compressive stress that has a significant magnitude even ten atom layers below the surface. One consequence of this is that any atom or defect that produces a dilitational stress in the system will be strongly attracted to this surface region. This definitely applies to the formation of surface vacancies and is one of the reasons for the strong adsorption of C to a Si surface. In this case, it is not only the fact that the Si–C bond is stronger than the Si–Si bond, but that the molecular volume of SiC (10 Å^3) is much less than that of Si (20 Å^3). When one extends the idea of the surface stress tensor to binary compounds like GaAs that have a strong piezoelectric coefficient, one realizes that the surface relaxation will give rise to an electric field effect that may extend $\sim 10^2$–10^3 Å from the interface.

Another important example to consider is water. In this system the molecular dipole strength is such that, although the bulk dipole–dipole interaction is randomized by thermal fluctuations, a region of ~ 5–10 molecular layers is strongly oriented at the free surface. At the ice–water interface, because of a strong interface electric contact potential, this oriented dipole layer is expected to be even thicker. In addition, even for perfectly pure water, H_2O dissociation into H^+ and OH^- will lead to ion relaxation to screen the interface potential. The net effect is the development of a space charge layer with its attendant electric field variation to a distance $\sim 1 \ \mu m$ on either side of the interface. Such a resultant electric field will strongly influence the interface adsorption of minor constituents. From the foregoing, we see that a short-range molecular rearrangement in the interface region can give rise to subsequent relaxation effects that extend the range of thermodynamic inhomogeneity to $\sim 0.1 - 1.0 \ \mu m$.

Because of these interface fields, we must expect the equilibrium solute distribution in each phase to be given by

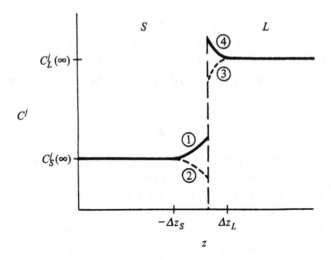

Fig. 4.10. Concentration profiles in the solid, C_S, and liquid, C_L, for a static interface which exhibits all four possible types of distributed binding potentials.

$$\frac{X_\beta^j(z)}{X_\beta^j(\infty)} = \left(\frac{\hat{\gamma}_\beta^j(\infty)}{\hat{\gamma}_\beta^j(z)} \right) \exp \left[\frac{-\delta \widetilde{\Delta G}_{0\beta}^j(z)}{\kappa T} \right] \tag{4.22a}$$

where we now take into account the exchange of Ω^j/Ω^0 molecules of solvent at location (z) with one molecule of species j from location (∞) by defining $\delta \widetilde{\Delta G}_{0\beta}$ as

$$\delta \widetilde{\Delta G}_{0\beta}^j = \widetilde{\Delta G}_{0\beta}^j - \frac{\Omega^j}{\Omega^0} \widetilde{\Delta G}_{0\beta}^0 + kT\ell n \left[\frac{\hat{a}^j(z)/\hat{a}^j(\infty)}{[\hat{a}^0(z)/\hat{a}^0(\infty)]^{\Omega^j/\Omega^0}} \right] \tag{4.22b}$$

Depending on the sign of $\delta \widetilde{\Delta G}_0^j$ in Eqs. (4.22), where $\hat{a}$ refers to activity, we have either an enhancement or a depletion of the jth species in the interface region. This interface adsorption or desorption increases as the temperature decreases and as $\delta \widetilde{\Delta G}_{0\beta}^j$ becomes more negative. Further, since the crystal surface density and the magnitude of the contact potential are both functions of crystallographic orientation, θ, both $\delta \widetilde{\Delta G}_{0\beta}^j$ and the interface excess quantity

$$\Gamma^j = n^j(\infty) \int_{-\infty}^{\infty} [X_\beta^j(z) - X_\beta^j(\infty)]\mathrm{d}z \tag{4.23}$$

are expected to be functions of θ. Thus, for a particular θ, at a crystal–nutrient interface with $V = 0$, the four different types of equilibrium solute profile are as illustrated in Fig. 4.10. Although we may expect the (1,4)-type of interface to be found most often, any combination of the four types is possible. For further discussions, we shall define the extent of the field-influence zone in the liquid to be $0 < z < +\Delta z_L$ and in the solid to be $-\Delta z_S < z < 0$.

Using the foregoing, we are now in a position to define the interface distribution coefficient, k_i^j, as

$$\tilde{k}_i^j = \frac{X_S^j(0)}{X_L^j(0)}; \quad k_i^j = \frac{C_S^j(0)}{C_L^j(0)} \tag{4.24}$$

so that

$$\tilde{k}_i^j = \tilde{k}_0^j \hat{\gamma}^{*j} \exp[-\delta^* \widetilde{\Delta G}_{0SL}^j(0)/\kappa T] \tag{4.25a}$$

with

$$\delta^* \widetilde{\Delta G}_{0SL}^j(0) = \delta \widetilde{\Delta G}_{0S}^j(0) - \delta \widetilde{\Delta G}_{0L}^j(0) \tag{4.25b}$$

and

$$\hat{\gamma}^{*j} = \frac{\hat{\gamma}_S^j(\infty)/\hat{\gamma}_S^j(0)}{\hat{\gamma}_L^j(\infty)/\hat{\gamma}_L^j(0)} \tag{4.25c}$$

This is the condition that leads to equality of $\tilde{\eta}^j$ for the j molecules in both phases at the interface. Neglecting activity coefficient effects, we can see that $k_i^j \sim k_0^j$ depending upon the relative magnitudes and signs of the various factors involved in $\widetilde{\Delta G}_{0\beta}^j$.

For use in numerical evaluation later, a reasonable and simple choice of interface fields is one that decays exponentially with distance from the interface so we shall assume the following forms

$$\delta \widetilde{\Delta G}_{0S}^j(z) = \bar{\beta}_S \exp(\alpha_S z); \quad z \lesssim 0 \tag{4.26a}$$

and

$$\delta \widetilde{\Delta G}_{0L}^j(z) = \bar{\beta}_L \exp(\alpha_L z); \quad z \gtrsim 0 \tag{4.26b}$$

Adsorption occurs when $\bar{\beta}_S$ or $\bar{\beta}_L$ are negative. Using Eqs. (4.26) we have

$$k_i^j/k_0^j = \hat{\gamma}^{*j} \exp[(\bar{\beta}_L - \bar{\beta}_S)/\kappa T] \tag{4.27}$$

with $\hat{\gamma}^{*j} \sim 1$. We also have $\Delta z_S \sim 5\alpha_S \ll \delta_C$ and $\Delta z_L \sim 5|\alpha_L| \ll \delta_C$ so that, in most crystal growth situations with stirring present, the interface field distortion of the macroscopic solute profile is confined to a microscopic region close to the interface. We shall see later, in the companion book[5], that the effect of the interface field on the solute profile during crystallization can extend orders of magnitude beyond the Δz_L distance for some special range of growth parameters. One is

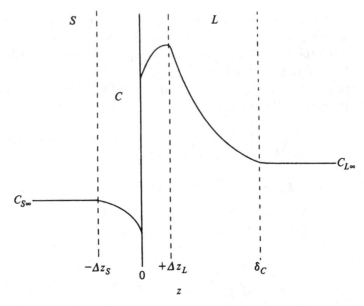

Fig. 4.11. Schematic representation of the solute distributions in solid and liquid during freezing from a convectively stirred liquid ($z \gtrsim \delta_C$) plus with an interface field present (2, 3)-type for $k_0 < 1$.

thus led to ask, "What are the consequences of this microscopic solute profile on the net solute incorporation into the crystal?".

4.4   Net interface distribution coefficient ($\hat{k}_i$)

In a real crystal growth situation where interface fields and stirring are present, as illustrated in Fig. 4.11 for the (2,3)-type interface field case, we see that our old definition of k is no longer appropriate and our new definition should be of the form

$$k^j = C_S^j(-\Delta z_S)/C_L^j(\infty) \tag{4.28}$$

Since transport by field-driven migration can occur in the region $-\Delta z_S < z < 0$, the experimentally viable domain that fits the negligible diffusion assumption is $z \lesssim -\Delta z_S$. For the majority of cases, where $\Delta z_S \ll \delta_C$, if we are interested only in the macroscopic solute profile in the crystal, we could think of the interface as having a width $\Delta z_L + \Delta z_S$, treat it like a "black box" and define a net interface distribution coefficient across it as

$$\hat{\tilde{k}}_i^j = \frac{X_S^j(-\Delta z_S)}{X_L^j(\Delta z_L)}; \quad \hat{k}_i^j = \frac{C_S^j(-\Delta z_S)}{C_L^j(\Delta z_L)} \tag{4.29}$$

Then we would find that the BPS equation (Eq. (4.16)) must be replaced by

$$k = \frac{\hat{k}_i}{\hat{k}_i + (1 - \hat{k}_i)\exp(-V\delta_C/D_C)} \tag{4.30}$$

since $\hat{k}_i$ replaces our former use of k_0. To see if $\hat{k}_i$ fits with the Hall[3] and other data, we need to understand its (V, θ) dependence. We do this by considering the effect of the interface field on transport rates.

4.4.1   Interface transport and up-hill diffusion

The solute conservation condition in the interface region for the case of a zero interface field is given by

$$V[C_L(0) - C_S(0)] = V(1 - k_0)C_i = -D_L\left(\frac{\partial C_L}{\partial z}\right)_i \tag{4.31a}$$

With a non-zero interface field present, the conservation condition becomes

$$
\begin{aligned}
V(1 - k_i)C_i &= -\hat{M}_L\left(\frac{\partial \tilde{\eta}_L}{\partial z}\right)_i + \hat{M}_S\left(\frac{\partial \tilde{\eta}_S}{\partial z}\right)_i \\
&= -\left[D_L\left(\frac{\partial C_L}{\partial z}\right)_i + \frac{D_L C_i}{\kappa T}\left(\frac{\partial}{\partial z}\delta\widetilde{\Delta G}_{0L}\right)_i\right] \\
&\quad + \left[D_S\left(\frac{\partial C_S}{\partial z}\right)_i + \frac{D_S k_i C_i}{\kappa T}\left(\frac{\partial}{\partial z}\delta\widetilde{\Delta G}_{0S}\right)_i\right]
\end{aligned} \tag{4.31b}
$$

where $\hat{M}$ is the atomic mobility. From Eq. (4.31b), a driving force now exists at the interface for "up-hill" diffusion if $(\partial/\partial z)\delta\widetilde{\Delta G}_0$ is of opposite sign to $(\partial/\partial z)C$. In addition, sufficient driving force now exists in the solid to produce a short-range solute distribution (within the distance Δz_S of the interface). This field effect is always such as to bring the actual solute distribution for the $V > 0$ case back towards the distribution for the $V = 0$ case of Fig. 4.10.

By comparing the solute flux, J, for the case of simple convection at velocity v ($J = vC$) with the Fickian diffusion flux ($J = -D\partial C/\partial z$) and with the field-driven transport flux ($J = -DC\partial/\partial z\,(\delta\widetilde{\Delta G}_0/\kappa T)$), one can write an effective convective velocity, v_{eff}, for the latter two contributions. Thus, in our present example we should write v_{eff} as

$$v_{eff} = -\frac{\partial}{\partial z}D\left(\ell n C + \frac{\delta\widetilde{\Delta G}_0}{\kappa T}\right) \tag{4.31c}$$

and compare it to the actual interface velocity, V. When $v_{eff} > V$, the equilibrium solute profile is maintained; when $v_{eff} < V$, departures

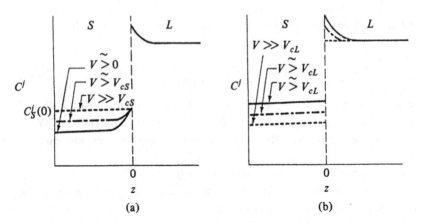

Fig. 4.12. Concentration profile changes due to interface field effects as V increases: (a) loss of transport communication in the solid as $V \to V_{cS}$ and (b) loss of transport communication in the liquid as $V \to V_{cL}$.

from this equilibrium profile must be expected. Finally, the relative importance of the two contributions can be assessed by comparing the magnitudes of the two terms in the bracket of Eq. (4.31c).

4.4.2   Transport communication and critical growth velocities

The (1,4) case of Fig. 4.10 will be used to illustrate the changes that occur as V increases. We shall neglect long-range solute redistribution effects in the liquid and shall assume a semiinfinite liquid in a state of complete mixing beyond $z = \Delta z_L$.

For conditions of slow freezing where $V \stackrel{\sim}{>} 0$, the bulk solid has sufficient time to communicate with the bulk liquid through the solute-enhanced layers so that the equilibrium profiles of Fig. 4.10 are maintained as the interface moves; i.e., in the solid even though a concentration $k_i C_L(0)$ freezes in at the interface, it does not stay at that level as that particular slice of solid recedes further behind the interface due to additional crystallization.

Up-hill diffusion towards the interface occurs to deplete the slice of solute and lower it to the concentration of the equilibrium condition. It is able to do this in step with the interface movement so that, at $z = -\Delta z_S$, the concentration of that slice has attained the bulk value where now transport occurs only by simple diffusion.

Since $D_S^j \ll D_L^j$, for communication to be maintained, the species j must be able to migrate the distance Δz_S in a time $\tau \leq \tau_S = \Delta z_S/V$,

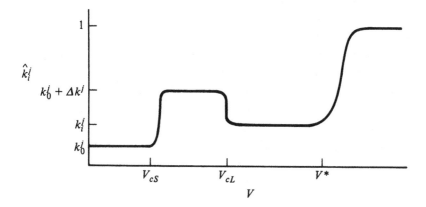

Fig. 4.13. Schematic variation of the net interface solute distribution coefficient, $\hat{k}_i^j$, as a function of interface velocity, V, when the interface field is of the $(1, 4)$-type.

the time needed for the interface to move a distance Δz_S. At some critical growth velocity, V_{cS}, transport communication begins to be lost across the interface in the solid where

$$V_{cS} = D_S/\Delta z_S, \text{for a negligible drift term} \qquad (4.32a)$$

and

$$V_{cS} = \frac{D_S}{\Delta z_S} \left| \frac{\delta \widetilde{\Delta G}_{0S}(0)}{\kappa T} \right|, \text{for a significant drift term} \qquad (4.32b)$$

Thus, for a small field effect, one uses Eq. (4.32a), for a large field effect, Eq. (4.32b) is used and one uses Eq. (4.31c) in the mid region. As transport communication is lost, the concentration profile in the solid for case (1,4) of Fig. 4.10 begins to rise until, at $V \gg V_{cS}$, the profile is flat so that the bulk solute content is now the same as that frozen in at the interface. This solute profile change is illustrated in Fig. 4.12(a). As V increases further, nothing additional happens until we begin to approach the condition where transport communication begins to be lost across Δz_L. Here, field-driven transport cannot pile up the solute at the interface fast enough to keep pace with the movement of the interface at these values of V. This occurs at V_{cL} which is given by Eqs. (4.32) with $D_L, \Delta z_L$ and $\widetilde{\Delta G}_{0L}$ replacing $D_S, \Delta z_S$ and $\widetilde{\Delta G}_{0S}$. Because interface equilibrium still holds at V_{cL}, k_i is constant and, as the profile in the liquid begins to flatten (see Fig. 4.12(b)), the profile in the solid is lowered as well.

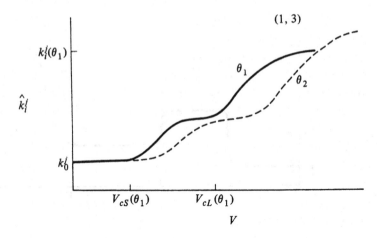

Fig. 4.14. $\hat{k}_i^j$ variation with V for two different orientations, θ, when the interface field is of the $(1,3)$-type.

Using the procedure of Fig. 4.12, one is able to evaluate $\hat{k}_i^j$ as a function of V and this result is shown in Fig. 4.13. We find that $\hat{k}_i^j = k_0^j$ at $0 < V < V_{cS}$, that $\hat{k}_i^j$ begins to rise at $V \gtrsim V_{cS}$ to a new plateau of $\hat{k}_i^j = k_0^j + \Delta k^j$ where

$$\Delta k^j = \exp(\delta \widetilde{\Delta G}_{0S}^j(0)/\kappa T) - 1 \qquad (4.33a)$$

$$= \exp(\bar{\beta}_S/\kappa T) - 1 \qquad (4.33b)$$

for the field profile of Eqs. (4.26) and that $\hat{k}_i^j$ begins to fall towards a third plateau of $\hat{k}_i^j = k_i^j$ when $V \gtrsim V_{cL}$. In general, we expect $V_{cS} \sim 10^{-8}$–10^{-1} cm s^{-1} and $V_{cL} \sim 10^{-4}$–10^{-1} cm s^{-1} for different materials.

Since the field characteristics of the interface are a function of crystal orientation θ, both the magnitude and the rate of change of $\hat{k}_i^j/k_0^j$ will also be functions of θ. This is illustrated in Fig. 4.14 for the $(1,3)$ case of Fig. 4.10. For some systems, V_{cS} and V_{cL} are sufficiently close together that the intermediate plateau will not be clearly resolved. For other systems, D_S may be so small that one is not able to obtain the k_0^j initial plateau because V_{cS} is less than any realistically controllable V. Of course, for the different cases in Fig. 4.10, a different type of $\hat{k}_i^j$ versus V plot obtains. Finally, if $\delta \widetilde{\Delta G}_{0S}$ or $\delta \widetilde{\Delta G}_{0L}$ are partially dependent on V and time, then the $\hat{k}_i^j$ versus V plot might look quite different.

As V is further increased, $\hat{k}_i^j = k_i^j$ so long as interface equilibrium is maintained and the interface fields remain constant. However, at extremely high velocities, $V \gtrsim V^*$, interface equilibrium will begin to be lost and $\hat{k}_i^j = k_i^j \to 1$. This condition is illustrated in Fig. 4.13 and will be returned to in the next section.

When one is concerned with the incorporation of two solutes, such as C and Si in GaAs, by crystallization from the melt, it is useful to consider what might happen as a result of an interface field effect. This is illustrated in Fig. 4.15 for the cases of $\Delta\phi_i < 0$ and $\Delta\phi_i > 0$ with donor and acceptor dopants present. In case (a) where $\Delta\phi_i < 0$, the equilibrium concentration profile of the donor in the solid will bend up while that for the acceptor will bend down. The results for $V < V_{cS}$ and $V > V_{cS}$ are shown in Fig. 4.15(c) where the crystal remains of donor electrical type but of higher uncompensated doping level. For case (b) where $\Delta\phi_i > 0$, the equilibrium donor profile in the crystal will bend down while that for the acceptor profile will bend up. Even when the equilibrium donor content is greater than the equilibrium acceptor content, the rate effect at $V \gg V_{cS}$ produces non-equilibrium concentration profiles wherein the acceptor content may be greater than the donor content so that the crystal becomes p-type (see Fig. 4.15(d)) due to the rate effect. A switch from case (a) to case (b) is possible if the surface state condition is sufficiently altered by doping. This is one possible explanation for a shift in dominant carrier type in a crystal as a function of V.

4.4.3   Non-equilibrium interface effects

In Chapter 2 we found that $V = 2haV_k/\lambda_\ell\lambda_k$ (see Eq. (2.20b)) so the interface velocity, V, is determined by the kink velocity, V_k. In turn, the value of V_k is given by $V_k \approx (a\nu/6)\exp(-\Delta G_A/\mathcal{R}T_i)[1 - \exp(-\Delta G_i/\mathcal{R}T_i)]$ which has a maximum value when there is no backward reaction $(\Delta G_i/\mathcal{R}T_i \gg 1)$. The condition of interface equilibrium occurs from a kinetic viewpoint when a large number, n, of state change cycles can occur for each molecule attaching to the crystal at a kink site. As $\Delta G_i/\mathcal{R}T_i$ increases beyond a critical value, n decreases below some critical value n^* and ΔG_E begins to increase because the molecules are no longer in the minimum free energy configuration. Thus, the interface velocity at which interface non-equilibrium effects begin to appear, $V = V^*$ in Fig. 4.13, is given by

$$V^* \approx \frac{haa'\nu}{12n^*\lambda_\ell\lambda_k}\exp\left(-\frac{\Delta G_A}{\mathcal{R}T_i}\right) \tag{4.34a}$$

$$\sim 5 \times 10^3 \exp(-\Delta G_A/\mathcal{R}T_i) \text{ cm s}^{-1} \tag{4.34b}$$

for $n^* \sim 10, \lambda_\ell \sim 10a, \lambda_k \sim 4a'$ and $h \sim a \sim 2 \times 10^{-8}$ cm. Thus, V^* is

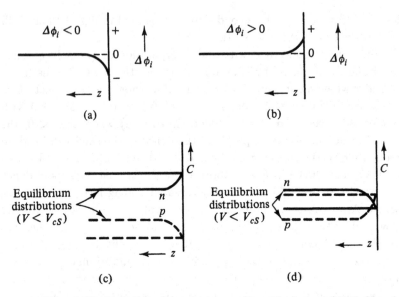

Fig. 4.15. Variation of net doping profile with change of both V and interface electrostatic potential, $\Delta\phi_i$ (a) leads to (c) and (b) leads to (d) which produces a net carrier-type change.

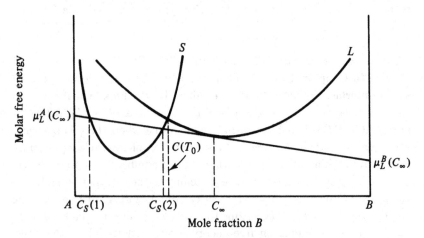

Fig. 4.16. Concentration range of solid $(C_S(1)-C_S(2))$ that may spontaneously form from a liquid of composition C_∞.

expected to lie in the 10–10^2 cm s^{-1} range for metals and semiconductors and $k_i^j \sim 1$ for V an order of magnitude larger ($n^* \to 1$). For even larger V, the crystallinity begins to be lost (ΔG_E increases greatly and an amorphous solid results).

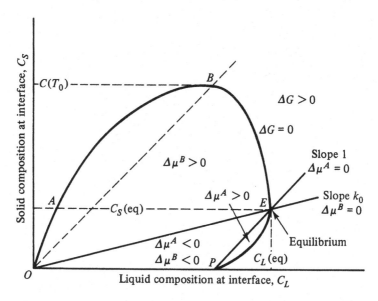

Fig. 4.17. Envelope $OABEP$ presents all possible solid compositions that can form for liquid of concentration C_L at a given temperature. The lines OE and PE represent conditions of equal chemical potential for solute and solvent, respectively, at dilute solution conditions. The line OB is for the diffusionless transformation condition (T_0 condition in Fig. 4.18).

From a thermodynamic viewpoint, if one wishes to inquire whether a particular solid can form from a single phase liquid of composition C_∞, a tangent is drawn to the liquid free energy curve, G_L, at C_∞ in Fig. 4.16 and one looks to see if any portion of the free energy curve for the solid phase, G_S, containing this particular solid composition lies below this tangent. The composition range over which G_S lies below the tangent provides the range of solid composition that is thermodynamically allowed to form. At a given temperature, T_0, the range of thermodynamically possible solid compositions, C_S, that can form from a liquid of varying composition, C_L, is shown in Fig. 4.17 as an area enclosed by the curve $OABEP$ where the free energy change associated with the concentration change is zero.[6] The locus of this curve is given by

$$(1 - C_S)\Delta\mu^A + C_S\Delta\mu^B = 0 \qquad (4.35)$$

The interior of $OABEP$ gives $\Delta G < 0$ while the exterior gives $\Delta G > 0$.

The right hand point, E, on the curve depicts the equilibrium interface condition for $T = T_0$. As the liquid composition moves to the left, the

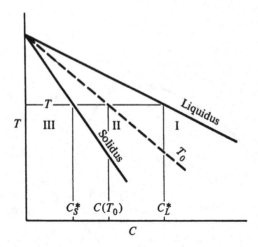

Fig. 4.18. Schematic phase diagram illustrating the diffusionless transformation condition (curve T_0).

range of possible solid compositions first increases and then decreases. The maximum solid composition, point B in Fig. 4.17, is the composition where the two free energy curves of Fig. 4.16 cross. Only one liquid composition, $C^*(T_0)$, can give this solid. Diffusionless transformations ($k_i = 1$) can occur only if the liquid composition is below this cross-over composition. On the phase diagram of Fig. 4.18, the T_0 line marks the upper limit to the composition and temperatures for diffusionless transformations ($k_i^B = 1$).

Two other curves in Fig. 4.17 are of interest. These are the lines where we have equal chemical potentials in the liquid and solid; for line OE we have $\Delta \mu^B = 0$ while for line PE we have $\Delta \mu^A = 0$. Where the two lines cross, we have equilibrium. These lines bound those regions in the figure where a particular component solidifies with a decrease in chemical potential. In the triangular region OEP, both components experience a decrease in chemical potential on solidifying. Outside of OEP but within the $\Delta G = 0$ locus, one of the components experiences an increase in chemical potential although the overall free energy decreases on solid formation. Solid formation can occur in this domain only if the two species do not solidify completely independently of each other. One species enters the solid with an increase in chemical potential (generally the B component) because it is either passively trapped by the advancing crystal front or because it is a required participant in a coupled reaction leading to an overall free energy decrease. We shall define "trapping" of a component when that species experiences an increase in chemical potential; i.e., $\Delta\mu_E^B > 0$ above the line OE while $\Delta\mu_1^A > 0$ below the line PE. Using the Gibbs equations for dilute solutions with zero interface field, we find that

$$\Delta\mu^B = \mu_S^B - \mu_L^B \approx \kappa T \ell n(k_i^B/k_0^B) \qquad (4.36)$$

Thus, when $\Delta\mu^B = \Delta\mu_E^B > 0$, $k_i^B > k_0^B$ and solute trapping occurs. This effect is accentuated if B strongly adsorbs to the interface kink sites and makes the attachment kinetics of A become very sluggish.

For the special case where the material under consideration is a binary compound, this interface adsorption/attachment kinetic limitation serves to require off-stoichiometric growth of the crystal. For a pure AB compound, we have for the A and B constituents at the interface

$$V = \beta_K^A[C_i^A - (C^{*A} + \Delta C_E^A)] \qquad (4.37a)$$
$$= \beta_K^B[C_i^B - (C^{*B} + \Delta C_E^B)] \qquad (4.37b)$$

where β_K^A and β_K^B are the attachment kinetic coefficients for A and B, respectively, while C^* is the equilibrium concentration of the liquid and ΔC_E is the departure from the equilibrium concentration associated with non-equilibrium solid. Thus, if $\beta_K^A \gg \beta_K^B$, Eqs. (4.37) require that $\Delta C_E^A \gg \Delta C_E^B$ which leads to excess A in the crystal given by $(\Delta C_E^A - \Delta C_E^B)$. This will probably be in the form of anti-site defects. Thus, one sees that k_i^A is a function of β_K^A, β_K^B and V.

If an impurity is present in this compound system and it adsorbs to the active kink sites then both β_K^A and β_K^B will be reduced in magnitude. For constant V, this requires $C_i - C^*$ to increase and yields a consequent increase in magnitude of ΔC_E. If the impurity prefers a particular type of kink site (A or B), then the corresponding kinetic coefficient will be decreased accordingly. Overall, this is bound to lead in most cases to a greatly enhanced departure from stoichiometry for the AB compound as well as to significant trapping of the impurity.

4.4.4   *Some experimental data on* $\hat{k}_i$

Using the analysis of the companion book,[5] Hall's Ge data for Sb has been analyzed[4] using the assumption that $\bar{\beta}_L/\kappa T \approx 0$ with the exponential fields of Eqs. (4.26) and one finds that $\bar{\beta}_S/\kappa T = -1.3$ and $\alpha_S = 2.0 \times 10^6$ cm^{-1}. As can be seen from the data match in Fig. 4.19, the values of $\bar{\beta}_S/\kappa T$ for the various crystal faces differ by ~ 0.1 with the (111) orientation having the largest magnitude. The sign of the binding energy indicates an Sb build-up at the interface probably due to a negative electrostatic potential binding the Sb$^+$ ion cores to the

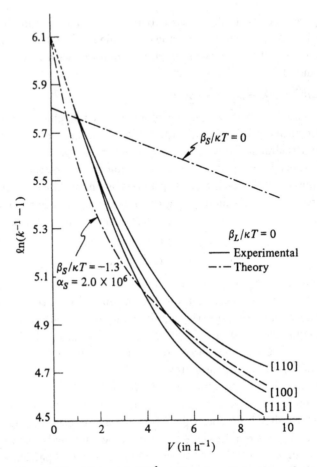

Fig. 4.19. Plot of $\ln(k^{-1} - 1)$ versus V to show the comparison between Hall's data for Sb in Ge and interface field-effect theory using $\bar{\beta}_L/\kappa T = 0$.

interface region. Of course, because we have neglected any $\bar{\beta}_L$ effect, there is some degree of uncertainty with this conclusion.

Using CsCoCℓ_3 as solute, CsCdCℓ_3 crystals were grown by Ahn[7] for a range of stirring conditions and pull rates with the Co distribution investigated using a laser scanning technique. He found distinct interface field effects for the $\langle 1000 \rangle$ and the $\langle 10\bar{1}0 \rangle$ orientations. Once again, $\bar{\beta}_L$ was neglected and the following results were obtained

$$\frac{\bar{\beta}_{S_{\langle 0001 \rangle}}}{\kappa T} = -1.50; \quad \alpha_{S_{\langle 0001 \rangle}} \bullet D_S = 1.6 \times 10^{-3}$$

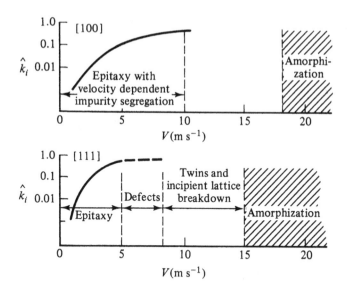

Fig. 4.20. Typical experimental variation of $\hat{k}_i$ with V for Si(100) and Si(111) as a result of high velocity freezing in the 1–20 m s^{-1} range.

and

$$\frac{\bar{\beta}_{S_{\langle 10\bar{1}0\rangle}}}{\kappa T} = -0.25; \quad \alpha_{S_{\langle 10\bar{1}0\rangle}} \bullet D_S = 4.0 \times 10^{-3}$$

Using the technique of pulsed laser annealing of Si, investigators have examined the regrowth regions that solidified at extremely high rates and have extracted values of $\hat{k}_i$ needed to fit the data. All of the data is in the range of $V > V^*$ with the general qualitative results being illustrated by Fig. 4.20 for the (100) and (111) orientations with velocities in the range 10^2–2×10^3 cm s^{-1}. At the highest velocities, amorphous Si is produced as expected. Some quantitative data for Bi and In as dopants in Si is illustrated in Figs. 4.21(a) and 4.21(b) respectively. The plateau at $\hat{k}_i \sim 0.3$ may be related to V_{cL} but it is more likely to signify a quantitative error in the data.

In 1968, Jindal and Tiller[8] calculated the variation of k_i with V by using a two-step partitioning process. In step 1, a layer of thickness λ freezes at V and the solute is incorporated into the solid with $k_i = 1$ while, in step 2, the solute back-diffuses towards the interface under the chemical potential gradient associated with the excess energy state, $\Delta\mu_E$, for the solid given from the difference between the $k_i = 1$ and the equilibrium k_i condition ($k_i = k_i^e$). The amount of excess solute retained in the solid is determined via solute conservation leading to an

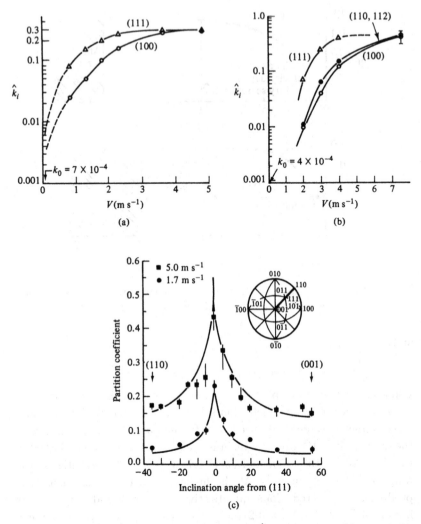

Fig. 4.21. Experimentally determined $\hat{k}_i$ variations for (a) Bi and (b) In in Si as a function of velocity V and (c) a comparison of Goldman's and Aziz's aperiodic stepwise growth model to the Bi in Si data for the orientation dependence of k_i at $V = 1.7$ m s^{-1} and $V = 5.0$ m s^{-1}.

expression for $k_i(V)$ of the following form.

$$\left(\frac{1+k_i}{1-k_i}\right)\ell n\left(\frac{1+k_i}{2k_i^e}\right) = \frac{2aV}{D_S} \tag{4.38a}$$

where D_S is the value in the first atomic layer of solid at the interface. This result has the general form exhibited by Fig. 4.21.

More recently, Goldman and Aziz[9] used a ledge flow model to calculate $k_i(V)$ with $V_\ell = V/\sin\theta$ and θ is the inclination angle of the interface from the (111). They assumed the same total solute trapping condition and the same back-flow diffusion conditions through the solid with a diffusive speed $v_D^L = D_i^L/\lambda$ where λ is a monolayer thickness and D_i is the diffusion coefficient in the solid in the ledge movement direction. They also included the effect of diffusion in the solid in the terrace direction yielding $v_D^T = D_i^T/\lambda$. Finally, they obtained a result for $k_i(V)$ as their aperiodic stepwise growth model (ledges randomly spaced) of the form

$$k_i = \frac{\{k_i^e + \beta_t[(\beta_\ell + k_i^e)/(\beta_\ell + 1)]\}}{(\beta_t + 1)} \tag{4.38b}$$

where

$$\beta_t = V_\ell/v_D^T, \ \beta_\ell = V_\ell/v_D^L \tag{4.38c}$$

They showed that there was indeed a measurement error in the Bi data of Fig. 4.21 and evaluated $k_i(\theta)$ for Bi in Si at $V = 1.7$ m s^{-1} and $V = 5.0$ m s^{-1}. These results are shown in Fig. 4.21(c) using $v_D^T = 6$ m s^{-1} and $v_D^L = 20$ m s^{-1} in the aperiodic stepwise growth model.

4.5   Extension to film formation

This case is dealt with more completely in Chapter 5 but here a mention should be made of the absolute need to apply the field effect concepts of this chapter. For film formation at some vicinal angle θ, because of the terrace reconstruction, the ledges will have a certain dynamic height $h_\ell = h_\ell(T)$ where T is the deposition temperature. This is governed to a significant degree by the kinetics of unreconstruction at the lower terrace/ledge junction and by the kinetics of reconstruction at the upper terrace/ledge junction. The net effect of this is that there will be a field effect acting at the lower terrace/ledge contact influencing solute (dopant) transport on the lower terrace and a different field effect acting at the upper terrace/ledge contact influencing transport at this location. In addition to these field effects which act in the plane of surface transport, a second field effect acts in the solid perpendicular to the surface. This latter field effect is very similar to those already discussed earlier in this chapter.

For the film growth case, k_i acts at the ledges and k_0 acts between the bulk solid and the adlayer. As the binding energy increases for the solute in the adlayer, perhaps because it greatly influences surface reconstruction or satisfies dangling bonds (see Fig. 5.10), k_0 decreases. For such cases, the sign of the field perpendicular to the surface will drive the solute back to the adlayer after it has been incorporated into the crystal at the ledges. From Eq. (4.31c) and a field like Eq. (4.26a), the effective field-driven transport velocity, $v_{eff} \approx -(\bar{\beta}_s/\kappa T)D_s\alpha_s$. Thus, if $\alpha_s \sim 10^6$ cm^{-1}, $D_s \sim 10^{-12}$ cm^2 s^{-1} and $\bar{\beta}_s/\kappa T \sim 5$, all reasonable possibilities, then for MBE ($V \sim 10^{-8}$–10^{-7} cm s^{-1}) and CVD, chemical vapour deposition, ($V \sim 10^{-6}$–10^{-5} cm s^{-1}) we find $v_{eff} \gtrsim V$ and the equilibrium profile should be present. Unfortunately, in these cases, as V is altered so is the terrace reconstruction effect that controls $\bar{\beta}_s$ and k_0. In addition, as θ and h_ℓ change, $\bar{\beta}_s$ and k_0 also change because the adlayer concentration changes. All of these factors make the doping of defect free epitaxial films a substantial challenge.

Problems

1. For the system of pure ice–water, use the dissociation/ionization reaction $H_2O \rightleftharpoons H^+ + OH^-$ for each phase plus the Gibb's equilibrium equation to prove Eq. (4.7). How will $k_0^{H^+}$ and $k_0^{OH^-}$ change when HCℓ is added to the water at a concentration of $C_\infty^{HC\ell}$?

2. For the four possible combinations of adsorption/desorption of solute illustrated in Fig. 4.10, draw the qualitative profiles for $\hat{k}_i^j/k_i^j$ versus V.

3. You are growing a crystal for which $V_{cS} = 10^{-3}$ cm s^{-1} and the level of the second plateau in a $\hat{k}_i/k_0$ plot is $k_i/2$ for a solute where $k_i = 5k_0$. Using the modified BPS equation, plot k versus $V\delta_C/D_C$ for $k_0 = 10^{-2}$ and all $V < 2V_{cL}$. Alternatively, plot $\ell n(k^{-1} - 1)$ versus $V\delta_C/D_C$.

4. You are growing an elemental semiconductor crystal from a stirred solution containing both n-type and p-type solutes where $C_\infty^n = 10\,C_\infty^p$, $k_0^n = 10^{-2}k_0^p$, $k_0^p < 1$ and no interface fields are present. Find the value of $V\delta_C/D_C$ for which the initial portion of the crystal to crystallize is totally charge compensated.

5. You are growing an AB crystal from a solution where complete dissociation and complete ionization have occurred to give a far-field concentration $C_\infty^{A^+} = C_\infty^{B^-}$ at $z > \delta_C$. The electrostatic potential difference between the interface and solution is $\phi_i - \phi_\infty > 0$ for steady state one-dimensional crystal growth at

velocity V so that a positive electric field $E \approx (\phi_i - \phi_\infty)/\delta_C$ resists the movement of A^+ to the interface and enhances the movement of B^-. Show that the interface concentrations, C_i^+ and C_i^- are given by

$$\left(\frac{C_i}{C_\infty}\right)^{\pm} = \frac{\dfrac{V}{\Omega(D_C C_\infty)^{\pm}} - \left(\dfrac{1}{\delta_C} \mp \dfrac{E}{2\kappa T}\right)}{\left(\mp \dfrac{E}{2\kappa T} - \dfrac{1}{\delta_C}\right)}$$

where Ω is the molecular volume of AB.

6. Using a two-step monolayer trapping and back-diffusion model, show that the dependence of k_i on V is given by Eq. (4.38a).

5

Thin film formation via vapor phase processing

In heteroepitaxial film formation, at least five distinct stages need to be distinguished:

(1) the time-dependent development of chemical species in the surface adsorption layer and the attendant adsorption layer isotherm development;

(2) cluster formation and nucleation in the adsorption layer as influenced by surface reconstruction processes;

(3) non-impinging crystallite growth of well-defined aspect ratio with attendant long-range stress, electrostatic and chemical fields;

(4) impinging crystallite growth where the aspect ratio increases because of the intervening fields so that a rumpled surface topography and residual small holes in the film are often developed; and

(5) subsequent homoepitaxial film growth on this rumpled surface leading to the development of film texture.

In the extensive literature on film growth, three different mechanisms have been listed to distinguish which of these stages dominates the process. The most general mechanism of three-dimensional nucleation, crystal growth and crystal coalescence, where a more or less connected network of islands containing empty channels is formed, is called the

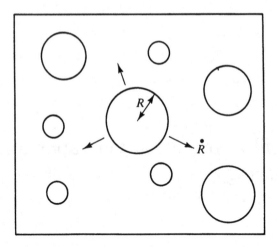

Fig. 5.1. Schematic illustration of circular crystallite growth on a substrate.

Volmer–Weber (VW) mechanism. A second more restrictive mechanism, the Stranski–Krastanov mechanism, relates to the adsorption of a monolayer and subsequent nucleation of the new phase on top of this buffer layer. The third mechanism involves the restriction of layer-by-layer dominated growth and is called the van der Merwe mechanism. The VW mechanism is one characterized by a high nucleation rate and a slow lateral spreading rate while the van der Merwe mechanism is characterized by a low nucleation rate and a rapid lateral spreading rate. The Stranski–Krastanov mechanism is characterized by an interfacial energy that increases with overlayer thickness and then saturates at a constant value (analogous to Fig. 5.7(c)). Thus, the nucleation rate is initially slow compared to the lateral spreading rate and then becomes much faster so the new film starts out with a van der Merwe-type of characteristic and then develops a VW-type of characteristic.

The surface area fraction coverage of the new phase, $\theta(t)$, due to both nucleation *and* growth depends upon the number of nuclei formed between time τ and $\tau + d\tau$ which is $I^* d\tau$ and their growth at rate $\dot{R}(t - \tau)$; i.e., assuming circular symmetry as in Fig. 5.1, we have

$$\theta(t) = 1 - \int_0^t I^*(\tau')\theta(\theta(\tau')\tau')g(t - \tau')d\tau' \qquad (5.1a)$$

where $g(t)$ is the area of the nucleus at t given by

$$g(t - \tau') = \pi \left[\int_{\tau'}^t \dot{R}(t')dt' \right]^2 \qquad (5.1b)$$

For the special case where the nuclei form at a constant rate, I_0^*, and grow by either a diffusion-controlled or an interface attachment-controlled mechanism, we have $g(t) = \alpha_1 t$ and $g(t) = \alpha_2 t^2$, respectively, so that

$$\theta(t) = \cos\left[(\alpha_1 I_0^*)^{\frac{1}{2}} t\right] \tag{5.1c}$$

and

$$\theta(t) = \frac{1}{3}\left[\exp(-\beta_1 t) + 2\exp(\beta_{\frac{1}{2}} t)\cos\left(\frac{\sqrt{3}}{2}\beta_1 t\right)\right] \tag{5.1d}$$

respectively, with $\beta_1 = (2\alpha_2 I_0^*)^{\frac{1}{3}}$.

In many cases, one is interested in knowing whether the deposited film should be amorphous versus polycrystalline and this involves an interplay between the deposition flux, J, and the surface diffusion coefficient D_s. When the time of transport between adjacent adatom sites, t_D, is more than the time, t_L, needed to deposit a monolayer, then the film is expected to be amorphous. When $t_L \gg t_D$, one expects to find a polycrystalline film. This can be made more quantitative by defining N_s as the number of surface sites per square centimeter per second so that $t_L = N_s/J$ and by assuming that the lifetime of the adatom on the surface before evaporation, $\tau_S \gg t_L$. Then, the mean diffusion distance, $\bar{X}_s^B$, of the adatom before being buried is given by $\bar{X}_s^B = (4D_s t_L)^{\frac{1}{2}}$ and we have

$$\text{Amorphous deposition:} \qquad D_s \lesssim J/4N_s^2 \tag{5.2a}$$

$$\text{Polycrystalline deposition:} \qquad D_s \gg J/4N_s^2 \tag{5.2b}$$

In this chapter, we will focus on the domain characterized by Eq. (5.2b) and restrict our attention to single crystal and polycrystal deposition.

In the crystalline film domain, the ratio of the mean diffusion distance before evaporation, $\bar{X}_s = (4D_s \tau_s)^{\frac{1}{2}}$, to the distance, L, between ledges is a very important consideration and, for $\bar{X}_s < L/2$ as illustrated in Fig. 5.2(a), we have

$$V \approx \frac{2\bar{X}_s}{L}\left(\frac{J}{N_s}\right) a \tag{5.3}$$

Only the material impinging on the surface within a distance $\bar{X}_s$ of the surface can reach the ledge and be incorporated into the crystal before evaporating. Thus, the capture efficiency can be less than unity. Eq. (5.3) assumes that the separation distance between active kinks along the ledge is much less than L. To ensure that a single crystal film retains its single crystal character, the growth conditions must be adjusted so that no two-dimensional nucleation of misoriented pill-boxes

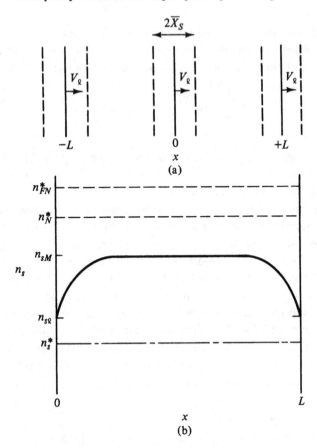

Fig. 5.2. (a) Illustration of effective atom collection zones of width $2\bar{X}_s$ about a parallel train of ledges and (b) solute profile $n_s(x)$ between ledges relative to the equilibrium ledge concentration $n^*_{s\ell}$ and the unfaulted and faulted terrace pill-box nucleation conditions, n^*_N and n^*_{NF}, respectively.

can occur on the terraces between the ledges. The relevant surface concentrations are illustrated in Fig. 5.2(b) where n^*_s is the equilibrium adsorbate concentration. Thus, so long as n_{sM}, the far field adsorbate concentration between ledges, is less than n^*_N no two-dimensional nucleation can occur. When n_{sM} exceeds n^*_{FN}, faulted two-dimensional nucleation can occur so polycrystalline film formation has a high probability of happening. For good quality epitaxial film formation, one wishes to operate in the domain where $n_{sM} < n^*_N$. We will see later how this depends on various growth conditions and material parameters.

The four main methods of film deposition of interest here are: (1) evaporation (electron beam, etc.); (2) sputtering; (3) chemical vapor deposition (CVD, MOCVD, etc); and (4) molecular beam (MBE, etc.). In the sputter deposition method, ions are accelerated over a mean free path length before striking the surface at some flux J and some excess energy, ΔE, beyond the thermal energy value. This method generally leads to compressively stressed films at low total pressure, P_T, and low substrate temperature, T_{sub}, probably due to ion implantation types of considerations. Increasing P_T reduces the mean free path so ΔE is reduced and the film stress can be changed from compressive to tensile. With a given film stress, the excess energy of the film increases with film thickness and a critical film thickness is reached where the free energy of the system is lowered by separation of the film from the substrate via a rumpling type of fracture process. For evaporative deposition at low T_{sub}, the depositing species strike the surface at flux J and energy between κT_{source} and κT depending upon the mean free path length in the system. In this case the films are usually deposited under a tensile stress (trapped vacant sites) and, at thicknesses above some critical thickness, fracture is expected to occur across the film. Thus, in what is to follow, we implicitly assume film stress levels sufficiently low that mechanical failure is not a consideration. However, it should be clear that, at low substrate temperatures, film stresses will generally be present because of excess interstitial or vacancy species, if for no other reason.

5.1   Key surface processes

5.1.1   *Surface adsorption and the sticking coefficient*

On collision with the surface, a gas molecule is first physisorbed to form a precursor state. It may then transform to the chemisorbed state at an unoccupied site of the adsorbent or it may acquire sufficient thermal energy from the lattice to be desorbed. In Fig. 5.3(a), possible potential energy plots for the process of chemisorption (ABC) and that of physisorption (DEF) to form the precursor state at E are shown. Although the cross-over point at G lies below the zero potential energy state, an activation energy, E_p, is necessary for transition into a chemisorbed state. This potential energy profile is similar in appearance to that found for the occurrence of dissociative chemisorption as illustrated in Figs. 5.3(b) and (c) for H on Cu and on Au, respectively.

The sticking coefficient, S_*, indicates the capture efficiency of an atom or molecule in the final chemisorbed state; i.e., the rate of capture in

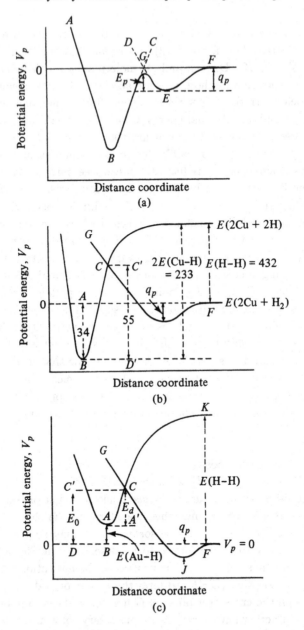

Fig. 5.3. (a) Potential energy as a function of distance coordinate for the physisorbed precursor state (DEF) and chemisorbed state (ABC) with heat of physisorption, q_p, and activation barrier, E_p; (b) same for H on Cu; and (c) same for H on Au.

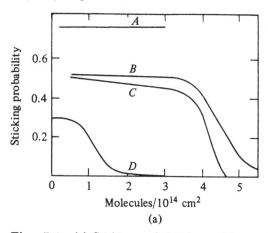

Fig. 5.4. (a) Sticking probabilities on W as a function of amount absorbed for different adsorbates (A ≡ Cs, B ≡ N_2 at 113 K, C ≡ CO at 340 K, D ≡ N_2 at 298 K.

the chemisorbed state to the rate of surface collisions. The sticking probabilities are almost invariably less than unity and possible reasons are: (i) the act of chemisorption may require an activation energy; (ii) the translational component of the kinetic energy of the gas molecule perpendicular to the surface may not be completely transferred to the adsorbate on impact and some desorption within the period of vibration in the pre-chemisorbed state takes place; and (iii) the incoming gas molecule collides with an occupied site and is back-scattered to the gas phase so that S_* is then proportional to the fraction of sites that are still unoccupied. This type of behavior for S_* is shown in Fig. 5.4(a) for curves B, C and D but not for A. It has been shown that S_* for Cs molecules at a W surface remains constant and approximately unity until an almost complete monolayer of Cs adatoms has been deposited. For such a case, it is thought that adsorption occurs on top of the primary chemisorbed layer with a sufficiently long lifetime in this precursor state that a high probability of transfer to the chemisorbed state occurs before desorption. Clearly, from Fig. 5.4(a), a variety of different S_* versus θ dependencies exists depending upon the particular chemical species involved. Fig. 5.4(b) shows that, as a function of temperature, two uniquely different adsorption states with different adsorption energies and adlayer thicknesses exist for Cd on W, while Fig. 5.4(c) indicates that S_* for As_2 and As_4 species on a GaAs surface depends on the flux of Ga atoms to the surface. Thus, very detailed modeling is needed to describe a specific case.

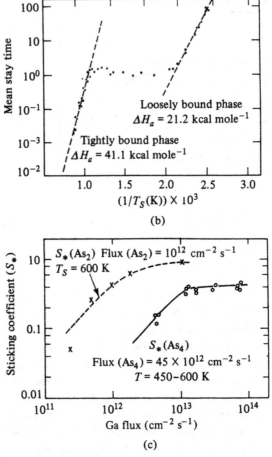

Fig. 5.4. (b) Surface lifetime for Cd adsorption on W; and (c) sticking coefficients of As₂ and As₄ on GaA(100) versus the flux of Ga.

The equilibrium surface population for a gas species of the j-type is given most generally by

$$\frac{n_s^j}{C_g^j h_a} = \exp\left(\frac{\Delta\mu_a^j}{\kappa T}\right) \tag{5.4a}$$

where n_s is the surface concentration, C_g is the gas concentration immediately adjacent to the surface, h_a is the adlayer thickness and $\Delta\mu_a$ is the chemical potential change between the gas and a particular type of adlayer state. The different adsorption isotherms available in the literature[1] make different approximations to $\Delta\mu_a$, make a few additional assumptions and then generate a specific relationship between

(P, T) and surface coverage, θ^j, of the adsorbate j. A few of these will be discussed below. An additional level of approximation, when one wishes to evaluate a material parameter, is to use the simple pair potential, dangling bond approach that is normalized to the heat of vaporization as in Eq. (2.1). Because of surface reconstruction and surface stress tensor effects, this is often an inadequate approximation.

People often approximate $\Delta\mu_a$ by the enthalpy of adsorption, ΔH_a, and neglect any entropy effects. As was seen in Chapter 3, this can lead to an overly packed adlayer at moderate to low temperatures and severely overestimates $\Delta\mu_a$. Here, the neglected entropy effects fall into two categories: (1) those involved in the transition from the gas phase to some type of condensed phase (liquid or solid) or a two-dimensional gas phase; and (2) those involved in entropy of mixing changes. Thus, to a better degree of approximation we can write $\Delta\mu_a$ in the form

$$\Delta\mu_a^j \approx \Delta H_a^j - \Delta\mu_{cond}^j - T\Delta S_{mix}^j \qquad (5.4b)$$

where

$$\Delta S_{mix}^j \approx -\kappa \left[\ell n \left(\frac{X_g^j}{1 - X_g^j} \right) - \ell n \left(\frac{X_a^j}{1 - X_a^j} \right) \right] \qquad (5.4c)$$

In Eq. (5.4b), ΔH_a is the enthalpy of adsorption at small coverage (just bonding to the substrate) while $\Delta\mu_{cond}$ is the chemical potential change associated with going from the gaseous state to some type of condensed state in the adlayer (largely bonding effects in the two-dimensional sheet). In the simple entropy of mixing expression, ΔS_{mix}, X_g and X_a refer to mole fractions in the gas and surface site fraction in the adlayer, respectively. Combining Eqs. (5.4) and rearranging leads to a useful expression for X_a; i.e.,

$$X_a^j \approx \left(\frac{C_T h_a}{N_s} \right)^{\frac{1}{2}} \left(\frac{1 - X_a^j}{1 - X_g^j} \right) \exp \left(\frac{\Delta H_a^j - \Delta\mu_{cond}^j}{2\kappa T} \right) \qquad (5.5)$$

In Eq. (5.5), $C_T = P_T/\kappa T$ where P_T is the total gas pressure and N_s is the number of possible adsorption sites on the surface. The main new feature here is the factor of 2 in the exponent which leads to a significant reduction in X_a^j over that predicted by the various standard adsorption isotherms which all provide some sort of approximation for $\Delta\mu_{cond}$. Although we shall treat a few of these standard isotherms below, a meaningful treatment really requires the full-blown effort of a computer calculation utilizing semiempirical PEFs.

The common postulates shared between the three simplest isotherms (Langmuir, Freundlich and Temkin)[1] are: (i) the adsorbent surface is planar and the total number of adsorption sites is constant under all experimental conditions; (ii) the adsorption is restricted to a monolayer; (iii) on impact with an unoccupied or free site, the adsorbate molecule is adsorbed with zero activation energy; (iv) an occupied site comprises one adsorbate molecule at a single site and collisions of gaseous molecules with an occupied site are perfectly elastic; (v) desorption of an adsorbed molecule occurs as soon as it has acquired sufficient thermal energy to equal the heat of adsorption; and (vi) at equilibrium, the rate of adsorption on the unoccupied sites equals that of desorption from the occupied sites.

For the Langmuir adsorption isotherm, two additional postulates were made: (1) the adsorptive properties of all surface sites are identical; and (2) there are no lateral interactions between neighboring adsorbed molecules. Thus, the adsorption rate and desorption rate, respectively, are given by

$$\left(\frac{dn_s}{dt}\right)_a = k_a P(N_s - n_s) \tag{5.6a}$$

and

$$\left(\frac{dn_s}{dt}\right)_d = k_d n_s \tag{5.6b}$$

Here, k_a is the rate constant per site at unit pressure. Eq. (5.6a) is linear in pressure because the collision flux to the surface, J, is proportional to P; i.e.,

$$J = P/(2\pi m \kappa T)^{\frac{1}{2}} \tag{5.7}$$

where m is the mass of species colliding with the surface. At equilibrium, Eqs. (5.6a) and (5.6b) are equated leading to

$$P = K\theta/(1 - \theta) \tag{5.8a}$$

where

$$K = k_d/k_a = K_0 \exp(-\Delta\mu_a/\kappa T) \tag{5.8b}$$

and $\theta = n_s/N_s$. This is the famous Langmuir adsorption isotherm where $\Delta\mu_a \approx \Delta H_a$ is generally used.

The time dependence of adlayer build-up can be gained by combining Eqs. (5.6) to give

$$\frac{dn_s}{dt} = k_a P N_s - (k_d + k_a P)n_s \tag{5.9a}$$

For constant values of $k_a P N_s$ and $k_d + k_a P$, $n_s(t)$ is given by

$$n_s(t) = \frac{k_a P N_s}{(k_d + k_a P)}\{1 - \exp[-(k_d + k_a P)t]\} \tag{5.9b}$$

We thus see that it is an exponentially rising curve with a well-defined saturation level that could have been gained directly by inverting Eq. (5.8) to give a relationship for θ.

For many real systems of interest, we can expect at least two factors to modify Eqs. (5.9). The first is that $\Delta\mu_a$, and thus k_d/k_a, will be a function of n_s because of a variety of surface relaxation and reconstruction events associated with the adsorption. The second is that there will be additional losses from the pool of monomer adspecies due to dimer, trimer, etc., or other chemical species formation plus losses by diffusion to any nearby ledges. Thus, Eq. (5.9a) should really be written as

$$\frac{dn_{s1}}{dt} = k_a\{PN_S - [K(n_{s1}) + P]n_{s1}\} - \sum_{m=2}^{\infty}\left(\frac{mdn_{sm}}{dt}\right) + \cdots (5.9c)$$

For many semiconductor systems, it is the $K(n_{s1})$ term that is most important. In the literature,[1] departures from Eqs. (5.8) have been developed to account for (i) geometric factor changes influencing the surface "uniformity" assumption and (ii) effective adsorption potential variation with θ.

The Freundlich equation was one of the earliest relationships used to describe adsorption isotherms. It may be derived by assuming that the heat of adsorption decreases logarithmically with increase of θ and it relates the amount adsorbed, v, to a fractional power of the pressure at constant temperature, i.e.,

$$v = \text{constant} \cdot P^{\frac{1}{n}} \tag{5.10a}$$

where n is usually greater then unity. The Temkin equation can be derived by assuming that the heat of adsorption decreases linearly with increase of θ and leads to the relationship

$$v = \text{constant} \cdot \ell n P + \text{constant}. \tag{5.10b}$$

These isotherms are largely considered to be isotherms of the Fowler-type where the adsorbed molecule is completely localized, each in its own potential well. However, it is also useful to consider the case where the adsorbed species is quite mobile along the substrate and behaves somewhat like a two-dimensional gas. Such a mobile isotherm is of the Hill–deBoer-type.[1,2] Let us briefly consider such a mobile adsorption isotherm (MAI).

We shall consider only a single component surface species, like $C\ell$ on Cu, to illustrate the new principles involved. We shall assume that the gas atoms adsorbed on the surface behave as a two-dimensional gas with a spreading pressure φ. Here, we have the analogue of a van der Waals' equation of state in two dimensions

$$\left(\varphi + \frac{a'}{b^2}\theta^2\right)(1 - \theta) = \frac{\theta\kappa T}{b} \tag{5.11a}$$

where $\theta = n'b/A$, A is the substrate area, n' is the number of adsorbed atoms, b is the area per adsorbed atom and a' is related to the molecule–molecule binding energy in the adlayer. Using the Gibbs adsorption equation

$$\frac{\mathrm{d}\varphi}{\mathrm{d}\ell n P} = -\frac{\kappa T n'}{A} = -\frac{\kappa T \theta}{b} \tag{5.11b}$$

Placing Eq. (5.11b) into Eq. (5.11a) and integrating leads to the MAI, i.e.,

$$\frac{P}{P^*} = K_2 \left(\frac{\theta}{1-\theta}\right) \exp\left(\frac{\theta}{1-\theta}\right) \exp(-K_1\theta) \tag{5.11c}$$

where $K_1 = 2a'/b\kappa T$ is the adatom–adatom binding energy, and K_2^{-1} is proportional to the adatom–substrate binding energy and P^* is the equilibrium vapor pressure of the adsorbed substance. Thus, as K_2 decreases, θ increases for a given increment of external pressure. For fixed K_2, Fig. 5.5(a) illustrates the variation of θ versus P/P^* as K_1 changes. For $K_1 <\sim 5.5$, only a single and probably somewhat immobile phase is present on the surface. However, for $K_1 >\sim 5.5$, a miscibility gap develops and the adlayer contains a mixture of immobile and condensed or liquid-like phases. Similar behavior is seen in Fig. 5.5(b) associated with K_2 variations for fixed K_1. Thus, for a given K_2 and K_1, we may find that, for $P/P^* < (P/P^*)_c$, only an immobile phase isotherm exists while, for $P/P^* > (P/P^*)_c$, only a liquid-like isotherm exists with the two phases coexisting at $P/P^* = (P/P^*)_c$ illustrated in Fig. 5.5(c). As K_1 and/or K_2^{-1} increase, $(P/P^*)_c$ decreases.

Considering Fig. 5.5 as a general phenomenon for adlayer behavior, one can anticipate three important domains for condensation:

(1) $K_2^{-1} \gg K_1$ (small K_2); here we have a high desorption energy relative to the sublimation energy so that no supersaturation is possible because condensation begins in the undersaturated state leading to layer-by-layer condensation.

(2) $K_1 \approx K_2^{-1}$; here we have the classical region of nucleation and growth where less than monolayer coverage occurs and moderate supersaturation is possible before layer condensation occurs.

(3) $K_2^{-1} \ll K_1$ (large K_1); here we have a low desorption energy compared to the heat of sublimation of the adsorbate so that very low surface coverages should occur ($\theta \sim 10^{-6}$) and very high supersatura-

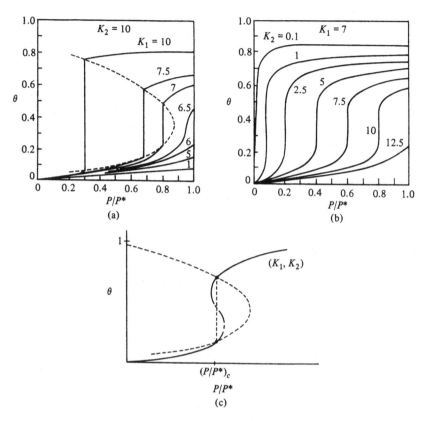

Fig. 5.5. (a) Adsorption isotherms for $K_2 = 10$ and various values of K_1; (b) adsorption isotherms for $K_1 = 7$ and various values of K_2; and (c) the general isotherm for the (K_1, K_2) set showing gas-like behavior at $P/P^* < (P/P^*)_c$ and liquid-like behavior at $P/P^* > (P/P^*)_c$.

tions are required to produce heterogeneous nucleation via very small clusters. One cannot apply classical nucleation theory in this regime.

The foregoing is simple in concept and considerable variation is possible due to the intermolecular PEFs involved. It also conceptually delineates a value of $(P/P^*)_c$ for a transition to an altered D_s state. Although only one adsorbate species was considered, in principle the general method could be extended to a multicomponent adsorbate. However, the method does assume a gas-like starting point for the two-dimensional adlayer and we have no clues to the types of PEF that should be consistent with this assumption. It will be necessary to perform Monte Carlo or molecular dynamics calculations using a many-body PEF before one can hope to understand the complete picture;

however, the present parametric scheme does provide some useful insight.

5.1.2   Surface diffusion

When one considers transport in the adlayer, simple models lead one to think that, as the surface adsorption of a foreign constituent increases, both D_s and $\bar{X}_s$ for the crystal-forming species should decrease. This is often true when the foreign constituent does not interact strongly with the substrate and adatoms to alter the surface reconstruction and thus the magnitude of the atomic potential wells. It is generally true for the surface self-diffusion of Cu as indicated in Fig. 5.6(a); however, it is not true when $C\ell$ is present on the surface.[3] The enhancement factor of $\sim 10^4$ in D_s^{Cu} is shown more fully in Fig. 5.6(b) and we see from Figs. 5.6(c) and (d) that such an enhancement also occurs for Pb, $T\ell$ and Bi as foreign species on copper as well as Pb and Bi on Au and S on Ag. These results indicate that we are dealing with a general phenomenon and that it occurs at a vapor pressure of the adsorbate that is below the saturation pressure for condensation of the adsorbate phase. How are we to understand this?

One possible rationale for the D_s^{Cu} behavior illustrated in Fig. 5.6 is that, as the surface coverage of adsorbate ($C\ell$, for example) increases so that a surface adsorbed Cu atom has an increasing number of $C\ell$ neighbours, the Cu–$C\ell$ interaction weakens the Cu–Cu binding so that, even though a larger activation energy exists for Cu surface self diffusion as indicated in Figs. 5.6(a) and (c), the vibrational frequency factor and the entropy of activation increase sufficiently to more than compensate. This is what one might expect if the adlayer exhibited two-dimensional liquid-like behaviour. If $(P/P^*)_c$ for $C\ell$ on Cu is $\sim$ 0.1–0.2, then the data of Fig. 5.6(d) would be associated with the formation of a two-dimensional liquid layer of $C\ell$ wherein the Cu can readily diffuse. One possible scenario is that the Cu–$C\ell$ interaction is such that both the Cu adatom concentration in the two-dimensional liquid-like layer and the kink density at the Cu ledges significantly increase. This would definitely increase the effective surface diffusion coefficient of Cu.

Most surface diffusion studies really monitor the flux of transporting species under some known driving force and implicitly use the relationship

$$J^j = \frac{-D_s^j C_s^j}{\kappa T} \nabla_s \mu_s^j \tag{5.12a}$$

to obtain D_s^j. The unstated assumption in an experiment with a foreign

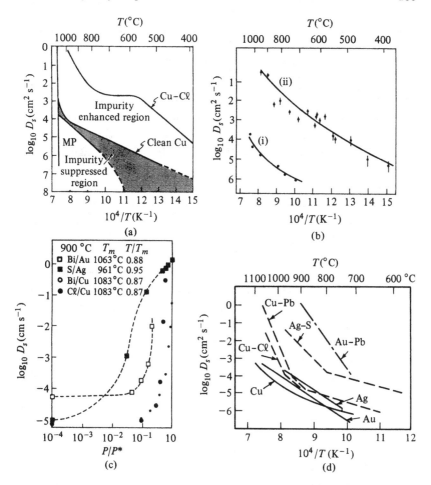

Fig. 5.6. (a) Surface diffusion of Cu on Cu showing significant enhancement in the presence of $C\ell$ and suppression by some unspecified impurity; (b) the same as (a) for pure Cu (i) and Cu in a $C\ell$ atmosphere (ii); (c) temperature variation of D_s^j for a variety of adsorbates; and (d) increase of D_s^j for j = Au, Ag and Cu with increasing adsorbate gas pressure (P^* is the adsorbate pressure above which condensation or compound formation occurs).

adsorbate species is that only D_s^j is changing whereas, in actual fact, both C_s^j and $\nabla_s \mu_s^j$ are also changing. Thus, if we neglect changes in $\nabla_s \mu_s^j$, the $D_{s\;eff}^j$ changes include the changes in both D_s^j and C_s^j and the activation energy for the process will be the sum of those pertaining to the two parameters.

For D_s^j, itself, an Einstein-type of relationship should hold yielding

$$D_s^j = X_{Vs}D_{Vs}^j = \frac{f}{4}\lambda_s^2\nu_s^j e^{(\Delta S_m^j/\kappa)} \exp\left(\frac{\Delta S_m^j}{\kappa}\right) \exp\left(-\frac{\Delta H_m^j}{\kappa T}\right) X_{Vs}$$

$$(5.12b)$$

where X_{Vs} is the surface fraction of vacant sites, f is the correlation factor, λ_s is the surface jump distance which may be more than an intersite distance, ν_s^j is the surface vibrational frequency while the subscript m relates to migration. We thus see that D_s^j, itself, can be strongly limited by the fraction X_{Vs} of available surface vacancies. As one goes to lower substrate temperatures, the degree of surface adsorption is expected to increase exponentially so that X_{Vs} strongly decreases and surface transport must also strongly decrease.

Overall, the foregoing discussion indicates that, at too high a temperature, $D_{s\ eff}^j$ is limited by reduced solubility in the adsorption layer because an immobile isotherm holds and, at too low a temperature, it is again limited by a greatly reduced population of surface vacancies. Between these two limits, for systems with the proper values of K_1, K_1', K_2 or K_2' the magnitude of $D_{s\ eff}^j$ can be enhanced (K_1' and K_2' refer to the strength of the Cu–Cℓ and Cu–substrate interaction, respectively, for the Cu + Cℓ vapor example).

The surface population of a species j is strongly determined by its surface lifetime, τ_s^j, given by

$$\tau_s^j = \frac{1}{\nu_s^j} \exp\left(\frac{\Delta\mu_a^j}{\kappa T}\right)$$

$$(5.13a)$$

so that the mean diffusion distance $\bar{X}_s^j$ of the j species before evaporation from the surface layer is

$$\bar{X}_s^j = (4D_s^j\tau_s^j)^{1/2}$$

$$(5.13b)$$

Thus, we see that the solubility of j in the surface adlayer can greatly increase τ_s^j and also greatly increase $\bar{X}_s^j$.

5.1.3 *Surface reconstruction, cluster formation and nucleation*

The *idealized surfaces* (dangling bond models), used theoretically by most investigators since the fifties, can relax only by adatom roughening and adatom clustering. *Real surfaces* relax first by ledge and surface reconstruction and only thereafter by adatom roughening and clustering. This is because the surface reconstruction process leads to a surface stress tensor which drives/influences *all* other subsequent relaxation. Sometimes ledges alter the terrace reconstruction so beneficially that $\gamma_\ell < 0$ as is found in PEF computations for GaAs($\bar{1}$00)/[$\bar{1}$10] ledges.

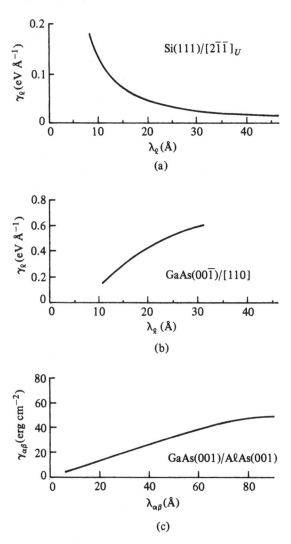

Fig. 5.7. Variation of calculated $[2\bar{1}\bar{1}]$ upper ledge energy, γ_ℓ, on Si(111) as a function of ledge spacing, λ_ℓ; (b) calculated [110] ledge energy on GaAs(00$\bar{1}$); and (c) calculated GaAs/AℓAs interfacial energy, $\gamma_{\alpha\beta}$, for a superlattice structure as a function of interlayer spacing, $\lambda_{\alpha\beta}$.

Sometimes surface point defect formation interacts with the reconstruction/stress tensor pattern in such a favorable way that $\Delta E_{PD} < 0$ and spontaneous surface Frenkel defect formation occurs up to some limit. This mechanism is observed to be a part of the Si(111) − (7 × 7) reconstruction pattern. These effects are discussed in more detail in Section

8.3.1 with the important point to be noted here being illustrated in Fig. 5.7.[4] We see that PEF-calculated γ_ℓ-values for particular ledges on Si(111) and on GaAs($\bar{1}$00) are still changing with ledge separation out to values of $\lambda_\ell \sim 50$ Å$\gtrsim 10$ times the atomic force cut-off distances. This can only have occurred through the agency of the terrace stress tensor. In Fig. 5.7(c), we see that a similar effect occurs in a GaAs/AℓAs multilayer system where the interfacial energy increases with layer spacing out to a spacing of ~ 100 Å. This long-range effect is thought to be associated with the interface stress tensor generated by interface reconstruction. A similar type of result is found for the formation of a GaAs layer on Si as a function of GaAs layer thickness; i.e., the excess energy for the Si/GaAs plus GaAs/vacuum interfaces increases with GaAs layer thickness up to some saturation value at a particular layer thickness. This means that a ready explanation exists for two-dimensional pill-box nucleation in the first few monolayers of a new phase followed by much more rapid nucleation (at fixed driving force) in subsequent layers which allows the transition from two-dimensional crystallite growth to three-dimensional crystallite growth. This explains the Stranski–Krastanov mechanism.

In addition to the foregoing, one expects a marked effect to develop with respect to surface cluster formation because some portion of the ledge on one side of a pill-box cluster is interacting with some portion of ledge on the other and the reconstruction on the top of the cluster should be a function of cluster size. Thus, the energetics of cluster formation, $\Delta E(i)$ for a cluster of size i, is given for a circular disk-shaped cluster of height h by

$$\Delta E(i) = ai(\gamma_{SC} + \gamma_{CV} - \gamma_{SV}) + (4\pi ai)^{1/2} h\gamma_\ell \qquad (5.14)$$

In Eq. (5.14), a is the area per molecule of the cluster while the subscripts S, C and V refer to substrate, cluster and vapor respectively. For homogeneous nucleation on a substrate (i.e., Si on Si), one generally neglects the first term of Eq. (5.14). However, from Fig. 5.7 we can see that this may not be such a good idea. Using PEF calculations, this idea was tested on both Si and GaAs homodeposition and the results are shown in Fig. 5.8. If a simple dangling bond model could be applied to this type of cluster formation, then the first term of Eq. (5.14) could be neglected and $\Delta E(i) \propto i^{1/2}$ with a proportionality constant depending on the geometry and ledge energy of the cluster. Thus, the data on $\Delta E(i)$ is plotted versus $i^{1/2}$ against a backdrop of straight lines predicted by idealized theory. We note that, for the Si case, it is very

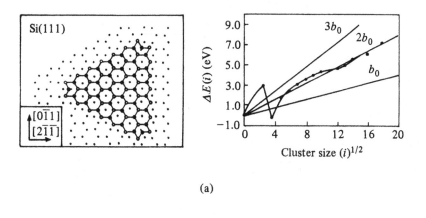

(a)

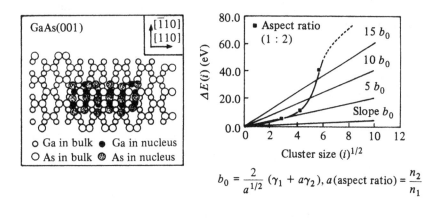

$$b_0 = \frac{2}{a^{1/2}} (\gamma_1 + a\gamma_2), \, a \, (\text{aspect ratio}) = \frac{n_2}{n_1}$$

(b)

Fig. 5.8. (a) Calculated Si cluster excess energy, $\Delta E(i)$ containing i atoms on the Si(111) plane as a function of cluster perimeter parameter $i^{1/2}$; and (b) calculated GaAs cluster excess energy containing $i/2$ molecules.

difficult to form hexamers but decamers should form spontaneously and only when the cluster has grown larger than 100 atoms does one attain the idealized model limit. For the GaAs case, the optimum cluster shape is rectangular with the long side being a [$\bar{1}$10] ledge (having negative energy in a parallel array of ledges) and $\Delta E(i)$ departs very strongly from the idealized model. These results for GaAs would give a cluster population $\sim 10^{10}$ greater than the idealized model ($\gamma_\ell \sim 20b_0$) at ~ 500 °C if the $5b_0$ line (small clusters) was used to determine γ_ℓ.

In the classical theory of nucleation (see Chapter 8), the steady state formation rate of critical sized nuclei, I, is given by

$$I = Z\beta S(i^*)n_s(i^*) \tag{5.15a}$$

where $n_s(i^*)$ is the equilibrium concentration of critical sized nuclei involving i^* molecules; i.e.,

$$n_s(i^*) = n_s(1)\exp(-\Delta G^*/\kappa T) \tag{5.15b}$$

In Eq. (5.15b), $n_s(1)$ is the concentration of monomer on the surface needed for cluster formation while ΔG^* is the free energy fluctuation needed for formation of the critical sized clusters involved in the surface nucleation process. In Eq. (5.15a), $S(i^*)$ is the peripheral area of the critical sized cluster involved in the growth to supercritical size, β is the rate of passage of monomer over the activation barrier to join the critical sized embryo and Z is the "Zeldovich" factor which numerically converts the equilibrium cluster population to the steady state cluster population.

The factor ΔG^* in Eq. (5.15b) depends upon the geometry of the clusters under consideration and is given, most generally, by

$$\Delta G^*(i^*) = \Delta E(i^*) - i^*|\Delta\mu_V| \tag{5.15c}$$

In Eq. (5.15c), $\Delta E(i^*)$ is the excess surface energy associated with forming a cluster of size i^* while $\Delta\mu_V$ is the bulk chemical potential change associated with the molecular change from the vapor state to the condensed state; i.e.,

$$\Delta\mu_V = -\kappa T\ell n\hat{\sigma} \tag{5.15d}$$

where $\hat{\sigma}$ is the molecular supersaturation given by

$$\hat{\sigma} = n_s(1)/n_s^*(1) \approx (P_i/P_i^*)t = (J_i/J_i^*)t \tag{5.15e}$$

In Eq. (5.15e), P_i is the vapor pressure at the gas–adlayer interface while P_i^* is the equilibrium vapor pressure, $n_s^*(1)$ is the equilibrium adlayer concentration of key monomer species, J_i is the surface flux of this species from the gas phase while J^* is the flux associated with the equilibrium vapor at P^*. The latter two relationships apply at small coverage where we are dealing with the initial slope region of Eq. (5.9b).

Having calculated $\Delta E(i^*)$, Eqs. (5.15c)–(5.15e) will tell us the supersaturation conditions needed for a uniform attachment type of deposition either on a face or on a ledge (generation of kinks). Thus, the considerations dealt with at length in Chapter 2 should be applied here. Because we have found that γ_ℓ is a function of both ledge spacing and ledge height, we need to realize that the γ-plot is no longer a function of just orientation but is a function of layer height, h, as well. In addition, it

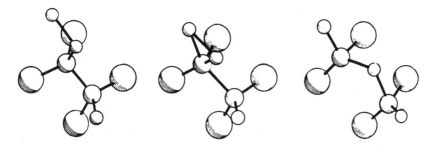

Fig. 5.9. Illustration of the predominant mechanism for stable dimer opening leading to unreconstruction on Si(100) and epitaxial growth.

has recently been found for Si that the surface reconstruction process is a thermally activated process so that γ_ℓ will also be a function of ledge velocity, V_ℓ. Thus, for dynamic processes, it is important to recognize that $\gamma = \gamma(\theta, \phi, h, V_\ell)$ and not just a function of the angles θ and ϕ as is generally thought.

Gossman and Feldman[5] investigated the effects that surface reconstruction had on the growth of Si single crystals via the use of atomic Si beams. Their studies have indicated that the minimum substrate temperature needed to grow epitaxial layers on Si(111) is 570 K while that needed for Si(100) is 790 K. This difference in epitaxial growth temperature is thought to be due to the differences in energy needed to reorder the different reconstruction patterns on these surfaces. It has also been found that, for Si beam deposition with energies in the 10–65 eV range for Si(100), good epitaxial growth is obtained at very low temperatures. However, for beam deposition energies above 100 eV, implantation damage is observed as far as 400 Å below the surface.

Using a PEF calculational approach, Garrison, Miller and Brenner[6] made molecular dynamics simulations for the Si(100) − (2 × 1) reconstructed surface (see Fig. 2.6) and found the predominant mechanism for stable dimer opening, which is necessary for epitaxial growth, to be that illustrated in Fig. 5.9. In the initial stage, two of the adatoms occupy the remaining dangling bonds of the dimer and one of them also becomes bonded to a third adatom. A rearrangement of the dimer is induced by this extra diffusing adatom so that one of the initial adatoms is inserted into the dimer while the third diffusing adatom occupies the vacated dangling bond. The initially dimerized atoms now occupy sites which are characteristic of a bulk terminated lattice. These authors also studied the effects of excess deposition energies in the 0.026–20 eV

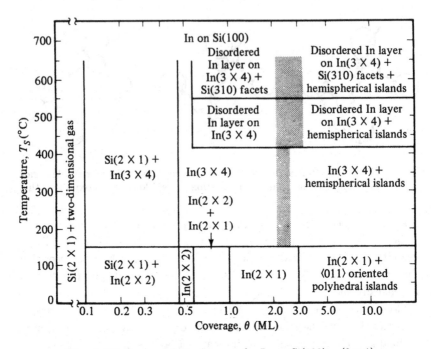

Fig. 5.10. Surface phase diagram for In on Si(100) − (2 × 1).

range. For kinetic energies of 0.026 eV and perpendicular incidence on a dimerized surface, no significant motion of the dimers occurred. However, for energies in the 5–10 eV range, significant motion of the dimers occurred which led to epitaxial atomic configurations. For incoming atoms with excess energies in the 20 eV range, implantation into the lattice was observed. Experimental work by Iyer, Heinz and Loy[7] showed the activation energy for surface unreconstruction and regrowth to be 1.2 ± 0.2 eV for the Si(111) − (7 × 7) faces. The effect of ledges and ledge spacing on the activation energy for unreconstruction has not yet been determined. The effects of various kinds of impurities on the unreconstruction activation energies is also unknown at present.

A careful study of In overlayers on clean Si(100)−(2×1) surfaces[8] has provided many valuable insights concerning the growth of films. Over the entire temperature range investigated, 30–600 °C, In was found to interact with clean Si(100) surfaces through a Stranski–Krastanov type of mechanism wherein the initially deposited In nucleated and grew two-dimensionally up to a coverage of between two and three monolayers (ML) depending upon the substrate temperatures, T_s, after which three-dimensional islands were formed. These results are illustrated in a type of phase diagram of the (T_s, θ)-type as shown in Fig. 5.10. All depositions were carried out at 0.01 ML s^{-1} < J^{In} < 1.0 ML s^{-1}. At 30 °C < T_s < 150 °C, In formed a two-dimensional gas at coverages, θ, below 0.1 ML and an In(2 × 2) structure with chains of In dimers along $\langle 011 \rangle$ directions at 0.1 ML < θ < 0.5 ML. Further increase in θ results in a transition to an In(2 × 1) pattern while polyhedral-shaped single crystal islands elongated along the [011] and [0$\bar{1}\bar{1}$] directions were observed at θ > 3 ML. For 150 °C < T_s < 420 °C, the In overlayer exhibited an In(3 × 4) structure and hemispherical In islands were observed at $\theta \gtrsim 2$ ML. For T_s > 420 °C, a disordered In layer formed on top of the In(3 × 4) phase at $\theta \gtrsim 0.5$ ML. The introduction of even minute amounts of contaminants (O$_2$ or C at ~ 0.01 ML) prevented the formation of the Si(100)/In(2 × 2) phase at low T_s and resulted in the growth of epitaxial hemispherical FCC In islands on Si(100)/In(2 × 1) at $\theta \gtrsim 2$ ML.

Modulated-beam desorption spectra showed that, for the θ < 0.07 ML and T_s > 450 °C domain, $\Delta H_a^{In} = 2.8 \pm 0.2$ eV and $\nu_s^{In} = 3 \times 10^{14 \pm 1}$ s^{-1}. First order desorption from a single phase was observed. For the three-dimensional In islands, the activation energy for desorption was 2.5 eV which is very close to the enthalpy of evaporation for bulk In ($\Delta H_a^{In} = 2.47$ eV). The binding energy of In in the (3 × 4) phase was found to be 2.85 eV while the disordered two-dimensional phase that adsorbed on top of the (3 × 4) phase had a binding energy of 2.45 eV.

For metal overlayers on Si, one needs to recognize that many three-dimensional metastable silicide phases are generally possible for each metal, M, so that a variety of two-dimensional reconstruction patterns (two-dimensional silicide phases) are also possible. The interface energetics for formation of these phases involves not only the quasi chemical contribution to γ as discussed earlier for the GaAs/AℓAs multilayers and as discussed at length in Chapter 7, but it also involves the other contributions to γ, particularly the electronic contribution, γ_e, which will also be a function of layer thickness (see Chapter 7). The point here is that, depending upon the variation of γ_c, γ_e and γ_σ with layer thickness and layer structure, one can readily account for Stranski–Krastanov versus VW types of film growth behavior.

5.2   Mathematical aspects of film growth at vicinal surfaces

By treating vicinal surfaces only, we can temporarily neglect any difficulties associated with nucleation. Here, we will consider only the

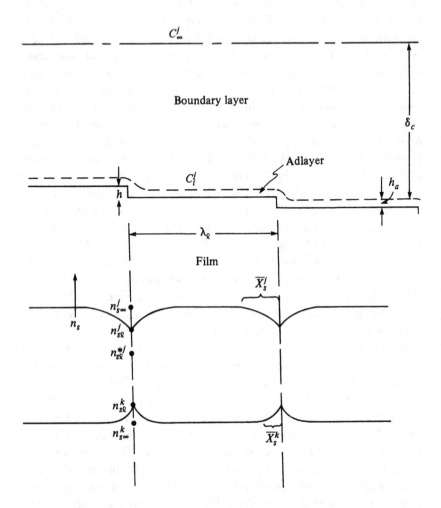

Fig. 5.11. Schematic illustration of bulk gas, surface boundary layer, adlayer and vicinal surface plus concentration profiles of the key Si-bearing adlayer species (j) and rejected adsorbant species at the layer edges (k).

steady state growth of ledges and will be satisfied with an approximate mathematical result that retains all the essential physics of the situation.

During film growth from the vapor phase, the surface terrace concentration of the j species far from a ledge, $n_{s\infty}^j$, is given approximately by

$$n_{s\infty}^j \approx J^j \tau_s^j \tag{5.16a}$$

where τ_s is given by Eq. (5.13a) and the exact equation comes from Eq. (5.9b). For a simple gas/surface reaction at vapor pressure, P, and this approximate type of formulation,

$$J^j = (P^j - P^{*j})/(2\pi m^j \kappa T)^{\frac{1}{2}} \tag{5.16b}$$

which is slightly different than Eq. (5.7), because we are dealing here with the net flux.

In Fig. 5.11, the growth of a vicinal surface of ledge spacing λ_ℓ and ledge height h is considered to be growing at velocity V due to a net flux J of the j species from the gas phase where

$$J_g^j = D_g^j \frac{(C_\infty^j - C_i^j)}{\delta_c} = \frac{D_g^j}{\delta_c} \frac{(P_\infty^j - P_i^j)}{\mathcal{R}T} \tag{5.16c}$$

where D_g is the gas phase diffusion coefficient and C_i^j is the average gas concentration at the edge of the adlayer. In the surface adsorption layer of thickness h_a, the lateral flux of species to the ledges (for $\bar{X}_s \ll \lambda$) is given by

$$h_a J_s^j \approx 2D_s^j \frac{(n_{s\infty}^j - n_{s\ell}^j)}{\bar{X}_s^j} = \frac{hV_\ell}{\Omega^j} \tag{5.16d}$$

$$= \frac{2aV_k}{\lambda_k} \left(\frac{h}{\Omega^j} \right) = \beta_0'(n_{s\ell}^j - n_{s\ell}^{*j}) \tag{5.16e}$$

using Eqs. (2.21) and (2.22) and the relationship

$$\Delta G_k = \mathcal{R}T\ell n(n_{s\ell}^j / n_{s\ell}^{*j}) \tag{5.16f}$$

with $\bar{X}_s$ being given by Eq. (5.13b). Thus, β_0' is an orientation-dependent, interface attachment coefficient. Here, β_0' can be used to account for kink site only attachment, uniform ledge attachment, terrace unreconstruction-limited attachment, etc. For this case, the excess energy, ΔG_E, is composed of the usual strain and defect parts but also a transport part due to the diffusion of species k (either an impurity or a component of the species) away from the ledges, i.e.,

$$\Delta G_E = \Delta G_\sigma + \Delta G_D + \Delta G_C^k \tag{5.17a}$$

where

$$\Delta G_C^k = \left[\mathcal{R}T\ell n \left(\frac{n_{s\ell}^k}{n_{s\infty}^k} \right) \right] \left(\frac{J_s^k}{J_s^j} \right) \tag{5.17b}$$

The ratio of fluxes is involved because the analysis is based on either a per atom or per mole of species j basis.

From Eqs. (5.16), we have

$$V = \frac{hV_\ell}{\lambda_\ell} = \frac{\Omega^j}{\lambda_\ell} \left(\frac{n_{s\infty}^j - n_{s\ell}^{*j}}{1/\beta_0' + \bar{X}_s^j/2D_s^j} \right) \tag{5.18a}$$

with

$$n_{s\infty}^j = (C_i^j h_a) \exp(\Delta G_{a\infty}^j / \mathcal{R}T) \tag{5.18b}$$

$$n_{s\ell}^{*j} = \left(\frac{P^{*j}}{\mathcal{R}T} h_a \right) \exp(\Delta G_{a\ell}^j / \mathcal{R}T) \tag{5.18c}$$

and

$$P^{*j} = P_0^{*j} \exp(\Delta G_E / \mathcal{R}T) \tag{5.18d}$$

In Eqs. (5.18), $\Delta G_{a\infty}$ and $\Delta G_{a\ell}$ are the adsorption free energies at the center of the terrace and at the ledge, respectively. The ΔG_E contribution enters the equilibrium vapor pressure because one is producing a film with effectively a higher free energy than the standard state film involved in the determination of ΔG_∞ and P_0^*. Our final step is to apply conservation between the gaseous flux (Eq. (5.16c)) and film formation rate; i.e.,

$$V = t\Omega^j J_g^j \tag{5.19}$$

Combining Eqs. (5.18) and (5.19), one obtains

$$V = \Omega^j \left\{ \frac{\epsilon C_\infty^j - (P_0^{*j}/\mathcal{R}T) \exp[(\Delta G_{a\ell}^j - \Delta G_{a\infty}^j + \Delta G_E)/\mathcal{R}T]}{\delta_c/D_g^j + \left(1/\beta_0' + \bar{X}_s^j/2D_s^j\right)(\lambda_\ell/h_a) \exp(-\Delta G_{a\infty}^j/\mathcal{R}T)} \right\} \tag{5.20}$$

where $0 < \epsilon < 2x_g^j / \lambda_l$ is the collection efficiency and thus V is given in terms of the key variables (C_∞^j, λ_l, δ_c, T and ΔG_E) and all the key materials parameters of the problem. From Eq. (5.20), one sees that, at a given (T, θ, δ_c), V is linear in C_∞^j and has an intercept proportional to P_0^{*j}. Although most people prefer to plot V versus $1/T$ for the analysis of their data (see Fig. 5.15(a)), it is much more preferable to evaluate plots of $(\partial V/\partial C_\infty^j)^{-1}$ versus $1/T$, λ_l and gas flow velocity if one wishes seriously to study the basic processes involved. For the MBE technique, the analogous plots would be with $(\partial V/\partial J^j)^{-1}$ or $(\partial V/\partial T_s)^{-1}$ where T_s refers to the source temperature. Of course, in a certain range of C_∞^j, ΔG_E may no longer be independent of the bulk gas composition and data evaluation becomes even more complex. When the film growth technique involves the formation of a variety of chemical intermediates on the surface then the data analysis complexity increases even further.

From Fig. 5.11 one can note that, as T decreases for fixed J^j, $n_{s\infty}^j$ will increase because τ_s^j increases, $\bar{X}_s^j$ decreases because D_s^j decreases and the supersaturation between ledges, $\hat{\sigma}^j = n_{s\infty}^j / n_{s\infty}^{*j}$, increases. Thus, V decreases and the probability of two-dimensional nucleation on the terraces between ledges increases. For small values of $\hat{\sigma}^j$, largely unfaulted two-dimensional nucleation occurs and only minor chemical segregation defects are generated as the layer edges grow together. For larger values of $\hat{\sigma}^j$, faulted two-dimensional nucleation should be anticipated and, for very large $\hat{\sigma}^j$, even amorphous film deposition is expected (see Fig. 5.2(b)).

When $\bar{X}_s \gtrsim \lambda_\ell/2$, movement of the ledge is neglected in solving the surface diffusion equation (small Péclet number domain) and a symmetrical surface flux is assumed about the mid-point of the terrace. The terrace concentration profile is given by

$$\frac{n_s(x) - n_{sM}}{n_{s\ell} - n_{sM}} = \frac{\cosh\left[(x - \lambda_\ell/2)/(\bar{X}_s/2)\right]}{\cosh(\lambda_\ell/\bar{X}_s)} \tag{5.21a}$$

where, here n_{sM} refers to the mid-point of the terrace at $x = \lambda_\ell/2$. This altered concentration profile changes Eq. (5.16d) yielding

$$\frac{hV_\ell}{\Omega^j} = \frac{4D_s}{\bar{X}_s}(n_{s\infty} - n_{s\ell})\tanh\left(\frac{\lambda_\ell}{\bar{X}_s}\right) \tag{5.21b}$$

so that Eq. (5.20) holds but with $\bar{X}_s^j/2D_s^j$ multiplied by $[2\tanh(\lambda_\ell/\bar{X}_s)]^{-1}$. For this case, we also need an expression for n_{sM} which is obtained by using Eq. (5.21a) at $x = \lambda_\ell/2$ so that

$$n_{sM} = n_{s\infty} + \frac{(n_{s\ell} - n_{s\infty})}{\cosh(\lambda_\ell/\bar{X}_s)} \tag{5.21c}$$

Since $n_{s\infty}$ is given by Eq. (5.16a) we see that n_{sM} decreases as $\lambda_\ell/\bar{X}_s$ decreases and has the limits $n_{s\ell} < n_{sM} < n_{s\infty}$.

As stated earlier, the cosh form of solution given by Eq. (5.21a) holds in the small Péclet number domain where $\hat{P} = V_\ell\lambda_\ell/2D_s$. Since $V = hV_\ell/\lambda_\ell$, $\hat{P} = V\lambda_\ell^2/2hD_s$; thus, provided $V \sim 1$ Å s^{-1}, such as one finds for MBE, $\hat{P} \ll 1$ and Eq. (5.21a) holds. However, for CVD with $V \sim 10$ μm min^{-1} and orientations close to the facet plane where $\lambda_\ell \sim 10^{-4}$ cm, we find that $\hat{P} \gtrsim 1$ if diffusion in the adlayer is slowed down by strong surface adsorption. As shown in Chapter 3 of the companion book,[9] interface field effects in the plane of the surface can be incorporated into the surface transport equation as a modified value of $V_\ell = V_\ell'$. For this case, if the new $\hat{P} = V_\ell'\lambda_\ell/2D_s \ll 1$, then Eq. (5.21$a$) can be used. This is important because, if $\hat{P} \gtrsim 1$, the in-plane field effect leads to an asymmetrical potential distribution around a ledge so that a more complicated concentration profile is required (includes a sinh term).

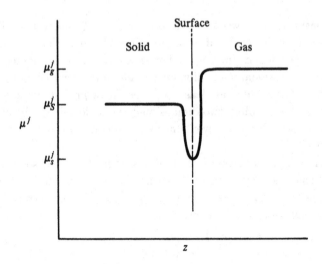

Fig. 5.12. Possible chemical potential variation with distance relative to the surface for a dopant species, j.

The second interface field contribution, that in the underlying solid, cannot be so easily dispensed with and can have very important consequences for the doping of semiconductor films. If one has a binding potential distribution for the species such as illustrated in Fig. 5.12, the force of the field will be driving the solute, built into the crystal at a ledge, back to the interface where it has a lower potential. This leads to a greatly reduced incorporation of dopant species into the film.

Returning to consideration of Eq. (5.20), it is interesting to note that it is independent of step height h. However, experimentally one finds a specific $h = h_\ell(T)$ for film growth at different temperatures, T. How are we to understand this and can we predict $h_\ell(T)$? The place to begin seems to be in recognizing that the terrace must unreconstruct before the ledge can pass over it and that, even though there is a supersaturation for solid formation existing at the ledge, unreconstruction of the lower terrace leads to an increase in free energy locally and it may not be fully recovered at the ledge of the upper terrace. Thus, not only is there a need for a ledge supersaturation to allow the usual attachment process to occur but an additional supersaturation is needed for the reconstruction change. For modeling purposes, we can force fit both effects into the β_0' term in Eq. (5.20) but we must recognize that β_0' will be a function of λ_ℓ and V_ℓ. In the domain where V is limited by β_0', increasing h_ℓ will increase λ_ℓ and decrease V_ℓ so more time is available for unreconstruction. Thus, we can presume that $h = h_\ell(T)$ is the dynamic value that maximizes V in Eq. (5.20) by the optimization of β_0' (seeking larger values) and λ_ℓ (seeking smaller values). For spontaneous unreconstruction to occur at the lower ledge, we must have

$$\kappa T \ell n \frac{n_{s\ell}}{n_{s\ell}^*} > \overline{\Delta \mu}_{rec} \tag{5.22}$$

where $\overline{\Delta \mu}_{rec}$ is the average free energy increase due to the unreconstruction process at the ledge. For such a situation to hold, this would mean that no ledge motion would occur, even at very slow velocities, until this minimum supersaturation had been exceeded and, of course, this minimum value of $\hat{\sigma}_s$ would be a function of orientation. For this case, the β_0' formulation of Eq. (5.20) would not hold because it neglects this onset value of supersaturation needed for ledge motion. It is also possible that the ledge movement process is a thermally activated one wherein, via a fluctuation type of process, lower ledge unreconstruction and upper ledge reconstruction plus intermediate ledge structural change can occur in a superkink-like fashion which then propogates down the multiatom high ledge without an increase in free energy. Such an event requires significant cooperative movement of atoms, especially for large $h_\ell(T)$.

In closing this section, some mention should be made of layer sources, two-dimensional nucleation and screw dislocations. In both cases the free energy coupling equation (see Eqs. (1.6)) holds. Here, the ΔG_C term can be expressed in terms of Bessel functions whose net result will be somewhat like that of Eq. (5.21a). The ΔG_E contribution will have a ΔG_γ contribution in terms of both the radius of curvature of the ledge, r, and the separation distance to adjacent ledges, Δr, since γ_ℓ is a function of ledge spacing and may also be a function of V_ℓ. For the screw-dislocation case, the ΔG_E contribution will also contain a ΔG_σ term because the strain field associated with the dislocation falls off as r^{-2}. For both cases, the ΔG_K term contains the unreconstruction factor plus the ledge attachment factor. The determination of layer flow velocities, layer height and surface hillock morphology due to the interplay of these factors is best left to the discussions of the companion book.[9]

5.3   Chemical vapor deposition of Si

Chemical vapor deposition (CVD) is a technique in which the substance to be deposited is transported, using a carrier gas, towards a heated substrate in the form of a volatile compound. Chemical reactions,

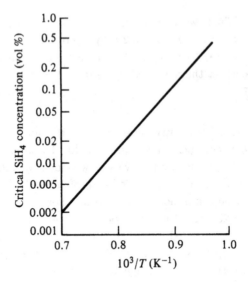

Fig. 5.13. Critical partial pressure for homogeneous gas phase nucleation of Si in a SiH_4/H_2 atmosphere as a function of temperature ($P_{tot} = 1$ atm).

occurring either heterogeneously at the substrate surface or homogeneously in the gas phase, then give rise to layer growth (see Fig. 5.13). The released constituent from the volatile compound generally acts as a strongly bound adsorbate which can significantly influence the layer growth depending upon the degree of binding between this species and the surface. Metal organic CVD (MOCVD) is the same type of process but with the organic becoming the adsorbate and it is usually less tightly bound to the surface than the inorganic species involved in CVD. The MOCVD process thus allows epitaxial film growth to occur at lower temperatures than does the CVD process. In either the CVD or MOCVD process, it is not a simple matter to deliver a uniform flux of reactant to a large diameter Si wafer because of the various non-linearities involved in the convection process. This aspect of the crystal growth will be not discussed here but is dealt with elsewhere.[9]

From early work on CVD Si via SiH_4 in H_2 carrier gas, it was found that high deposition rates were impeded by the formation of Si clusters in the gas phase via the homogeneous nucleation process. Fortunately, the addition of some $HC\ell$ to the gas impeded this gas phase cluster formation (by etching the clusters) and higher film deposition rates became possible. Fig. 5.14(a) shows that, at 1150 °C with $P_{SiH_4}/P_{HC\ell} = 1$, V

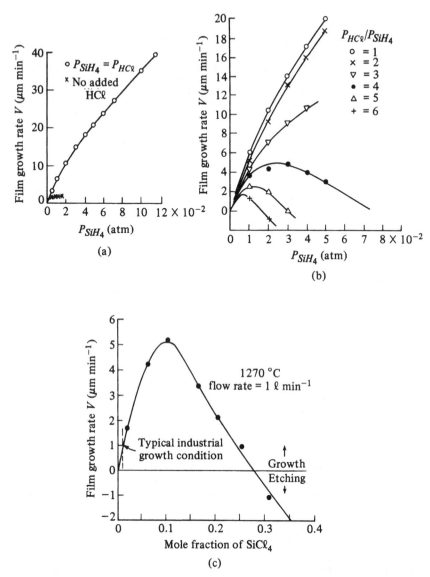

Fig. 5.14. (a) Si growth rate from mixtures of SiH_4 and HCℓ in H_2 at a substrate temperature of 1150 °C; (b) the same as (a) but for a range of $P_{HC\ell}/P_{SiH_4}$ ratios; and (c) epitaxial film growth rate at 1270 °C versus SiCℓ_4 concentration showing the growth and etching domains.

was increased to 6.7×10^{-5} cm s^{-1} (40 μm min^{-1}) at $P_{SiH_4} = 0.12$ atm using extremely high carrier gas flow rates. These data result from a

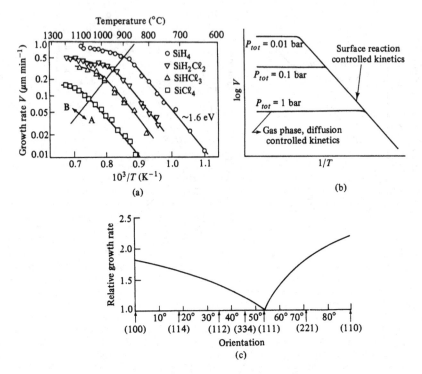

Fig. 5.15. (a) Temperature dependence of Si film growth rate from different gas sources; (b) general pressure dependence of V for Si; and (c) orientation dependence of V for Si.

combination of deposition and etching reactions where the Si etching rate increases with added $HC\ell$ as indicated in Fig. 5.14(b). Si film growth from $SiC\ell_4$ in H_2 carrier gas is illustrated in Fig. 5.14(c) where one can see that $V \propto C_\infty$ at small values of C_∞ but that the etching reaction dominates at large values of C_∞. Fig. 5.15(a) represents data on V from a variety of assorted Si sources revealing a fairly constant activation energy (~ 1.6 eV) in the low temperature regime for all sources. The effect of total gas pressure, P_{tot}, is qualitatively indicated in Fig. 5.15(b) while the relative orientation dependence is illustrated in Fig. 5.15(c).

Bloem and Giling[10] carried out a more careful study of the SiH_4 source and found the results shown in Figs. 5.16(a) and (b). At low temperatures (600–700 °C), a third process dominates the growth. They attribute this to the formation of a hydrogenated surface layer because they detected an abrupt change in surface emissivity at a given temper-

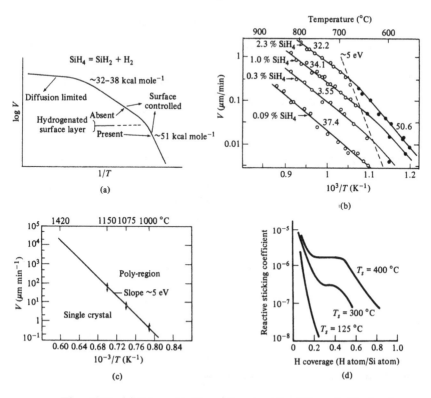

Fig. 5.16. (a) Schematic Si growth rate at fixed $P_{\mathrm{SiH_4}}$ in H_2 showing a new rate-controlling step at low temperature; (b) V versus T^{-1} at low temperature with the filled dots representing the presence of surplus emissivity from the Si films (indicating trapped H in the films with an activation energy for trapping ~ 5 eV); (c) the dividing line between regions of stable single crystal film formation and polycrystalline film formation on $\{111\}$ oriented substrates; and (d) the reactive sticking coefficient of SiH$_4$ on Si(111) $-$ (7 × 7) at different temperatures.

ature for a given SiH$_4$ concentration in the bulk gas. They report an activation energy of ~ 5 eV for the onset of this emissivity change which correlates in an interesting way with the much earlier finding, shown in Fig. 5.16(c), that the stability line between good quality Si(111) film and poor quality Si(111) film as a function of V versus T^{-1} has an activation energy of 5 eV. It is interesting to note that, even at high T, this process occurs so it must be intimately related to the magnitude of $n_{s\ell}^k$ built up at the ledges due to the large V_ℓ.

Returning to Eq. (5.20) and Figs. 5.14–5.16, one can see the following: (i) the high temperature domain is clearly dominated by δ/D_g^j with D_g^j

Table 5.1. *Dissociation reaction and equilibrium constant for each species of the Si–H–Cℓ system*

Dissociation	Equilibrium constant
$SiCℓ_4 + 2H_2 \rightleftharpoons Si + 4HCℓ$	$K_{SiCℓ_4} = \dfrac{P_{HCℓ}^4}{P_{SiCℓ_4}P_{H_2}^2}$
$SiHCℓ_3 + H_2 \rightleftharpoons Si + 3HCℓ$	$K_{SiHCℓ_3} = \dfrac{P_{HCℓ}^3}{P_{SiHCℓ_3}P_{H_2}}$
$SiH_2Cℓ_2 \rightleftharpoons Si + 2HCℓ$	$K_{SiH_2Cℓ_2} = \dfrac{P_{HCℓ}^2}{P_{SiH_2Cℓ_2}}$
$SiH_3Cℓ \rightleftharpoons Si + HCℓ + H_2$	$K_{SiH_3Cℓ} = \dfrac{P_{H_2}P_{HCℓ}}{P_{SiH_3Cℓ}}$
$SiCℓ_2 + H_2 \rightleftharpoons Si + 2HCℓ$	$K_{SiCℓ_2} = \dfrac{P_{HCℓ}^2}{P_{SiCℓ_2}P_{H_2}}$
$SiCℓ + \frac{1}{2}H_2 \rightleftharpoons Si + HCℓ$	$K_{SiCℓ} = \dfrac{P_{HCℓ}}{P_{SiCℓ} + P_{H_2}^{1/2}}$

decreasing as the total pressure increases; (ii) the intermediate temperature range is dominated by a combination of the second term in the numerator and/or the second term in the denominator; and (iii) the low temperature range is dominated by the ΔG_E term in the numerator due to the abrupt increase of ΔG_D associated with the trapping of H or some other adsorbate species into the film at the layer edges at some critical ledge concentration condition. If this were the proper explanation, one would expect that, by going to lower H_2 pressures, less H or adsorbate adsorption would occur and one could grow good quality films at still lower temperatures. This is exactly what one finds from low pressure CVD (LPCVD) studies; thus, it appears that the general perspective is correct; however, some uncertainty exists with respect to details.

The reactive sticking coefficient of SiH_4 on the $Si(111) - (7 \times 7)$ surface has been studied as a function of H coverage, θ_H, in the surface temperature range $80\ °C < T_s < 500\ °C$. At $400\ °C$, evidence is seen for two adsorption regimes proposed to correspond to minority and majority sites. On the minority sites, thought to be the corner holes of the (7×7) structure, $0 < \theta_H < 0.08$, the sticking coefficient is independent of T_s and has a value $S_* \sim 10^{-5}$. On the majority sites $0.08 < \theta_H < 1.0$, S_*

Table 5.2. *Values of the change in entropy and the change in enthalpy at 1500 K for the dissociation reactions given in Table 5.1*

Species	ΔS^0_{1500} (cal mole^{-1} K^{-1})	ΔH^0_{1500} (kcal mole^{-1})
$SiCl_4$	35.9	59.8
$SiHCl_3$	29.9	49.3
SiH_2Cl_2	26.0	31.1
SiH_3Cl	23.7	12.9
SiH_4	23.6	-5.2
$SiCl_2$	- 5.1	-7.7
$SiCl$	-21.5	-61.4

is a more complicated function of T_s and θ_H and falls off in the fashion illustrated by Fig. 5.16(d). After SiH_4 exposure, SiH (chemisorbed H) is observed to be the dominant stable decomposition intermediate from 80 °C to 500 °C (~ 1 SiH per Si) with detectable amounts of SiH_2 and SiH_3 (~ 0.08 SiH_2 per Si and ~ 0.04 SiH_3 per Si).

5.3.1   *Thermodynamics of the Si–H–Cl system*

Tables 5.1 and 5.2 provide a listing of the chemical reactions, equilibrium constants, change of entropy and change of enthalpy (at 1500 K) for this system. Fig. 5.17(a) gives the formation free energies as a function of temperature. Using the phase rule; we see that three degrees of freedom exist because we have three components and two phases to be considered. These degrees of freedom are T, P and $J_{Cl}/J_H = P_{Cl}/P_H = X$ with conservation being given by

$$P = P^2_{HCl}\left(K^{-1}_{SiCl_4}X^2 + K^{-1}_{SiHCl_3}X + K^{-1}_{SiH_2Cl_2} + \frac{K^{-1}_{SiH_3Cl}}{X} + \frac{K^{-1}_{SiH_4}}{X^2}\right)$$

$$+ P_{HCl}\left(K_{SiCl_2}X + 1 + \frac{1}{X}\right) + P^{1/2}_{HCl}K_{SiCl}X^{1/2} \qquad (5.23a)$$

This equation neglects free Cl_2, Cl plus H species and is of the form

$$AP^2_{HCl} + BP_{HCl} - P = 0 \qquad (5.23b)$$

with $P_{HCl} \ll P$. This quadratic equation can be readily solved for P_{HCl} with fixed X so that the partial pressure for each species in Table 5.2 can be given as a function of temperature. An example of such thermo-

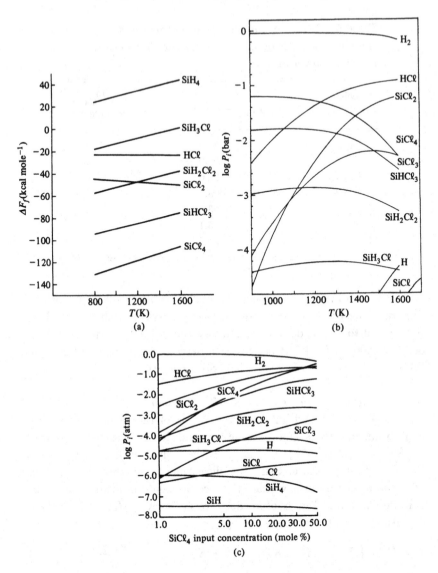

Fig. 5.17. (a) Free energies of formation for the important gas species in the Si–H–Cℓ system; (b) composition of the gas phase in equilibrium with Si as a function of T for a Cℓ/H ratio of 0.162 and $P_{tot} = 1$ atm; and (c) equilibrium partial pressures of the key species as a function of SiCℓ$_4$ input concentration at 1500 K.

dynamic equilibrium data in the gas phase is presented in Fig. 5.17(b) for $X = 0.162$ and $P = 1$ bar. Data as a function of SiCℓ$_4$ input con-

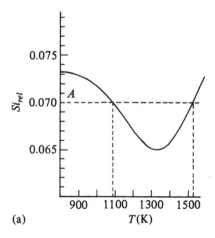

(a)

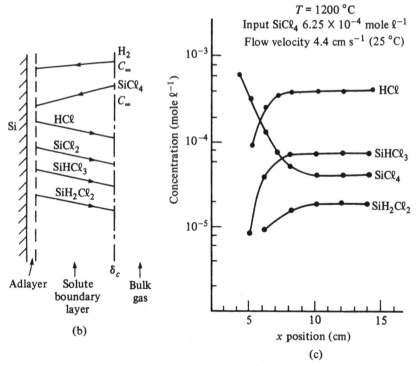

(b)

(c)

Fig. 5.18. (a) Relative Si solubility in the gas phase for a gas mixture *A*, growth is only possible between ∼ 1100 K and ∼ 1500 K; (b) possible gas species transferred across the boundary layer for a surface-catalysed reaction; and (c) species detected downstream from a Si wafer by IR spectroscopy in a horizontal reactor when using $SiCl_4 + H_2$ gases.

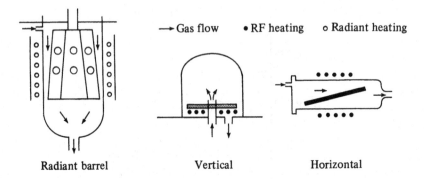

Fig. 5.19. Schematic illustration of three common CVD reactors.

centration at $T = 1500$ K are given in Fig. 5.17(c). It is interesting to note that $SiC\ell_4$ is the dominant species at low temperatures and decreases with increasing temperature while $SiC\ell_2$ becomes the dominant Si-bearing species at the higher temperatures. The relative Si solubility in the gas is denoted by

$$Si_{rel} = \frac{\Sigma P_{Si}}{\Sigma P_{H_2} + \Sigma P_{C\ell_2}} \qquad (5.24a)$$

where

$$\Sigma P_{Si} = P_{SiC\ell_4} + P_{SiHC\ell_3} + P_{SiH_2C\ell_2} + P_{SiC\ell_2} + \cdots \qquad (5.24b)$$

and

$$\Sigma P_{C\ell_2} = \frac{1}{2}P_{HC\ell} + P_{SiH_2C\ell_2} + \frac{3}{2}P_{SiHC\ell_3} + 2P_{SiC\ell_4} + \cdots \qquad (5.24c)$$

Fig. 5.18(a) shows that, for a starting gas composition of $Si_{rel} = 0.70$, $X = 0.162$ and $P = 1$ bar, growth of a Si film is possible only between 1100 K and 1500 K.

The foregoing thermodynamics has assumed a homogeneous gas phase reaction but perhaps it is a heterogeneous surface reaction that is really involved. If so, should we expect anything different for the downstream gas composition? When one actually grows a Si film on a substrate using a standard CVD reactor (see Fig. 5.19), the net chemical transport across the boundary layer should be like that indicated in Fig. 5.18(b). However, the species profiles actually detected by Nishigawa[11] are as shown in Fig. 5.18(c) and the key point to notice is that $SiC\ell_2$ is not detected in this concentration range although $HC\ell$ is abundantly present as is $SiHC\ell_3$ and $SiH_2C\ell_2$. Clearly, the chemical intermediate reactions are far from equilibrium and we should ask why that might be so. Perhaps the most obvious answer is that, whereas $HC\ell$, $SiHC\ell_3$ and $SiH_2C\ell_2$

are all valence saturated species so they should only be physisorbed at the surface (τ_s is small), SiCℓ_2 is valence undersaturated and should be chemisorbed at the surface (τ_s is large). It is thought that the large value of $\tau_s^{\text{SiC}\ell_2}$ increases its surface population sufficiently that SiCℓ_2 becomes the major source of Si for ledge growth. Its effective lifetime is determined by the reactions

$$\text{SiC}\ell_2 * + 2\text{C}\ell * \rightleftarrows \text{SiC}\ell_4 \qquad (5.25a)$$

$$\text{C}\ell * + \text{H} * \rightleftarrows \text{HC}\ell \qquad (5.25b)$$

where the $*$ state refers to the singly bonded chemisorption state and the states without a $*$ refer to the physisorption state. From Eq. (5.25a), when the surface coverage of Cℓ, $\theta_{C\ell *}$, is small, desorption is in the form of SiCℓ_2 but when it is large, desorption is in the form of SiCℓ_4. Thus, the effective surface lifetime of SiCℓ_2 is influenced by the Cℓ/H ratio and decreases as this ratio increases.

To make a simple approximation to the surface population of SiCℓ_2*, let us consider the coverage of all species determined via a Langmuir adsorption isotherm for this multicomponent adlayer; i.e.,

$$\theta^j = K^{*j} P_i^j / \left(1 + \sum_{m=1}^{N} K^{*m} P_i^m\right) \qquad (5.26a)$$

where N is the total number of components, $\theta^j = n_{s\infty}^j/N_s$ where $N_s \approx 8 \times 10^{14}$ cm^{-2} on Si(111), $K^{*j} =$ the equilibrium constant for the adsorption of j while P_i^j is the partial pressure of j in the gas phase at the interface. From a balance of incident and departing fluxes at the surface

$$K^{*j} = \frac{\exp(\Delta G_a^j/\kappa T)}{N_s \nu_s^j (2\pi m^j \kappa T)^{\frac{1}{2}}} \qquad (5.26b)$$

where the adsorption free energy change, ΔG_a^j, is generally approximated by the adsorption energy, ΔE_a^j, and $\nu_s \approx 10^{13}$ s^{-1}. At $T = 1500$ K and Cℓ/H $= 0.01$, and $P = 1$ atm, Chernov[12] calculated the results shown in Table 5.3 for Si(111); i.e., the order of 2% vacancies are present in the adlayer under these conditions so this allows significant surface mobility of the SiCℓ_2 to obtain. As one tries to grow films under these conditions at lower T, the degree of Cℓ and H adsorption is expected to increase so that θ^V decreases, the adlayer loses its mobility and poor film growth should result unless one lowers the total pressure, P. Although the assumed values of ΔE_a^j used in Table 5.3 may not be completely appropriate in a quantitative sense, the qualitative conclusions are sound. To illustrate the significant quantitative differences

Table 5.3. *Calculated species population in the adlayer for the Si–H–Cl system* ($T = 1500K, Cl/H = 0.01, Si(111)$)

Parameter	Component						Vacancy		
	H_2	HCl	$SiCl_4$	$SiHCl_3$	$SiCl_2$	Si	Cl	H	
p^j dyne cm^{-1}	10^6	2.10^4	4	3.10	7.10^2	4.10^{-2}	2.10^{-1}	10	
ΔE_a^j kcal mole^{-1}	8	10	11	12	65	54	92	74	
K^{*j} cm^2 dyne^{-1}	9×10^{-9}	4×10^{-10}	3×10^{-10}	4×10^{-10}	2×10^{-2}	6×10^{-4}	75	5	
θ^j	10^{-4}	10^{-7}	10^{-11}	10^{-10}	0.15	3.10^{-7}	0.2	0.63	0.02

that can arise from small changes in ΔH_a^j and ΔS_a^j, Figs. 5.20(a) and (b) gives results for H adsorption on Si as a function of T.

For growth in chlorosilanes, one finds from Fig. 5.15(a) that the activation energy in the intermediate temperature range is ~ 40 kcal mole^{-1}. Let us speculate with Giling that the following is the important sequence of reactions on the surface for the film forming species, $SiCl_2$

$$\text{Terraces:} \quad SiCl_2 + * \quad \rightleftharpoons SiCl_2* \qquad ; \Delta H_1 = -51.5 \text{ kcal mole}^{-1} \tag{5.27a}$$

$$\text{Ledges:} \quad SiCl_2* + * \quad \rightleftharpoons SiCl_2** \qquad ; \Delta H_2 = -51.5 \text{ kcal mole}^{-1} \tag{5.27b}$$

$$\text{Kinks:} \quad SiCl_2** + H_2 \rightleftharpoons 2HCl + Si + ** \ ; \Delta H_3 = +\Delta E \text{ kcal mole}^{-1} \tag{5.27c}$$

$$\Delta H_{tot} = 40 \text{ kcal mole}^{-1}$$

In these equations, $*$ refers to a substrate dangling bond. This requires that ΔE be ~ 143 kcal mole^{-1} which is just twice the Si–Cl bond strength (~ 77 kcal mole^{-1}). Thus, the rate limiting step in this temperature range may be the breaking of Si–Cl bonds of chemisorbed $SiCl_2$ species. We can apply the same approach to the silane reaction where SiH_2 would be the key Si bearing adlayer species. For this case, the important sequence of reactions is

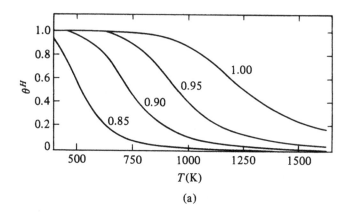

(a)

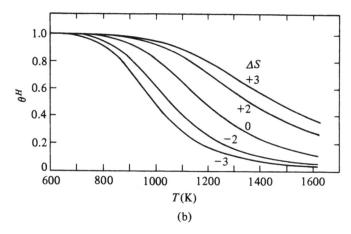

(b)

Fig. 5.20. (a) Si surface coverage, θ^H, versus T with Si–H bond strength as parameter ($\Delta H_{Si-H} = 71.3$ kcal mole^{-1} $\equiv 1.0$) at $P_{tot} = 1$ atm and (b) changes in θ^H with ΔS_{ads} as parameter for $\Delta H_{Si-H} = 71.3$ kcal mole^{-1}.

$$\text{Terraces: } SiH_2 + * \; \underset{\leftarrow}{\rightarrow} SiH_2* \qquad ; \Delta H_1 = -51.5 \text{ kcal mole}^{-1}$$

$$(5.28a)$$

$$\text{Ledges: } SiH_2 * + * \; \underset{\leftarrow}{\rightarrow} SiH_2 * * \qquad ; \Delta H_2 = -51.5 \text{ kcal mole}^{-1}$$

$$(5.28b)$$

$$\text{Kinks: } SiH_2 * * + C\ell_2 \; \underset{\leftarrow}{\rightarrow} 2HC\ell + Si + ** \; ; \Delta H_3 = +\Delta E \text{ kcal mole}^{-1}$$

$$(5.28c)$$

$$\overline{\Delta H_{tot} = 35 \text{ kcal mole}^{-1}}$$

This requires ΔE to be $\sim$ 138 kcal mole^{-1} $\sim$ twice the Si–H bond strength.

5.3.2   *Extension of concepts to diamond film formation*

Early experimental work in the diamond film CVD case was mainly focused on adjustments to the gas phase chemistry via plasma adjustments in order to control the diamond film growth. This was not particularly successful, probably because it is the thermodynamics and kinetics of reactions for intermediate species generated in the adlayer that are controlling rather than the bulk phase thermodynamics. The key film-forming precursors appear to be created in the adlayer rather than in the bulk gas phase. Current experiments show that the incorporated H content is quite high (several atomic % in some cases) as is the density of internal twin lamellae and that these levels increase as the CH_4 and H_2 partial pressures increase. It is interesting to note that the equilibrium gas phase population for $CH_4 + H_2$, analogous to Fig. 5.17(b) for Si species, shows C_2H and C_2H_2 as the most populous high temperature C species. However, for photon and electron-enhanced etching of diamond in H at 500 °C, the desorbing species are found to be greatly varied with large non-H populations in the CH_2^+ and $(CH_2)_n^+$ categories (with $n = 2$ and 3).

Additional experiments on crystallite morphology in the films indicate that sustained columnar growth does not usually occur and that, more often, repeated sphere-like or boulder-like crystal shapes are present in the films; i.e., the growth of an existing diamond crystallite slows down and is stopped by the nucleation of a graphite layer. This implies a time-dependent build-up of adsorbate (probably H, CH_2, CH_3 and $(CH_2)_n$, etc.) on the crystal terraces and ledges. The $(CH_2)_n$ species are strongly bound to the surface and are very unlikely to convert to diamond. However, these short chain polymers can readily form a graphite-like raft nucleus when the polymer concentration reaches a critical supersaturation level. After the growth of the graphite layer, the surface polymer species have been consumed and conditions once again exist for the nucleation and growth of diamond crystallites. Thus, if this speculation is correct, good quality diamond films will only be grown when we learn to control either the break-up or desorption of $(CH_2)_n$ species in the adlayer. This may involve specific photon energies, bombarding atom beams, etc.

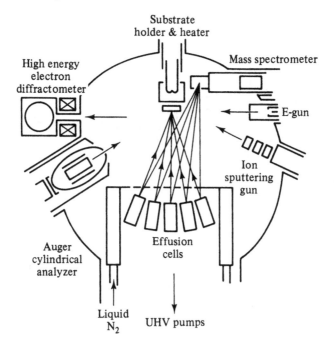

Fig. 5.21. Schematic diagram of a MBE system.

5.4   MBE of Si and GaAs

MBE is a process of epitaxial deposition from molecular or atomic beams in an ultra high vacuum system. The beams are usually generated thermally in Knudsen-type cells where quasi equilibrium is maintained so that both the beam compositions and intensities are constant and predictable from thermodynamics. The effused beams (fluxes), guided by orifices and controlled by shutters, travel in rectilinear paths to the substrate where they build up in the adlayer and provide a local environment for clustering, nucleation and crystal growth under well-defined conditions. A schematic illustration of a typical MBE system is shown in Fig. 5.21. The system is pumped via a combination of techniques to achieve a base pressure of $\sim 10^{-10}$ torr after an overnight bakeout at temperatures up to 250 °C. Using liquid N_2 cooling, residual gases are drastically reduced leaving CO as the major contaminant.

In Fig. 5.21, the mass spectrometer has the advantage of detecting not only the deposition beams but also the residual molecule environment. The ion sputtering gun is used primarily for cleaning the substrate prior to deposition while the Auger electron spectroscopy (AES) attachment

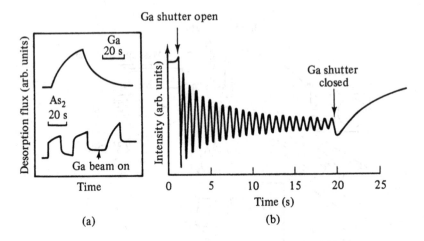

Fig. 5.22. (a) Desorbed fluxes from pulsed beams incident on GaAs(111) for Ga at 885 K and As₂ at room temperature demonstrating total reflection unless a prior Ga coverage exists; and (b) illustration of RHEED intensity oscillations after the Ga shutter opening.

provides information on the chemical composition of the surface so as to ensure its cleanliness. The reflection high-energy electron diffraction (RHEED) attachment operates at glancing angles to provide information on surface reconstruction, surface micro-structure and surface smoothness.

One can gain important information concerning the adsorption and film growth processes by modulating either the adsorption beam or the desorption beam from a substrate surface and by correlating the mass spectrometer results with the modulation frequency. If one begins with the source shutter in the closed position, then opens the shutter and waits for equilibrium to be achieved and then closes the shutter again, the results on an oscilloscope trace of the mass spectrometer response look like rising and decaying exponentials from which time constants can be obtained and correlated with Eqs. (5.9). Thus, information concerning k_d and k_a plus their activation energies can, in principle, be deconvolved from such data provided one can properly separate out the clustering, nucleation and layer growth processes. Some results from Arthur[13] are shown in Fig. 5.22(a).

Oscillations in both the intensity and width of RHEED beams, measured during MBE growth of either Si or GaAs, have been observed to depend upon most growth parameters. A typical example is shown in Fig. 5.22(b) where the period of oscillation exactly corresponds to the

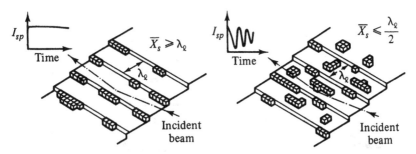

Fig. 5.23. Illustration of the connection between RHEED oscillations and cluster formation on terraces. J. H. Neave, B. A. Joyce, P. T. Dobson and N. Norton, *Appl. Phys.*, **A31**, 1 (1983).

growth of one monolayer of film so these oscillations are used to provide an absolute measurement of V. For GaAs, growth is initiated by the impinging flux of Ga atoms onto a substrate maintained at temperature, T_s, in a flux of either As_2 or As_4. If the Ga flux is temporarily terminated by closing the Ga source shutter for periods as short as 1 s and then reopened, the oscillations recommence at a larger amplitude rather than at the value present immediately prior to shutter closure. Equivalent oscillations occur for growth of $A\ell_x Ga_{1-x}As$ ($x \lesssim 1$) and, if an $A\ell$ beam is injected during the growth of GaAs at a stage when the oscillations had become completely damped, they reappear very strongly.

The principle on which the RHEED oscillation technique is based is quite simple and is illustrated in Fig. 5.23 for growth on a vicinal surface of step spacing, λ_ℓ. If $\bar{X}_s > \lambda_\ell$, the ledges will act as major sinks for the diffusing adatoms, $n_{sM} < n_N^*$ in Fig. 5.2, so no two-dimensional nucleation will occur on the terraces. In this case, RHEED oscillations will be absent. If, however, the growth conditions are varied to the point where $\bar{X}_s < \lambda_\ell/2$, n_{sM} can rise above n_N^* and two-dimensional nucleation will occur on the terraces as indicated in Fig. 5.23. For such growth conditions, RHEED oscillations are observed due to the dynamic change in cluster population on the terraces with time. The development of RHEED oscillations in conjunction with the formation of two-dimensional nucleation and growth on a perfect terrace plane is illustrated in Fig. 5.24. As indicated, the RHEED intensity decreases due to the surface roughening due to new ledge formation until half a monolayer has formed. Then, the intensity increases again as the islands fill in to remove the ledges until a completed monolayer has formed. The process repeats with the second, third, etc., monolayer formation. As the number of monolayers present on the surface increases, the net surface

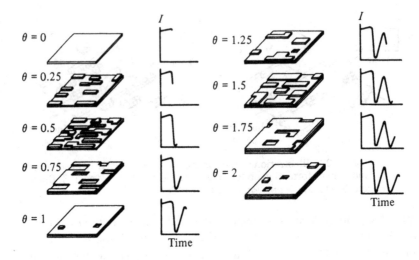

Fig. 5.24. Illustration of the temporal development of facet plane cluster formation and growth fill-in correlated with RHEED intensity variation with time for the growth of two monolayers (θ = number of monolayers deposited). J. H. Neave, B. A. Joyce, P. T. Dobson and N. Norton, *Appl. Phys.*, **A31**, 1 (1983).

roughness at the fill-in of the lowest monolayer increases so the maximum RHEED intensity will decrease with time. For a sufficiently large number of layers, the interface roughness may reach a steady state at some multilevel character. For such a condition, the oscillations should disappear and the intensity level should be a measure of the dynamic ledge site fraction on the multilevel surface.

Before discussing some results on Si and GaAs, it is important to point out that Eq. (5.9c), in its fully expanded form, operates here and that the dynamic changes associated with adatom and cluster interactions with the initial surface reconstruction pattern of the terrace plane can significantly influence the RHEED intensity. For the present application, we write Eq. (5.9c) in a slightly different form to illustrate the factors involved; i.e.,

$$\frac{dn_{s1}^A}{dt} = D_{s1}^A \nabla_s^2 n_{s1}^A + [J_1^A - (n_{s1}^A - n_{s1}^{*A})J_1^{*A}] - (J_{s2}^A + J_{s2}^{AB}) \qquad (5.29)$$

The middle bracketed term replaces the RHS of Eq. (5.9a), the D_s term is driven by the surface flux to ledges (this is an effective "D_s" value since it includes any field effects) and the other bracketed term represents the net loss to surface clusters. Similar equations will exist for the B monomer, the AA dimer, the AB dimer, etc., because these dimers, in turn, grow to form trimers, tetramers, and n-mers which all add to the ledge population on the surface. Thus, imbedded in the RHEED oscillation data is important information concerning not only surface adsorption but terrace diffusion and cluster formation kinetics. Learning how to solve this equation accurately and correlate the results with both mass spectrometer and RHEED oscillation data can provide us with key fundamental information that is not readily accessible by other procedures. The final piece of information that one needs to begin to appreciate film growth by MBE is that J/J^* is generally very large ($J/J^* \sim 10^4$–10^6) so that growth occurs far from equilibrium. In fact, the normal MBE flux conditions may be close to those needed for the *uniform attachment* growth mechanism to be operative (see Chapter 2). Certainly, if no two-dimensional nucleation occurred on a terrace plane surface until after the steady state adlayer population had been deposited, the supersaturation would be so high that a single molecule would be the critical nucleus. However, here, the actual nucleation process depends more upon the kinetics of cluster collisions than upon the thermodynamic driving force for n-mer formation.

Because the adlayer supersaturation is increasing in an almost linear fashion with time after the onset of beam flux, the critical size cluster for two-dimensional nucleation is decreasing strongly with time ($r^* \propto \gamma_\ell/T\ell n\hat{\sigma}(t)$). The average adatom separation distance on the surface is given by $(Jt)^{-1/2}$ and the average adatom diffusion distance on the surface is $(4D_s t)^{1/2}$ so the average time for adatom collision and perhaps dimer formation is $\tau_2 = (4D_s J)^{-1/2}$. For $J \sim 10^{14}$ cm^{-2} s^{-1} and $D_s < 10^{-14}$ cm^2 s^{-1}, $\tau_2 > 1$ s and $\tau_2 < 10^{-3}$ s when $D_s > 10^{-8}$ cm^2 s^{-1}. Since typical cluster formation times in melts is $\tilde{<} 10^{-6}$ s, it is reasonable to expect that, here, cluster formation is indeed limited, on the average, by the surface diffusion process leading to bimolecular collisions. Thus, if surface diffusion is very rapid at the substrate temperatures, T_s, an n-mer cluster may form long before the surface supersaturation reaches its maximum possible value. However, if surface diffusion is slow, uniform attachment conditions may obtain. In any event, it is likely that r^* will be quite small for the typical MBE conditions. Fig. 5.25 illustrates the interaction between these two factors in the nucleus-forming process. For a given J/J^* condition, $\hat{\sigma}$ increases linearly with t so the number of molecules, n, in the critical nucleus decreases logarithmically with t while the collision number, n', wherein the collision probability of an n-mer event , P_ν^n, is unity, increases with t. Only in the hatched region of Fig. 5.25 do these collisions form with unit sticking probability so that a critical sized cluster for nucleation is formed.

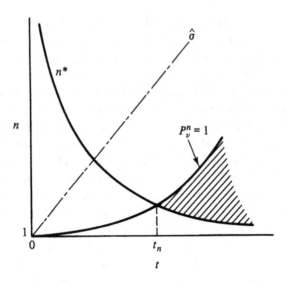

Fig. 5.25. Qualitative variation of surface supersaturation, $\hat{\sigma}$, equilibrium cluster size, n^*, and the unit probability $P_\nu^n = 1$, of n-mer collisions via surface diffusion with time.

When diffusion is infinitely rapid, the statistical probability of finding an n-mer cluster is simply

$$P_\nu'^n = N_s X \cdot \alpha_1 X \cdot \alpha_2 X \ldots \ldots \alpha_{n-1} X \qquad (5.30a)$$

$$= N_s \left(\prod_1^{n-1} \alpha_i\right) X^n = \left(\prod_1^{n-1} \alpha_i\right) (Jt/N_s)^{n-1} Jt \qquad (5.30b)$$

where X is the adatom site fraction coverage ($X = Jt/N_s$), N_s is the number of surface sites per square centimeter and α_i is the number of surface sites available adjacent to the central molecule. For a system of finite D_s, this probability must deal with an effective site fraction $X' = X$ times the average diffusion area/the average area per surface molecule; i.e., $X'/X = (4D_s t)/(Jt)^{-1} = 4D_s Jt^2$ so long as this ratio is less than unity ($t_c = (4D_s J)^{-1/2}$). Thus, for times $0 < t < t_c$, we have

$$P_\nu^n = \left(\prod_1^{n-1} \alpha_i\right) \left(\frac{4D_s Jt^3}{N_s}\right)^{n-1} Jt \qquad (5.30c)$$

Setting $P_\nu^n = P_\nu'^n = 1$, we can obtain the time for n-mer collision, i.e., for $0 < t < t_c$

$$t_n \approx \left[\frac{1}{\alpha_g J} \left(\frac{N_s}{D_s J}\right)^{n-1}\right]^{\frac{1}{3n-1}} \qquad (5.31a)$$

for $t_c < t$

$$t_n \approx J^{-1} \left[\frac{N_s^{n-1}}{\alpha_g} \right]^{\frac{1}{n}} \tag{5.31b}$$

where

$$\alpha_g = \prod_1^{n-1} \alpha_i \tag{5.31c}$$

Given any value of t_n, one can simply evaluate n_N^* and, when $n_N^* < n$, the sticking probability is unity so that the cluster is stable as the nucleus. When $n_N^* > n$, $\Delta G < \Delta G^*$ in Eq. (5.15c) so the sticking probability is less than unity and the cluster is not stable.

5.4.1 Si equilibrium and growth surfaces

Scanning tunneling microscopy has been used to investigate nucleation and growth phenomena during MBE growth of Si on Si(111) − (7 × 7) surfaces from the submonolayer range up to a few monolayers. In all cases, the initial reconstruction played a major role at temperatures below 530 °C. Even at room temperature, where the formation of amorphous clusters is found, the (7×7) substrate determines the location where the overlayer Si nucleate. In most cases there are two clusters of amorphous Si in each (7 × 7) unit cell (deposition rate ∼ 0.3 ML min⁻¹) with one cluster sitting on each triangular subunit of the (7 × 7) unit cell (see Fig. 5.26). The position of these deposited atoms is on top of a "rest atom", a dangling bond atom in the substrate layer that is not covered by an adatom. With this initial nucleation behavior for individual atoms, the existence of two clusters of amorphous Si in each (7 × 7) unit cell becomes understandable. Each triangle of the (7 × 7) unit cell has three rest atoms and, if these sites are occupied first, a basis is formed for two clusters separated by a boundary between the faulted and unfaulted halves where the atoms form dimers. This boundary has no dangling bonds for Si adsorption and thus inhibits the two clusters from growing together in the early stages of nucleation.

Increasing the substrate temperature to 250 °C yields larger islands, larger than the (7 × 7) unit cell size, that still look amorphous. Raising the temperature or annealing at this temperature causes more and more of these clusters to convert to the epitaxial state. In this early stage of epitaxial growth, all the islands are bilayers (3.1 Å high) and these ordered islands show a (2×2) adatom arrangement on top (the same local unit as in the triangular subunit of the (7 × 7) reconstruction pattern). When the deposition temperature is increased to 350 °C, all the islands

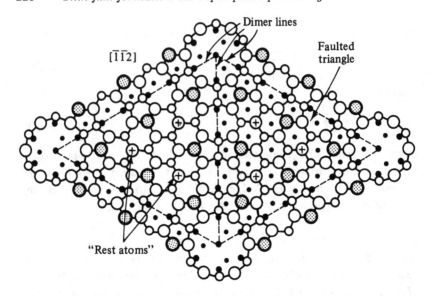

Fig. 5.26. Top view of Si(111) – (7 × 7) DAS (dimer–adatom stacking fault) model. Atoms on (111) layers of decreasing heights are indicated by circles of decreasing sizes. Heavily outlined dotted circles represent adatoms sitting at topsites. Larger open circles represent atoms in the stacking fault layer. Smaller open circles represent atoms in the dimer layer. Small circles and dots represent atoms in the unreconstructed layers beneath the reconstructed surface. J. E. Demuth, private communication (1989).

have grown epitaxially and the tops of the islands look (7 × 7)-like even though there is no long-range order. The shape of these epitaxial islands is quite random although some already show the tendency to be more triangular than round. This trend becomes the rule when the deposition temperature is raised to 450 °C.

Above 475 °C, superstructures of the triangular (7 × 7) islands and the substrate are in nearly perfect registry but often with a lateral shift of $1/\sqrt{3}$ of the Si lattice constant along the $[\bar{1}\bar{1}2]$ direction. This nearly ideal registry differs completely from the behavior one finds on a surface prepared by high-temperature annealing. After high-temperature annealing, the step edges show a random registration between (7 × 7) domains on the upper and lower terraces. Annealing above 1100 K, the specimen undergoes a surface phase transition from (7 × 7) to (1 × 1) and, on lowering the temperature again, the (7 × 7) superstructure nucleates at the upper edge of the substrate steps. Because the process starts independently at different steps, the lateral registration of these (7 × 7) superstructures between upper and lower terraces is random.

For the triangular island, the preferred step direction is the one with the outward normal along the [2$\bar{1}\bar{1}$] directions. This is consistent with early electron microscopy results which showed that [2$\bar{1}\bar{1}$] steps are straight on a macroscopic scale on the growing surface while [$\bar{2}$11] steps tend to roughen (see Figs. 2.6 and 2.7 for kink density results). Preferred nucleation of Si islands is found to occur along boundaries between (7×7) superstructure translational domains on the substrate and this preferred nucleation, which arises from defects in the epilayer, accounts for the formation of a second epitaxial layer long before the first layer is completed. It is also interesting to note that a variety of metastable reconstructions $((2n+1) \times (2n+1)$ with $n = 1, 2, 4$ and 7) which differ from (7×7) have also been found in these epitaxial islands.

Nucleation and growth of Si on Si(100) has been found to be far simpler than on Si(111) – (7×7) and involves predominantly dimer-derived structures. Nucleation involves the formation of linear and square dimer clusters. Growth at 200–450 °C proceeds by the preferential addition of dimers along dimer rows and produces relatively defect-free extended columns of symmetric and asymmetric dimers. Anti-phase boundaries between these columnar islands are the most predominant defects. Above 450 °C, the anti-phase boundaries begin to anneal and the columns coalesce to form more raft-like islands. At these temperatures the rafts tend to have defect structures similar to those of the substrate.

If a monatomic step occurs on a (001) surface, its height will be one-quarter of the bulk lattice constant and the atom rows on the upper terrace will run at 90° relative to the rows on the lower terrace. This DC lattice may be thought of as two interpenetrating FCC lattices turned 90° relative to each other; thus, the alternation of the row direction on the surface is a direct consequence of this feature of the diamond lattice. If the atomic rows of the upper terrace are parallel to the step edge, the step and the terrace have been called type A and one such runs along the [1$\bar{1}$0] direction. When the atom rows are perpendicular to the step edge, the step and terrace are called type B (see Fig. 5.27). This notation applies to steps of any height.

Vicinal Si(001) is known to exhibit unusual ordering when it is cut with a tilt about the [1$\bar{1}$0] direction. Such a surface forms type B double steps that can completely exclude type A terraces. For the double step structure, both a π-bonded chain model, similar to that seen in the Si(111) – (2×1) reconstruction and AA dimerlike rebonded geometry, in which the lower half of the step face tends to resemble a type A single

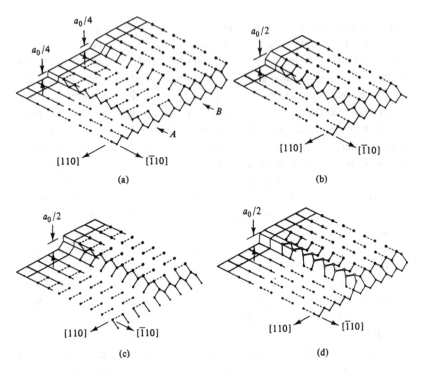

Fig. 5.27. Perspective view of a Si(100) surface showing single layer high steps and the terrace atom dimer orientation shift with alternate layers.

step, have been proposed. Experimental observations of an equilibrium surface tend to favor the latter model. Although a type A single step has the lowest calculated energy per unit length, on a vicinal surface it must alternate with the type B single step which has a relatively high calculated energy. The type B double step has a lower calculated energy per unit length than the combined single steps so it is favored for a Ge(001) surface cut with a small tilt about a line rotated from the [010]. Here, the relative area of the two terrace types, A and B, is different than expected. The type B terraces are found to dominate over the type A terraces; i.e., the areas with atom rows running along the [110] are about twice as prevalent as those with atom rows running along the [1$\bar{1}$0]. Practically no double steps are seen in this case. Although these steps exhibit a very high density of kinks, the entropy does not appear to be as high as expected so the kinks are correlated. A vicinal surface cut with a tilt more precisely aligned about the [010] direction might be expected to show an even greater tendency to form correlated kinks. As with Si, more of these Ge surfaces exhibited step heights greater than 2 ML. The spacing of the terraces between double steps was found to be relatively uniform while the spacing between monatomic steps was erratic. Monatomic type *A* and type *B* double steps are straight and, when double steps form kinks, they usually break up into single steps. The structure of the biatomic step edge is identical for Ge and Si.

During Si(001) film growth, a competition between adatom build-up on the surface, adatom nucleation into stable clusters plus cluster growth and coalescence determines the dynamic *microstructure* of the surface. RHEED oscillations, corresponding to an alternating roughening and smoothing of the surface, will be observed when the rate of nucleation of stable clusters is less than the lateral growth rate of these clusters. For some surface flux conditions, new clusters will occasionally nucleate on other clusters so that the surface gradually becomes distributed over many levels and the RHEED oscillations weaken and perhaps die out entirely. However, under some flux conditions, persistent, strong RHEED oscillations have been observed over *thousands of layers* of homoepitaxy on a single domain Si(001). This latter observation might be understood in the following way.

Consider deposition to be initiated on a perfect Si(001) surface with two-fold symmetry; i.e., reconstructed dimer rows that are orthogonal in adjacent exposed layers. Assume that islands of the new layer nucleate randomly on the surface and then grow preferentially as dimer rows orthogonal to those in the substrate (also note that the long dimension of the island is *along* the dimer bond breaking direction of the underlying layer: this is consistent with the notion that breaking of surface dimers (see Fig. 5.27) is a crucial step in homoepitaxy on Si(001)). Once such elongated islands form, atoms can land on top to nucleate a third layer island with rows orthogonal to those in the second layer (and parallel to those in the first layer) and so on. Because the second layer islands may not be very wide, the third layer islands cannot become very long. Third layer islands cannot bridge the vacant channels between second layer islands until this layer becomes interconnected. Thus, although growth is dominated by nucleation and coalescence, a large number of incomplete layers is topologically prohibited by the inability of the lower layers to form a continuous film. These are precisely the conditions needed for the formation of persistent RHEED oscillations. Such persistent RHEED oscillations are not observed when double steps, aligned very nearly along one of the ⟨110⟩ directions, flow on the surface. Only when the average alignment of the steps is away from the $\langle 110 \rangle$ direction are the persistent RHEED oscillations seen (for a type A double step, the dimer strings on both upper and lower terraces are orthogonal to the step direction while, for type B double steps, the dimer strings are parallel). Perhaps it is the kinks in these double steps that aid the unreconstruction of the dimers on the lower terrace and lead to the requisite lateral flow rate needed for the sustained oscillations.

5.4.2   GaAs and AℓGaAs multilayers

For the GaAs system (see Fig. 6.34), the vapor pressure of As_2 is above that of Ga over most of the GaAs stability temperature range so that it always evaporates from a GaAs surface first leaving a Ga-rich surface solution. If GaAs is used as the source in a Knudsen cell, the major species are Ga monomer and As dimer with their pressures remaining constant at a given temperature so long as a small amount of As is maintained in the liquid. At low temperatures, where P_{As_2} is no longer greater than P_{Ga}, congruent evaporation may occur. At 900 °C with a source to substrate distance of 5 cm and a source orifice area of 0.1 cm^2, the arrival rate of Ga and As_2 at the substrate are 6.4×10^{13} cm^{-2} s^{-1} and 8.4×10^{14} cm^{-2} s^{-1}, respectively. This leads to a film growth velocity of $V \sim 0.3$ Å s^{-1} at a surface ratio of As_2 to Ga of ~ 10. Generally, V is determined by the arrival rate of Ga which has $S_* = 1$.

The beam modulation technique was used long ago to investigate the interactions of Ga and As_2 beams with GaAs(111). At $T_s < 750$ °C, no Ga beam was found to be reflected indicating that $S_*^{Ga} = 1$. At $T_s > 750$ °C, Ga followed a first order desorption process with a well-defined τ_s as a function of T_s ($\tau_s^{Ga} = 10$ s at 885 °C). On the other hand, As_2 has a negligible value of S_* unless Ga atoms are present on the surface (see Fig. 5.4(c)). Such results show that the growth of GaAs is kinetically limited by the adsorption of As_2 while the magnitude of V is determined by the flux of Ga.

Studies on GaAs(100) and GaAs($\bar{1}\bar{1}\bar{1}$) show two distinct surface reconstruction patterns depending upon T_s and the arrival rates of Ga and As_2. These two surfaces are termed As-stabilized and Ga-stabilized and have been shown to be associated with the gain or loss of As_2 to the extent of about 0.5 ML on the surface. Because of this, it is felt that As_2 is adsorbed into a weakly-bound mobile precursor state which then dissociates upon encountering pairs of empty As sites adjacent to filled Ga sites. As_4 molecules appear to behave similarly to As_2 molecules

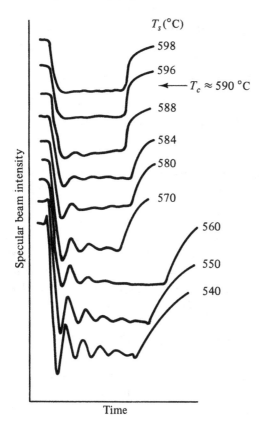

Fig. 5.28. Typical result for cut-off temperature, T_c, test at $J_{Ga} = 2.3 \times 10^{14}$ atom cm^{-2} s^{-1}.

in that they are adsorbed and form a weakly bound state. Subsequent chemisorption into suitable sites, however, has been found to be dissociative only at $T_s > 450\,°C$ where GaAs can be formed.

The influence of the various growth parameters (T_s, J_{Ga}, J_{As_x} and x) on film growth have been determined by maintaining J_{Ga}, J_{As_x} and x at some known values and changing T_s until the RHEED oscillations just disappear. Studies have been carried out on substrates oriented a small angle, θ, from the (001) towards the (011). For each sequence, either T_{Ga} or J_{As_x} was changed systematically and the cut-off temperature, T_c, was measured. Typical results are given in Fig. 5.28. For $\theta = 2.25°$ ($\lambda_\ell \approx 72$ Å), the terraces always displayed a (2×4) reconstruction, $\bar{X}_s$ increases with T_s increasing and J_{Ga} decreasing and is shorter with As$_2$ than with As$_4$ for the case where $J_{As_4} = 2J_{As_2}$. For $\theta = 0.4°$ ($\lambda_\ell \approx$

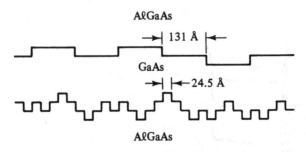

Fig. 5.29. Illustration of interface roughness, via calculated terrace widths during MBE formation of a GaAs quantum well in $A\ell_{\frac{1}{2}}Ga_{\frac{1}{2}}As$ at 650 °C.

400 Å), a much higher T_s gave the terrace a (3×1) reconstruction pattern but similar trends are observed. From such studies, values of $\bar{X}_s = 72$ Å and 400 Å at 580 °C and 650 °C, respectively, were found with $J_{Ga} = 1.2 \times 10^{14}$ cm^{-2} s^{-1} and a flux ratio of 3. From such studies, D_s was evaluated (from $\bar{X}_s^2 = D_s \tau_s$) giving $D_0 \approx 3.3 \times 10^{-5}$ cm^2 s^{-1} and an activation energy of 1.3 ± 0.1 eV for As$_4$ surface diffusion in a weakly bound precursor state during MBE growth of GaAs at low temperatures.

Since we generally expect that the interface equilibrium roughness will increase for alloy solutions, one should expect the $A\ell_x Ga_{1-x} As$ surfaces during growth to be dynamically rougher than GaAs surfaces. This effect is illustrated in Fig. 5.29 for a quantum well structure of $A\ell_{\frac{1}{2}}Ga_{\frac{1}{2}}As/GaAs$ where the analysis of surface diffusion data gave these predicted terrace widths. In some quantum well studies for this chemical system, the microscopic chemical/crystallographic disorder of the interface was found to have a period of much less than 17 nm for normal MBE growth at 700 °C and this increased to several microns for a growth interruption of 90 s. From what has been said earlier, one expects that, by periodically closing the shutter, the interface will smoothen and one will be able to grow smoother layers with enhanced control over layer thickness for thin layers. Because $A\ell$ diffuses more slowly than Ga on the surface, the time required for layer smoothness recovery is much greater for $A\ell_{\frac{1}{2}}Ga_{\frac{1}{2}}As$ than for GaAs.

To evaluate the effect of growth interruption versus no growth interruption, comparative photoluminescence (PL) studies of GaAs/$A\ell_{\frac{1}{3}}Ga_{\frac{2}{3}}As$ single quantum well structures were grown via MBE under identical conditions. Fig. 5.30 shows the RHEED specular beam

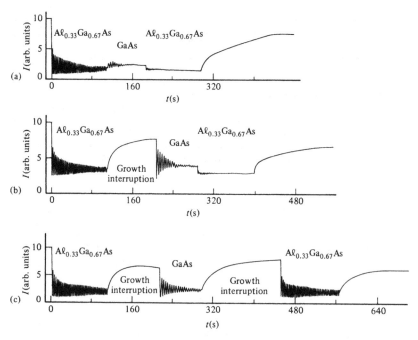

Fig. 5.30. Illustration of growth interruption effects via RHEED oscillation studies during GaAs quantum well formation in $A\ell_{\frac{1}{3}}Ga_{\frac{2}{3}}As$. ($P_{As_4} = 3.2 \times 10^{-6}$ torr, $T_s = 625$ °C, $V_{GaAs} = 0.25$ ML s^{-1}, $W_{GaAs} = 20$ ML (56.52 Å), $V_{Al_xGa_{1-x}As} = 0.315$ ML s^{-1} and $W_{AlxGa_{1-x}As} = 40$ ML).

oscillation behavior for the growth of a single 20 ML GaAs well sandwiched between 40 ML $A\ell_{\frac{1}{3}}Ga_{\frac{2}{3}}As$ layers for the case of (a) no growth interruption, (b) growth interruption only at the inverted interface and (c) growth interruption at both interfaces. This interruption time is seen to be just sufficient to allow full recovery of the specular beam intensity. The observation of a narrow single PL peak for this quantum well indicated an ideal well of thickness 20 ± 1 ML.

Problems

1. Using Eq. (5.7) for Si, calculate the time needed for 1 ML of atoms to collide with a Si surface of average orientation ($\Omega_{Si} = 20$ Å^3) as a function of vapor pressure between $P = 1$ atm and $P = 10^{-10}$ torr at room temperature, 500 °C and 1000 °C.

2. Using Eq. (5.4a) for Si(111), calculate the equilibrium surface coverage, θ_s^{*j}, in an adlayer of height $h_a = 2$ Å as a function of pressure 10^{-8} torr $< P^j < 10$ torr at $T = 200\ ^\circ$C and $700\ ^\circ$C when $\Delta\mu_a^j = 0.5$ eV, 1 eV and 2 eV.

3. For the six cases of question 2, using $\nu = 10^{13}$ s^{-1}, calculate the surface lifetime, τ_s^j.

4. For $D_{0s}^j = 10^{-1}$ cm^2 s^{-1} and $\Delta H_s^j = 1$ eV, calculate D_s^j for the six $(T,\ \Delta\mu_a)$ states of question 2. Assuming that j is Si and that $\Delta\mu_a^{Si} = 1$ eV, calculate the flux range of J^{Si} needed for the deposited film to be amorphous at a deposition temperature of $200\ ^\circ$C.

5. Knowing that $n_{s\infty}$ is given by Eq. (5.16a) and n_{sM} is given by Eq. (5.21c), use the data on $\Delta E(i)$ from Fig. 5.8 plus Eqs. (5.15) to make a plot of i^* versus J/J^* for both Si and GaAs homoepitaxy on the terraces between ledges. Assume that $\lambda_\ell/2 \gg \bar{X}_s$ and that $n_{s\ell} = n_s^*$.

6. Using Eqs (5.7) and (5.8), show that Eq. (5.16a) results when $N_s \gg n_{s\infty}$. What value of ν_s do you find?

7. For a vicinal surface orientation of spacing λ_ℓ, develop the average time dependence for adlayer build-up on the terraces between ledges by modifying the procedure of Eqs. (5.9) to account for the loss to the ledges. Assume that $\bar{X}_s \approx \lambda_\ell/2$ and that $n_{s\ell} = n_s^*$. Make a plot of $\bar{n}_s(t)$ versus time for three widely separated values of D_s/λ_ℓ in order to illustrate the effect of adatom loss to ledges on the rate of build-up to the steady state adlayer condition.

BACKGROUND SCIENCE

6

Thermodynamics of bulk phases

6.1   Thermodynamic functions

For phase changes such as sublimation, fusion and vaporization, the thermodynamics of homogeneous systems tells us how changes in such functions as the Gibbs free energy, volume, internal energy, etc., are related during a reversible process, i.e.,

$$dG = vdP - SdT + \Sigma \, \mu^j dN^j \qquad (6.1)$$

when the external variables are P, T and μ^j with N^j moles present. The P–v–T surface is illustrated in Fig. 6.1 and some features of this diagram will be discussed later when we deal with the "phase rule". For an isothermal, isobaric process involving only one component, $dG = 0$ or G = constant (a minimum). This means that the equations for sublimation, fusion and vaporization curves (v-constant intersections of Fig. 6.1) are given by setting equal the molar Gibbs functions for the appropriate phases. For a one component system, the molar Gibbs function is just the chemical potential, μ, for that species. For a constant pressure, plots of μ as a function of T for several possible solid phases α, β, γ and δ and a single liquid phase are given in Fig. 6.2(a) while, for a single solid phase, plots of μ versus T are given for several pressures P_1, P_2, P_3 in Fig. 6.2(b). Since the system, at equilibrium, takes up that state which has the minimum G ($\min \mu$, here), the melting temperature, T_M, is given by the intersection point between the liquid curve and a particular solid curve. The δ solid phase has no melting point and can never

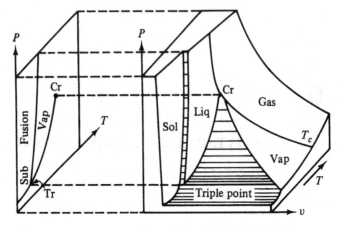

Fig. 6.1. Schematic diagram of a typical P–v–T surface.

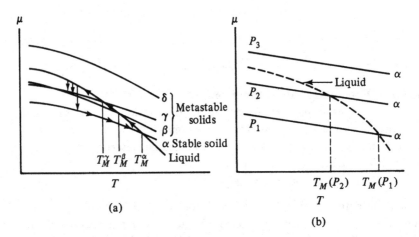

Fig. 6.2. Chemical potential versus temperature for a one component system illustrating (a) possible free energy changes during heating and cooling cycles for several possible solids; and (b) similar changes for one solid but three different pressures.

form from the liquid. In Fig. 6.2(b), $P_3 > P_2 > P_1$ provided the molar volume of the solid is greater than that for the liquid.

The free energy change, $\Delta G \equiv \Delta \mu$, in going from the liquid phase to one of these solid phases at temperature T is related to the entropy of fusion, $\Delta S_F(T)$; i.e.,

$$\Delta G(T) = -\int_{T_M}^{T} \Delta S_F(\tau) d\tau = \frac{\Delta H_F}{T_M} \Delta T + \int_{T_M}^{T} \frac{(T-\tau)}{\tau} \Delta c_P d\tau \quad (6.2a)$$

since

$$\Delta S_F(T) = \frac{\Delta H_F}{T_M} + \int_{T_M}^{T} \frac{\Delta c_P}{\tau} d\tau \tag{6.2b}$$

where ΔH_F is the latent heat and c_P is the heat capacity at constant pressure. For small undercoolings, the second term is approximately $\Delta c_P (\Delta T)^2 / 2 T_M$ and is negligible compared to the first so that

$$\Delta G \approx \frac{\Delta H_F}{T_M} \Delta T \tag{6.2c}$$

6.1.1 Bulk solutions

Very rarely are we concerned with completely pure materials. Rather, we are interested in the properties of solutions. Thus, we are more generally concerned with the thermodynamic properties of a particular component in solution called *partial molar quantities*. A given component will have a partial molar volume, $\bar{v}$ energy, $\bar{E}$, entropy, $\bar{S}$, enthalpy, $\bar{H}$, and free energy, $\bar{G}$, associated with it. For example, in a binary solution of A and B molecules, the volume may be thought to consist solely of the volume occupied by A molecules and the volume occupied by B molecules. The volume occupied by A is designated the partial molar volume of A and is referred to that for one mole of A for the purpose of standardization.

In terms of mole fractions, X^j, for an N-component solution, the key thermodynamic functions v, G, H and S are given in terms of the partial molar quantities $\bar{v}^j, \bar{G}^j, \bar{H}^j$ and $\bar{S}^j$ by

$$v = \bar{v}^1 X^1 + \bar{v}^2 X^2 + \cdots \tag{6.3a}$$

$$G = \bar{G}^1 X^1 + \bar{G}^2 X^2 + \cdots \tag{6.3b}$$

$$H = \bar{H}^1 X^1 + \bar{H}^2 X^2 + \cdots \tag{6.3c}$$

$$S = \bar{S}^1 X^1 + \bar{S}^2 X^2 + \cdots \tag{6.3d}$$

and

$$X^j = N^j / (N^1 + N^2 + \cdots) \tag{6.3e}$$

where N^j is the number of moles of component j in the solution. Since G is the potential function that determines the process of change in heterogeneous systems, it is given in terms of the other quantities by,

$$G = E + Pv - TS \tag{6.4a}$$

$$= H \qquad - TS \tag{6.4b}$$

In terms of the partial quantities

$$\bar{G}^j = \bar{H}^j - T\bar{S}^j \tag{6.5a}$$

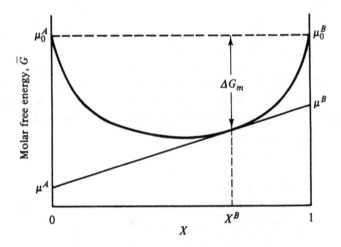

Fig. 6.3. Graphical method for obtaining the chemical potentials (μ^A, μ^B) from the molar free energy curve at the composition X^B.

This quantity $\bar{G}^j$ is actually the chemical potential μ^j of species j in the solution, which is more commonly written

$$\mu^j = \bar{G}^j = \mu_0^j + \mathcal{R}T\ell\mathrm{n}\hat{a}^j \tag{6.5b}$$

Here, μ_0^j is the standard state chemical potential of j and represents the intermolecular potential interaction of the j molecule with its neighbours in pure j of the same structure. Also, $\hat{a}^j$ is the chemical activity of j species in the solution and is given by

$$\hat{a}^j = \hat{\gamma}^j X^j \tag{6.5c}$$

where $\hat{\gamma}^j$ is the activity coefficient for j which represents all the enthalpy and entropy changes that occur to the state of the j species because of all the new interactions occurring between j and the solvent and between j and all the other solute species. For neutral species, in a field-free region, it is the spatial gradient of μ^j that drives any possible redistribution of the species in the solution. In the most general case, it is the expanded electrochemical potential, η^j, given by Eq. (7.49), which drives the process of change. In Fig. 6.3, the partial free energies and integral free energy at mole fraction X are illustrated for a binary system of A and B.

Now that the thermodynamic basis of the key functions has been defined, we wish to consider the free energy of formation of a binary solution from the pure components in their standard states, which is usually defined as the most stable form of the molecule at 1 atm pressure and the temperature specified. Common practice is to use 298 K as the reference temperature. For the reaction

$$X^A A(s) + X^B B(s) = (X^A, X^B) \tag{6.6}$$

the free energy of mixing, ΔG_m, in Fig. 6.3 is given by

$$\Delta G_m = G - X_A G_0^A - X^B G_0^B \tag{6.7a}$$
$$= X^A(\bar{G}^A - G_0^A) + X^B(\bar{G}^B - G_0^B) \tag{6.7b}$$
$$= \mathcal{R}T(X^A \ell n \hat{\gamma}^A X^A + X^B \ell n \hat{\gamma}^B X^B) \tag{6.7c}$$

where the subscript 0 refers to molecule j in its standard state. Thus, the heat of mixing, ΔH_m, and entropy of mixing, ΔS_m, are given by

$$\Delta H_m = \sum_{j=1}^{n} X^j (\bar{H}^j - H_0^j) \tag{6.8a}$$

$$\Delta S_m = \sum_{j=1}^{n} X^j (\bar{S}^j - S_0^i) \tag{6.8b}$$

At this point it is useful to classify solutions into a number of different categories according to their behavior and to specify what this means in terms of unique values of $\Delta H_m, \Delta S_m$, etc., (see Table 6.1).

(1) Ideal solution: In this case, no change occurs in the intermolecular potential interaction or in the vibrational spectrum on mixing. This case is characterized, for the partial pressure, P^j, by

$$P^j = X^j P_0^j \quad \text{or} \quad \hat{\gamma}^j = 1 \quad \text{Raoult's law} \tag{6.9a}$$
$$\Delta H_m = 0 \tag{6.9b}$$

(2) Non-ideal dilute solution: Here, the solute species are far enough apart, on average, not to interact with each other, but solute–solvent interaction occurs and it is different than for the solute in its standard state. The solvent behavior follows Raoult's law. We find

$$P^j = \tilde{b}X^j \quad \text{or} \quad \hat{\gamma}^j = b/P_0^j \quad \text{Henry's law} \tag{6.10a}$$

for solvent A : $S^A - S_0^A = -\mathcal{R}\ell n X^A; H^A - H_0^A = 0;$
$$v^A - v_0^A = 0 \tag{6.10b}$$
for solute j : $H^j - H_0^j = \tilde{f}_1(T) \neq 0; v^j - v_0^j = \tilde{f}_2(T) \neq 0 \tag{6.10c}$

where the functions f_1 and f_2 depend only on temperature and are independent of mole fraction X. Positive and negative deviations from

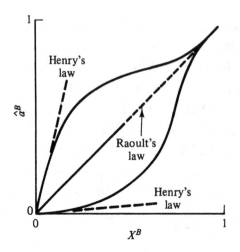

Fig. 6.4. Plots of activity, $\hat{a}^B$, versus composition, $X^2 = X^B$, illustrating positive and negative deviations from Raoult's law.

Raoult's law are illustrated in Fig. 6.4 in the plot of activity versus mole fraction, X^2, for components in a binary solution.

(3) Regular solutions: In this type of solution, ΔS_m takes on the ideal solution value and ΔH_m is different from zero; i.e.,

$$\Delta S_m = -\mathcal{R}(X^A \ell n X^A + \sum_j X^j \ell n X^j)   \tag{6.11}$$

Most solutions do not behave in this "regular" fashion; however, regular solutions are a step away from ideality and serve as a useful model.

(4) General solutions: Most solutions do not conform to the types of simple relationships characterized in (1), (2) and (3) above. Some attempt at further classification is given in Table 6.1 where (i) in associated solutions, departures from regularity are caused primarily by bonding between the molecules of one component in the solution whereas, in solvated solutions, the principal bonding is between the solute and solvent molecules and (ii) athermal solutions represent a particular case in which the enthalpy of mixing is zero but the activity does not have the ideal value.

Departures from ideal behavior may also be displayed by plotting ΔG versus activity as in Fig. 6.5. A positive deviation from Raoult's law ($\hat{\gamma}^2 > 1$) corresponds to a smaller solubility while a negative deviation ($\hat{\gamma}^2 < 1$) corresponds to an increased solubility of components. Ideal

Table *6.1.Classification of solutions*

Solution type	Enthalpy of mixing $\bar{H}^1 - H_0^1$	Entropy of mixing $\bar{S}^1 - S_0^1$	Remarks
Ideal	0	$-\kappa \ell n X^1$	$\hat{a}^1 = X^1$ $v^1 \approx v^2$
Regular	+	$-\kappa \ell n X^1$	$\hat{a}^1 > X^1$ $v^1 \approx v^2$
Non-ideal, athermal	0	$< -\kappa \ell n X^1$	$\hat{a}^1 < X^1$ $v^1 < v^2$
Irregular, associated	+	$> -\kappa \ell n X^1$	$\hat{a}^1 > X^1$
Irregular, solvated	−	$< -\kappa \ell n X^1$	$\hat{a}^1 < X^1$

solutions and solutions with $\hat{\gamma}^2 > 1$ will show a decrease in solubility at higher pressures whereas the $\hat{\gamma}^2 < 1$ case is frequently associated with an increase in solubility with pressure increase. An obvious and well-known fact is the higher solubility at higher temperatures for most systems (according to Eq. (6.5b)) for an endothermic dissolving reaction. The rare phenomenon of retrograde solubility (solubility decreasing with increased T) is connected with an exothermic dissolving process. In Fig. 6.6, free energy curves for ideal and regular solutions are given so that we may see the types of curves that arise for different heats of mixing, ΔH_m.

Our general goal is to gain some simple representative form for $\hat{\gamma}^j$ because, if $\hat{\gamma}^j$ can be characterized for all constituents in the solution, everything else follows. A very useful procedure can be obtained by considering the α-parameter formalism for a binary system.[1] This has been derived from general power series expansions for the partial molar enthalpy and the excess partial molar entropy (beyond ideal solution value) for the solvent and from the use of the Gibbs–Duhem equation.

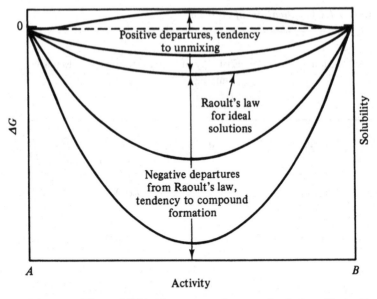

Fig. 6.5. Plots of Gibbs free energy for several solutions illustrating departures from ideal behavior.

The results of analysis for an (A, j) binary system yield

$$\ln\hat{\gamma}^A = \alpha^{Aj} X^{j^2} \tag{6.12a}$$

where

$$\alpha^{Aj} = \ell^{Aj} + a^{Aj} X^j + \frac{b^{Aj}}{T} + c^{Aj}\frac{X^j}{T} \tag{6.12b}$$

and

$$\ln\hat{\gamma}^j = -\alpha^{Aj} X^j (2 - X^j) + \ell^{Aj} + \frac{b^{Aj}}{T} + \frac{1}{2}\left[a^{Aj} + \frac{c^{Aj}}{T}\right][1 + X^{j^2}] \tag{6.12c}$$

where $\ell^{Aj}, a^{Aj}, b^{Aj}$ and c^{Aj} are constants for the particular phase. At infinite dilution

$$\ln\hat{\gamma}^j = \left(\ell^{Aj} + \frac{a^{Aj}}{2}\right) + \frac{1}{T}\left(b^{Aj} + \frac{c^{Aj}}{2}\right) \tag{6.12d}$$

For a strictly regular solution, the excess entropy is zero and one has $\ell^{Aj} = a^{Aj} = c^{Aj} = 0$.

Thurmond and Kowalchik[2] conducted an exhaustive study of the Ge and Si systems using a limited series expansion that amounted to choosing the following:

$$\alpha^* = \mathcal{R}T\ln\hat{\gamma}^A/(X^j)^2 = b + aT \tag{6.13}$$

which is obtained by setting $a^{Aj} = c^{Aj} = 0$ in Eq. (6.12b) and multiply-

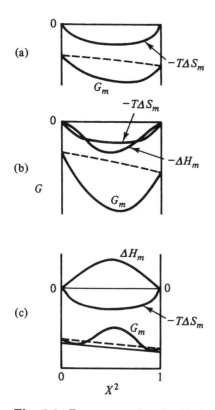

Fig. 6.6. Free energy plots for ideal and regular solutions: (a) ideal solution ($\Delta H_m = 0$); (b) regular solution ($\Delta H_m < 0$); (c) regular solution ($\Delta H_m > 0$).

ing by T. Their excellent compilation of data was used to test the utility of Eq. (6.12) with $c^{Aj} = 0$ and a good fit was obtained to the data for all systems. The coefficients obtained from a least squares analysis of their data are listed in Tables 6.2 and 6.3. In some cases, the fit was strikingly better using α^{Aj} rather than α^*, as can be seen from Figs. 6.7 and 6.8.

A variety of representations of excess free energy of solution are to be found in the literature. The particular advantage to using the α-parameter approach lies in the fact that this parameter generally has simple predictable forms as a function of temperature and composition. The magnitude and sign of α are direct indications of the direction and degree of deviation from ideality, with a value of zero indicating ideality. A form $\alpha = f(1/T)$ would indicate a strictly regular solution and the

Table 6.2. *Power series coefficients for the excess free energy in Ge alloy systems (1 ≡ Ge, 2 ≡ alloying element)*
$$\alpha_{12} = -A_2 - A_3 N_2 + B_2/T$$

Element '2' in Ge	A_2	A_3	B_2	Expected range of good reliability (N_{Ge})
Au	−0.6499	0.03995	−2515.1	0–1.0
Ag	−1.0352	2.16559	1201.3	0–1.0
Cu	−1.75575	0.367989	−1548.2	0–1.0
Ga	−0.07663	0.363630	57.99	0–0.50
In	−0.06404	−2.42964	1257.5	0–0.5
Tℓ	+1.75109	−0.38449	3367.68	0–1.0
Aℓ	−0.57877	+1.16948	−1270.8	0–1.0
Bi	+0.82808	−0.52725	2398.4	0–0.35
Sb	+0.85088	+0.03465	1189.6	0–1.0
As	−6.4920	−2.60552	−8323.7	0.45–1.0
Cd	+0.8957	−0.71869	1527.5	0–0.45
Zn	+2.2110	+3.68280	4121.6	0–1.0

absence of any temperature dependence indicates an athermal solution. For a regular solution, we can make an easy bridge to various quasi chemical bond models and find that

$$\alpha^{Aj} = b^{Aj}/T \tag{6.14a}$$

$$b^{Aj} = \frac{\bar{z}_1 N}{\mathcal{R}} \left(E^{Aj} - \frac{E^{AA}}{2} - \frac{E^{jj}}{2} \right) \tag{6.14b}$$

$$\Delta H_m = \bar{z}_1 N X^A X^j \left(E^{Aj} - \frac{E^{AA}}{2} - \frac{E^{jj}}{2} \right) \tag{6.14c}$$

where the Es are the respective bond energies, NX^j is the number of j atoms and ZX^j is the number of first nearest neighbor j atoms. Thus, we have a relationship between α and the bond energies for a regular solution. For a real solution, therefore, we have a model for the heat part of α^{Aj} and the entropic effects can be incorporated in a variety of ways. Basically, in Eq. (6.14b), b^{Aj} will need to be a function of both

Table 6.3. *Power series coefficients for the excess free energy in Si alloy systems ($1 \equiv Si$, $2 \equiv$ alloying element)* $\alpha_{12} = -A_2 - A_3 N_2 + B_2/T$

Element '2' in Si	A_2	A_3	B_2	Expected range of good reliability (N_{Si})
Au	+11.747	+10.220	−24290	0–1.0
Ag	−3.7174	−2.6631	−6725.3	0–1.0
Cu	−5.7915	−4.3019	−11751.6	0–1.0
Ga	+0.46958	+0.3560	+1938.8	0–0.5
In	+2.8389	−2.2214	4501.4	0–0.2
Aℓ	+0.2358	+0.63708	−620.85	0–1.0
Bi	–	–	–	*
Sb	−0.18475	−1.26158	931.88	0–0.25
As	−21.08	−10.310	+37370	†
Pb	–	–	–	*
Sn	+5.1473	−7.0262	1073.7	0–0.10
Zn	+8.8408	−10.078	548.9	0–0.10

* Data covers $N_{Si} < 0.01$; coefficients very unreliable and, hence, not listed.
† Only three data points available, in the range 0.60–0.70.

composition and temperature which, in turn, implies that the individual Es are expected to be composition and temperature dependent. It is not difficult to extend the α-parameter formalism to ternary, quaternary, etc., solutions.[3] One word of caution is needed here concerning the complete applicability of Eq. (6.12) to all systems. This approach basically rests on the Margules' expansion that is truncated after the third order term. Some error can be expected to enter when association as well as ionization reactions become important in the solutions.

Specific thermodynamic effects, such as those of relative sizes of the atoms and their relative electronegativity, are expected to be considerably smaller in the liquid solution than in the corresponding solid solutions. The size effect can be illustrated by considering the variation of α, at low solute concentrations in liquid solutions, with "size factor"

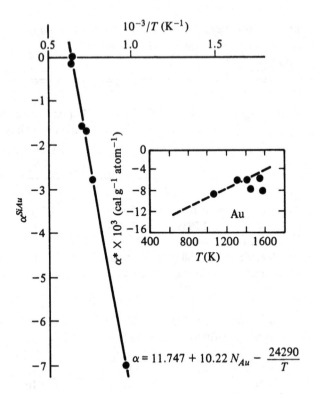

Fig. 6.7. Comparison of α_{SiAu} as a function of T^{-1} with α^*_{SiAu} as a function of T.

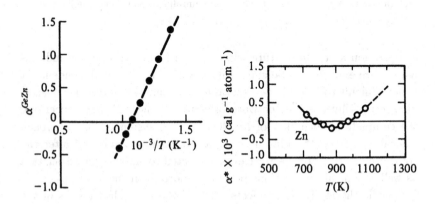

Fig. 6.8. Comparison of α^{GeZn} as a function of T^{-1} with α^* for Ge–Zn as a function of T.

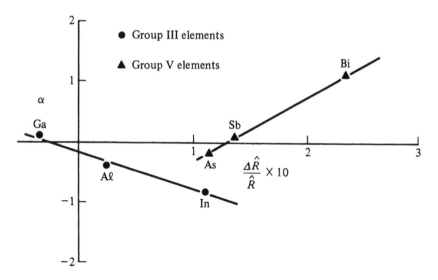

Fig. 6.9. Illustration of the size factor, $\Delta \hat{R}/\hat{R}$, dependence of α for Group III and Group V elements in Ge.

defined as $(\hat{R}^0 - \hat{R})$ where $\hat{R}^0$ and $\hat{R}$ are the metallic radii of solvent and solute atoms, respectively. Solutions of some Group III and Group V elements in Ge at $X_L^j = 0.02$ are presented in Fig. 6.9. The Goldschmidt CN_{12} atomic radii were used in this case and $T = 1333$ K for Ga, In and Aℓ whereas $T = 1234$ K for As, Sb and Bi. Good correlation has also been gained for other solvents.

Eqs. (6.5b) and (6.5c) can be applied to both neutral and charged molecules; however, this is not the preferable representation for the charged species case where μ^j depends upon the macro-potential ϕ. It is preferable to utilize the electrochemical potential η^j where we have

$$\eta^j = \mu_0^j + \mathcal{R}T\ell\mathrm{n}\hat{\gamma}^j X^j + \hat{z}^j e\phi \tag{6.15}$$

Here, $\hat{z}^j$ is the valence of the ion while e is the electronic charge. All that has been said earlier concerning $\hat{\gamma}^j$ still applies for this case.

6.2   Phase equilibria

6.2.1   The phase rule

The phase rule, first enunciated by Gibbs,[4] tells us the number of degrees of freedom, $\tilde{f}$, available to a chemical system of $\tilde{c}$ components when there are $\tilde{p}$ coexisting phases present. The rule is primarily a topological statement where all the terms are integers and where the relative

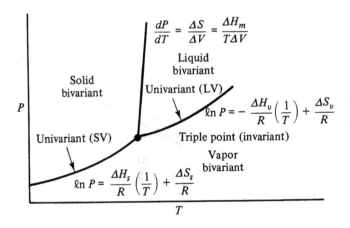

Fig. 6.10. Typical P–T phase diagram for a one component system.

quantities of the coexisting phases of their compositions do not enter. It generally assumes that the system is closed (mass conservation) and that the only extrinsic variables are temperature and pressure. However, it can be readily extended to include the electrostatic potential, ϕ, as an additional extrinsic variable. The rule statement for two extrinsic variables is

$$\tilde{f} = \tilde{c} - \tilde{p} + 2 \tag{6.16}$$

where $\tilde{p}$ is the number of coexisting phases, $\tilde{c}$ is the number of chemical components and $\tilde{f}$ is the number of degrees of freedom or the variance available for the system. It should be noted that it is the composition of the phases rather than the bulk composition that is considered variable in the phase rule sense.

The concept of variance can be illustrated most simply with the pressure–temperature phase diagram for a one component system as illustrated in Figs. 6.1 and 6.10. A one component system with one phase present has a variance of 2 from Eq. (6.16). This means that the pressure and temperature can be varied independently without changing the phase assemblage. These conditions appear on the phase diagram as *bivariant areas*, one for each state of matter: solid, liquid and vapor. When there are two coexisting phases, there is only one degree of freedom and the relationship appears on the diagram as a *univariant line*. Only for T and P described by a univariant line can two phases coexist at equilibrium. Any attempt to change pressure or temperature away from the line will result in the eventual disappearance of one of the phases via a loss or gain of heat. The univariant curve separating the bi-variant fields for solid and liquid is described by the Clapyron equation; i.e.,

$$\frac{dP}{dT} = \frac{\Delta S_F}{\Delta v} = \frac{\Delta H_F}{T \Delta v} \tag{6.17a}$$

where Δv is the volume change on melting. The univariant curves for solid/vapor (SV) and liquid/vapor (LV) equilibria in the low pressure region, where the ideal gas law is obeyed, are described by the Van't Hoff equation; i.e.,

(i) SV

$$\ell nP = \frac{\Delta H_{s*}}{\mathcal{R}} \left(\frac{1}{T}\right) + \frac{\Delta S_{s*}}{\mathcal{R}} \tag{6.17b}$$

and

(ii) LV

$$\ell nP = -\frac{\Delta H_{v*}}{\mathcal{R}} \left(\frac{1}{T}\right) + \frac{\Delta S_{v*}}{\mathcal{R}} \tag{6.17c}$$

where ΔH_{s*} and ΔH_{v*} are the heats of sublimation and vaporization respectively. These thermodynamic functions describe melting, subli-mation and boiling and are the analytical expressions between P and T defining the univariant equilibria lines expressed graphically in Fig. 6.10.

When three phases coexist in a one component system, the degrees of freedom are all used up and the system is *invariant*. Both P and T are fixed at the triple point in Fig. 6.10. Addition or subtraction of heat from the system will merely change the proportions of the three phases. The pressure and temperature remain stable against all heat losses or gains until one of the phases disappears. As a practical example, the system $SiO_2 - H_2O$ in Fig. 6.11 is a classic one and is relevant to the hydrothermal growth of quartz. A logarithmic scale has been used with the result that the solid–solid equilibria appear distorted. The horizontal dashed line shows the isobar at 1 bar with the traditional phase transition temperatures of the SiO_2 polymorphs.

Extending our considerations into the composition domain, in an iso-baric section of a P–T–X diagram at 1 atm pressure, we have the usual temperature–composition diagram illustrated in Fig. 6.12. Again, if there is only one phase, both T and X can be set independently and these are mapped on the diagram as the bivariant areas: a liquid phase, a solid solution composed mainly of component A, and a solid solution composed mainly of component B.

The expression of univariant equilibria in a T–X diagram is different from its expression in a P–T diagram although the governing principles

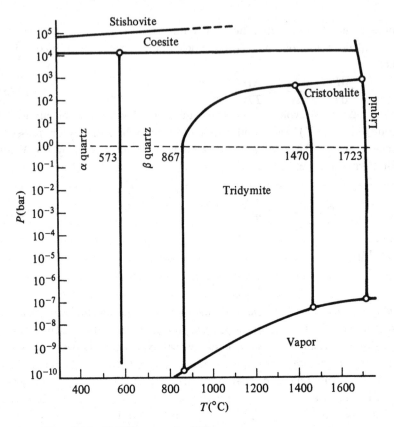

Fig. 6.11. P–T diagram for SiO_2.

are the same. If $P =$ constant and two phases coexist, the phase rule
dictates that there is only one degree of freedom so that the temperature and composition of the phases must be linked. However, there is no
requirement that the two phases have the same composition. Therefore,
the T–X diagram has univariant areas rather than univariant lines and
any bulk composition that falls within these areas will react, at equilibrium, to form two phases, the compositions of which are given by the
points on the curves that bound the univariant areas. Corresponding
points for the two phases are connected by isothermal lines called tie-
lines. If the temperature is altered, the compositions and proportions of
the coexisting phases will change in accordance with mass conservation
and the shape of the lines bounding the univariant areas. Likewise an
attempt to change the composition of the phases generates a change in
temperature. However, changing the bulk composition within the two

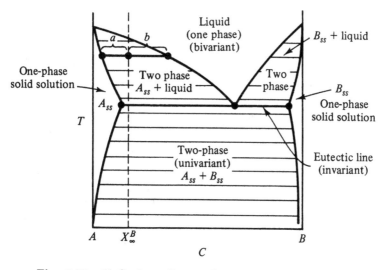

Fig. 6.12. T–C phase diagram for a two component system at a constant pressure.

phase area merely changes the proportions of the phases and not their composition. The requirement that the proportions of coexisting phases balance out to the original bulk composition leads to the lever rule; i.e., for liquid/A_{ss} equilibria in Fig. 6.12,

$$X_\infty = \frac{a}{a+b}X_{liquid} + \frac{b}{a+b}X_{A_{ss}} \qquad (6.18)$$

where the proportions a and b are shown in the figure.

The coexistence of the phases defines the *invariant* condition in this case as well. Here, it is described by an invariant line which connects A_{ss}, liquid and B_{ss}. The temperature and composition at which these three phases coexist are fixed. Of course, the bulk composition can be varied anywhere along the line without changing the three phase assemblage or the invariant condition. Only the proportions of the three phases change.

As a third example, an isothermal and isobaric section of a P–T–Φ–X diagram leads to a Φ–X diagram. A form of this diagram that is strongly utilized in the corrosion area is the Φ–pH diagram. The Φ–pH diagram for Cu shown in Fig. 6.13 (here, the reference potential is Φ_{H_2}) defines regions where Cu is soluble in an aqueous solution as Cu^{2+}, $HCuO_1^-$ or CuO_2^{2-} ions and where it exists as condensed phases such as the pure metal or as compounds. If Φ and pH are such that Cu^{2+} is stable, then Cu metal may dissolve until an equilibrium Cu^{2+} concentration is

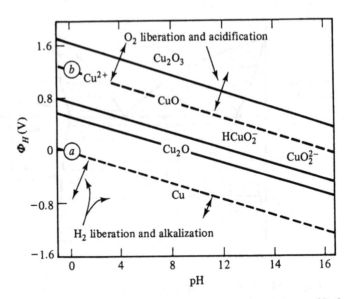

Fig. 6.13. Φ–pH diagram illustrating Cu–H$_2$O equilibrium (CuO was neglected). Dashed line a borders the H$_2$ liberation domain while dashed line b borders the O$_2$ liberation domain.

attained and the metal will corrode until this condition is reached. On the other hand, if the (Φ, pH) condition is such that a solid product like Cu$_2$O is stable, then the corrosion rate may be expected to be minimal because of a covering of the metal by a film of this solid product and the metal surface will be passivated.

To see the relationship between pH and chemical activity of one of these key species, we need only consider a chemical reaction of the form

$$A^+ + H_2O \rightleftharpoons B + H^+ \tag{6.19a}$$

and the application of the equilibrium relation leads to

$$\log \frac{\hat{a}^B}{\hat{a}^{A^+}} = \log K + \text{pH} \tag{6.19b}$$

where K is the equilibrium constant. A specific example might be

$$NH_4^+ + H_2O \rightleftharpoons NH_4OH + H^+ \tag{6.20a}$$

$$\log \frac{[NH_4OH]}{[NH_4^+]} = -9.27 + \text{pH} \tag{6.20b}$$

where [ ] refers to chemical activity. Thus, we see that, for a given reaction (determines K), the pH variation yields $\hat{a}^B$ for a given $\hat{a}^{A^+}$.

During crystal growth, of whatever kind, we are always working with the appropriate phase rule example in the limit of complete equilibrium obtaining. The appropriate phase diagram is our standard state according to which everything is referred. It behooves us to be clear concerning the constancy of the P and Φ coordinates in the system's description by a $(T–X)$ diagram. Neglect of vapor pressure effects or electrostatic potential effects can shift one, unknowingly, from a univariant domain in the $(T–X)$ diagram to a bivariant or trivariant domain in the (P, Φ, T, X) domain. Such an event would lead to anomalous crystal growth results. When the vapor pressure in the system becomes large enough that a significant portion of the system mass is in the vapor phase or when the vapor is of variable composition binary diagrams are no longer adequate to represent the phase relations and the chemical behavior of the system must be described by a $P–T–X$ diagram.

6.2.2 Phase diagrams

Two component phase equilibrium in a binary system is given by the equalization of the chemical potentials for the two species. The two chemical potentials are generally related by the Gibbs–Duhem equation which, for constant T and P, is

$$(1 - X)\mathrm{d}\mu^A + X\mathrm{d}\mu^B = 0 \tag{6.21}$$

when X is the mole fraction of B. The molar Gibbs free energy $\bar{G} = G/(n^A + n^B)$, is given from Eqs. (6.3b) and (6.5b), by

$$\bar{G}(X) = (1 - X)\mu^A + X\mu^B \tag{6.22a}$$

and

$$\mu^A = \bar{G} - X(\partial\bar{G}/\partial X) \tag{6.22b}$$

$$\mu^B = \bar{G} + (1 - X)(\partial\bar{G}/\partial X) \tag{6.22c}$$

Eqs. (6.22) form the basis for the popular graphical method of tangents to determine equilibrium conditions. They express the fact that the tangent to the molar free energy curve as in Fig. 6.14 at the composition of interest intercepts the $X = 0$ and $X = 1$ vertical axes at a value of $\bar{G}$ equal to the chemical potential of that component. Hence, a common tangent among two or more phases implies equality of chemical potentials and equilibrium (see Fig. 6.14). Let us now see how this simple technique allows us to determine the various types of phase diagram that we encounter in nature.

The simplest case is a system which is ideally miscible in both the liquid and solid phases. The $G_S(X)$ and $G_L(X)$ curves are both parabolas whose relative positions change with temperature as illustrated in Fig. 6.15. If the melting points of A and B are different, the G_S parabola

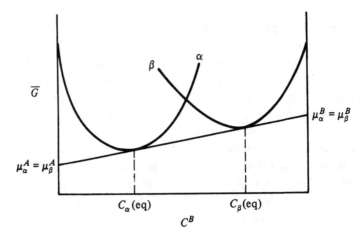

Fig. 6.14. Graphical construction for obtaining the equilibrium μs and Cs from the Gibbs free energy curves for a binary system.

must be tilted so that intersections and tangent constructions become possible yielding the well-known complete miscibility phase diagram of Fig. 6.15(f). The diagram changes as the heat of mixing, ΔH_m, becomes increasingly positive (See Eq. (6.15c)). First a solubility gap appears, then a congruently solidifying alloy and finally a eutectic system as in Fig. 6.16. If the melting points of A and B are very different, a completely different phase diagram is obtained as ΔH_m increases as shown in Fig. 6.17; i.e., a peritectic reaction ensues.

For the eutectic reaction, the eutectic temperature T_E is characterized by coincidence of the tangents to the SL two-phase regions, $L + \alpha_1$ and $L + \alpha_2$, at the temperature T_4 in Fig. 6.18. A similar construction is obtained if the $G_S(X)$ curve is replaced by two curves, $G_S^{\alpha_1}(X)$ and $G_S^{\alpha_2}(X)$ parabolas corresponding to solids of different structures. The peritectic reaction takes place when the minimum $G_S(X)$ corresponding to α_2 intersects the tangent joining $G_S^{\alpha_1}(X)$ and $G_L(X)$ (instead of the $G_L(X)$ curve itself). This is illustrated in Fig. 6.19.

For the case where $\Delta H_m < 0$ so that unlike nearest neighbors are energetically favorable, phase diagrams exhibiting intermetallic compound formation of more or less stoichiometric composition A_xB_y result. Fig. 6.20 shows how, with increasingly negative heat of mixing, ΔH_m, the phase diagrams change until the intermetallic compound A_xB_y forms from the melt. In this case, the liquidus and solidus form a common maximum and the phase diagram divides into two diagrams of the usual

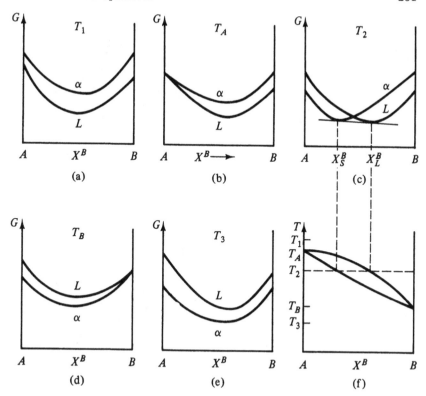

Fig. 6.15. Illustration of phase diagram derivation from free energy curves via the tangent method at different T $(T_1 > T_A > T_2 > T_B > T_3)$.

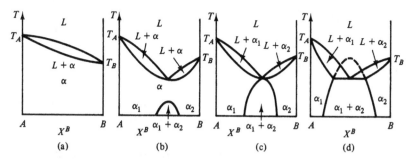

Fig. 6.16. Illustration of phase diagram changes with increasing enthalpy of mixing $(\Delta H_m > 0)$ from (a) to (d).

type, one extending between A and A_xB_y while the other extends between A_xB_y and B as illustrated in Fig. 6.21.

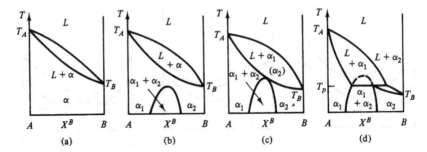

Fig. 6.17. Illustration of phase diagram changes with changes of the melting points for the two constituents ($T_A - T_B$ increasing from (a) to (d)).

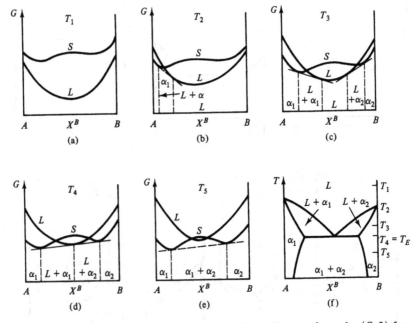

Fig. 6.18. Derivation of a eutectic phase diagram from the (S, L) free energy curves at successively lower temperatures.

The foregoing illustrates how basic G_S and G_L plots as a function of X and T can be used to generate the type of phase diagram displayed by the system. Let us now see how these diagrams might be utilized to evaluate the interface driving force, ΔG^j, available for crystallization from an alloy liquid. Suppose we have an alloy melt at (C_∞, T_∞) as in Fig. 6.22(a) so that it is supercooled an amount ΔT, what is the com-

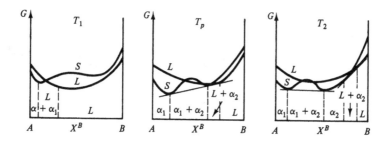

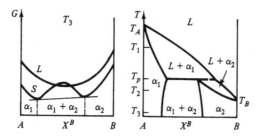

Fig. 6.19. Derivation of a peritectic phase diagram from temperature-dependent free energy curves ($T_1 > T_P > T_2 > T_3$).

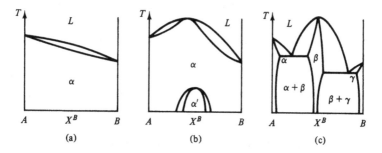

Fig. 6.20. Illustration of how a phase diagram changes from (a) to (c) as ΔH_m becomes increasingly negative.

position of solid that would provide the largest driving force, ΔG_∞, for solid formation from the liquid? With the aid of the tangent construction as in Fig. 6.22(b), it can be shown that, for infinitely rapid attachment kinetics, the solid composition C_S yields the maximum molar driving force given by

$$\Delta G_i = \left[\mu_L^A(C_i) - \mu_S^A(C_S)\right] = \left[\mu_L^B(C_i) - \mu_S^B(C_S)\right] \tag{6.23}$$

per mole of solution. The combined situation is illustrated pictori-

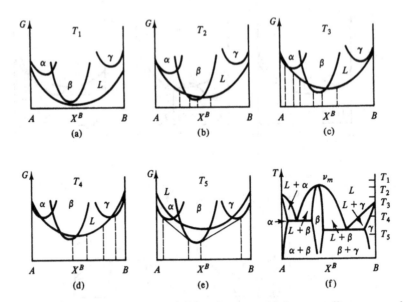

Fig. 6.21. Derivation of phase diagram with intermetallic compound β from the temperature-dependent free energy curves for α, β, γ and L ($T_1 > T_2 > T_3 > T_4 > T_5$).

ally in Fig. 6.22(c) where T_M^A is the melting temperature of pure A. In Fig. 6.22(c), $G(b) - G(a) = \Delta G_{sv}$ which should actually be less than ΔG_∞; the scale has been distorted so that the T_i and T_∞ curves could be clearly separated.

Many important thin film reactions encountered in semiconductor processing involve three chemical elements distributed in various phases and therefore must be described using ternary phase diagrams. Common examples are metals, M, in contact with SiO_2 films, Si_3N_4 films or GaAs films. One generally wants to know under what conditions the different films are stable and under what conditions they will react and break down. To answer such questions, binary phase diagram information is insufficient and ternary information is essential.

At fixed T and P, with three components, the Gibbs phase rule (see Eq. (6.16)) predicts a maximum of three phases in equilibrium in any portion of the phase diagram and regions of three-phase equilibrium form triangles in isothermal sections. The phases at the corners of a given triangle are thermodynamically stable in contact with each other; i.e., for the M–O–Si system, this means that μ^M is the same in all three phases (likewise μ^{Si} and μ^O). Regions of three-phase equilibrium are

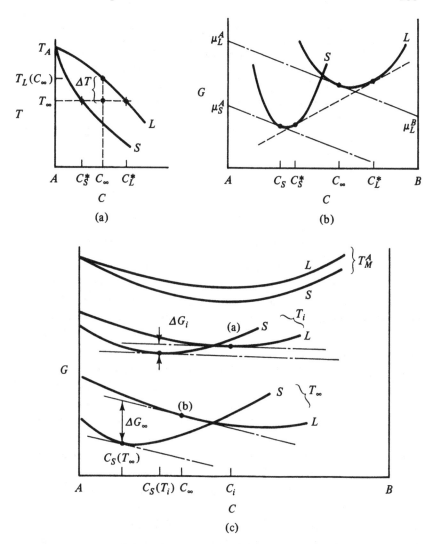

Fig. 6.22. (a) Typical binary phase diagram with melt at (T_∞, C_∞); (b) free energy curves for solid and liquid at T_∞ illustrating the maximum driving force available for solid formation from liquid at C_∞; and (c) similar free energy curve comparisons for the temperatures T_M^A, T_i and T_∞.

found by determining the stable two-phase tie-lines which are, in turn, established by calculating the Gibbs free energy of reaction at the point where two possible tie-lines would cross. For example, the existence of a stable tie-line between $TiSi_2$ and SiO_2 at 700 °C is determined by

Table 6.4. *Tie-line reaction energetics in (T$_i$, O, Si)*. R. B. Beyers, private communication (1983); *J. Appl. Phys.*, **56**, 147 (1984); R. B. Beyers, R. Sinclair and M. E. Thomas, *J. Vac. Sci. Technol.*, **B2**, 781 (1984); R. B. Beyers, K. B. Kim and R. Sinclair. *J. Appl. Phys.*, **61**, 2195 (1987).

Reactions	ΔG(kcal mole^{-1})
$2TiO + 5Si = 2TiSi_2 + 1SiO_2$	-25
$2Ti_2O_3 + 11Si = 4TiSi_2 + 3SiO_2$	-51
$2Ti_3O_5 + 17Si = 6TiSi_2 + 5SiO_2$	-94
$1TiO_2 + 3Si = 1TiSi_2 + 1SiO_2$	-23

considering all possible reactions between Ti oxide–Si couples and the TiSi$_2$–SiO$_2$ couple (see Table 6.4 and Fig. 6.23). Since the free energy change is negative in all cases, the reactions as written will go to the right and a stable tie-line runs between TiSi$_2$ and SiO$_2$ at this temperature. The remaining stable tie-lines in the Ti–Si–O phase diagram can be evaluated in a similar manner.

To evaluate metal reactivity with the underlying films, knowledge concerning tie-line configurations in the ternary phase field is needed. Elemental metals will not react with SiO$_2$ if a stable tie-line exists between the metal and the SiO$_2$, hence Mo and W are chemically stable on SiO$_2$ substrates. Although the metal and SiO$_2$ will interdiffuse until the solubility limit is reached, no chemical reaction between the two phases will occur. Conversely, if no tie-line exists between the metal and SiO$_2$, then a chemical reaction will occur between the two phases when sufficient activation energy is supplied to the system. Thus, a thin film of Ti or Ta on a SiO$_2$ substrate reacts to form a metal oxide and a metal-rich silicide at sufficiently high temperature. From the same line of reasoning, the native oxide found on Si is a chemically stable diffusion barrier to Mo and W but not to Ta or Ti. In addition, the same approach can be used to determine the stability of electrical contacts on GaAs where, from the ternary diagrams given in Fig. 6.24, one sees that only Si or Ge will produce stable films on GaAs.

6.2.3   Key parameter evaluation (zero interface field case)

For crystallization considerations, one is interested in more than the qualitative shape of the phase diagram of interest. However, one

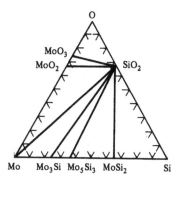
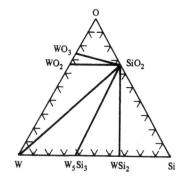
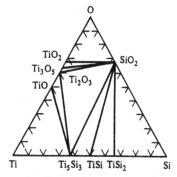
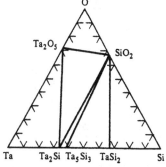

Fig. 6.23. Ternary phase diagrams with tie-lines for four M–O–Si systems. R. B. Beyers, private communication (1983); *J. Appl. Phys.*, **56**, 147 (1984); R. B. Beyers, R. Sinclair and M. E. Thomas, *J. Vac. Sci. Technol.*, **B2**, 781 (1984); R. B. Beyers, K. B. Kim and R. Sinclair. *J. Appl. Phys.*, **61**, 2195 (1987).

generally does not have available free energy curves for these solid and liquid phases although it is possible to obtain this information by computer analysis of the existing phase diagram and any additional thermodynamic information that is available. Thus, rather than using exact data, one makes recourse to the various models for thermodynamic solutions discussed earlier. Using such models, it is possible to calculate the phase diagrams, the slope of the liquidus and solidus lines, the equilibrium solute distribution coefficient, the eutectic concentration, the eutectic temperature, etc. These numbers are important for practical crystal growing so procedures for obtaining them will be developed in this section.

The prime characteristic of a solute element, j, that allows its manipulation during freezing is that, under equilibrium conditions, the atom fractions of solute in the bulk solid, X_S^j, and the bulk liquid, X_L^j, are dif-

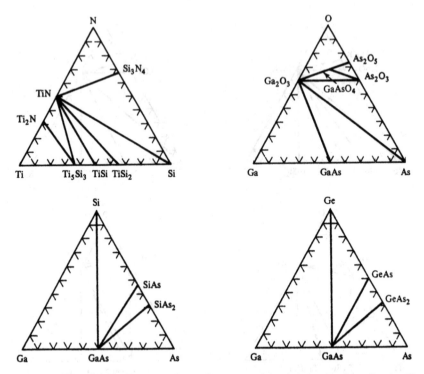

Fig. 6.24. Ternary phase diagrams with tie-lines for four Ga–As–X systems. R. B. Beyers, private communication (1983); *J. Appl. Phys.*, **56**, 147 (1984); R. B. Beyers, R. Sinclair and M. E. Thomas, *J. Vac. Sci. Technol.*, **B2**, 781 (1984); R. B. Beyers, K. B. Kim and R. Sinclair. *J. Appl. Phys.*, **61**, 2195 (1987).

ferent. The parameter used to represent this difference is the equilibrium phase diagram distribution coefficient, $\tilde{k}_0^j$, defined by

$$\tilde{k}_0^j = X_S^j / X_L^j \tag{6.24}$$

By convention, the ratio of the concentration in the solid phase, C_S, to the concentration in the liquid phase, C_L, is more commonly used and is called the equilibrium distribution coefficient

$$k_0^j = C_S^j / C_L^j \tag{6.25}$$

The solutes fall into two classes depending upon whether k_0 is greater or less than unity. Solutes with $k_0 < 1$ lower and solutes with $k_0 > 1$ raise the melting point of the solvent. Examples of k_0 are shown in Fig. 6.25 for the two cases. In both cases, for liquids of solute concentration C_∞, equilibrium freezing begins when the temperature of the liquid has been decreased to $T_L(C_\infty)$. At this point, solid of concentration $C_S = k_0 C_\infty$ begins to form (provided all interface fields are zero).

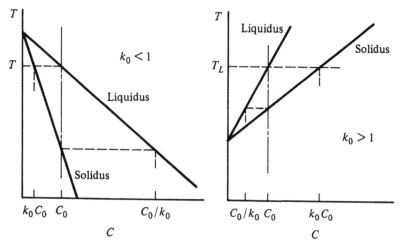

Fig. 6.25. Portions of phase diagrams for the cases where the solute either lowers ($k_0 < 1$) or raises ($k_0 > 1$) the melting point of the alloy relative to the pure solvent.

(i) Ideal binary liquid and solid solutions Equilibrium between any two phases requires equating the extended electrochemical potentials of any species in each phase. Provided all interface fields are zero, this becomes in our case,

$$\mu_S^j = \mu_L^j \tag{6.26}$$

Following Thurmond,[5] for ideal solutions in both liquid and solid phases, $\tilde{k}_0^j$ becomes very simply

$$\ln \tilde{k}_0^A = \ln \frac{X_S^A}{X_L^A} = \frac{\Delta H_F^A}{\mathcal{R}} \left(\frac{1}{T} - \frac{1}{T_M^A} \right) \tag{6.27a}$$

$$\ln \tilde{k}_0^j = \ln \frac{X_S^j}{X_L^j} = \frac{\Delta H_F^j}{\mathcal{R}} \left(\frac{1}{T} - \frac{1}{T_M^j} \right) \tag{6.27b}$$

In these equations, T_M^A and T_M^j are the respective melting points of pure solvent, A, and pure solute, j. The approximation of equal solid and liquid heat capacities for A and for j has been made (which is good to first order) so that ΔH_F^A and ΔH_F^j are independent of temperature. Fig. 6.26 illustrates Eqs. (6.27) for the case where $T_M^A > T_M^j$ and $\Delta H_F^A > H_F^j$. Note that, as $T \to T_M^A$, $\tilde{k}_0^A \to 1$ and $\tilde{k}_0^j \to a < 1$. Likewise, as $T \to T_M^j$, $\tilde{k}_0^j \to 1$ and $\tilde{k}_0^A \to b > 1$.

Since the sum of the mole fractions of A and j in each phase is unity, Eqs. (6.27) may be used to obtain any of the mole fractions as a function of temperature. This leads to Fig. 6.27 for the representative

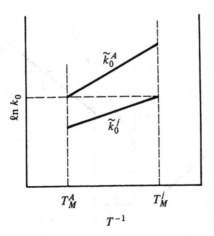

Fig. 6.26. Calculated equilibrium distribution coefficients for an ideal binary liquid and solid as a function of temperature.

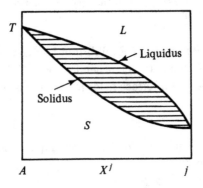

Fig. 6.27. Schematic phase diagram for ideal binary liquid and solid.

phase diagram illustrating complete solid and liquid solubility of the two constituents plus qualitative agreement with the $\tilde{k}_0$ values presented in Fig. 6.26.

(ii) Dilute solid and ideal liquid solutions In this case, since the solid solution is dilute, the thermodynamic properties of A in the solution are essentially those of pure A. Thus, the $\tilde{k}_0$ values are given by

$$\ell n \tilde{k}_0^A = \ell n \frac{1}{X_L^A} = \frac{\Delta H_F^A}{\mathcal{R}} \left(\frac{1}{T} - \frac{1}{T_M^A} \right) \tag{6.28a}$$

$$\ell n \tilde{k}_0^j = \frac{\Delta H_F^j - \Delta \bar{H}_S^j}{\mathcal{R}T} - \frac{\Delta H_F^j}{\mathcal{R}T_M^j} \tag{6.28b}$$

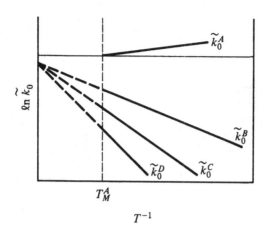

Fig. 6.28. Distribution coefficients for the major component, A, and various impurities, j, in an ideal dilute solid solution as a function of temperature.

These equations show that the liquidus curve is independent of X_S^j. However, $\tilde{k}_0^j \lesssim 10^{-2}$ is required to have a dilute solid solution which requires that the partial molar heat of mixing, $\Delta \bar{H}_S^j$, is a positive number. The smaller the value of $\tilde{k}_0^j$ at T_M^A, the larger is $\Delta \bar{H}_S^j$.

Fig. 6.28 illustrates the variation of $\tilde{k}_0^A$ and $\tilde{k}_0^j$ with temperature and the three choices of $\tilde{k}_0^j$ have different values of $\Delta \bar{H}_S^j$ but the same value of the entropy of fusion, $\Delta H_F^j / T_M^j$ (leads to a common intercept at $T^{-1} = 0$). The liquidus curve for this case is shown schematically in Fig. 6.29(a). Since the composition of the solid solution is too small to be represented on the same plot, Fig. 6.29(b) shows one of the solidus curves of Fig. 6.28 on an expanded scale. The solid solubility of j in A increases as T increases reaching a maximum value and then decreases again until it reaches zero at T_M^A. This phenomenon is called "retrograde solid solubility" and we will discuss it further below. The complete phase diagram representing liquid/solid equilibria between A and j would include another liquidus and solidus curve on the j side of the diagram and a eutectic minimum between the j and the A liquidus lines.

Using Eq. (6.28a), the equations for the liquidus line, T_L, and the liquidus slope, m_L^B, on the A side of the diagram are

$$1 - X_L^B = \exp\left[-\frac{\Delta H_F^A}{\mathcal{R} T_M^A} \left(\frac{T_M^A}{T_L} - 1 \right) \right] \tag{6.29a}$$

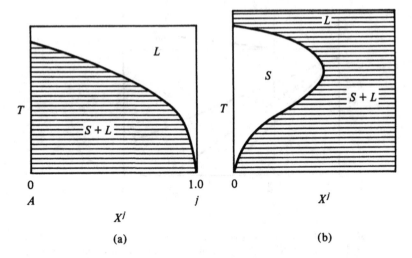

Fig. 6.29. (a) Schematic illustration of a phase diagram of a pure solid in equilibrium with and ideal binary liquid and (b) illustration of a retrograde solidus curve.

and

$$m_L^B = \frac{dT_L}{dX_L^B} = -\frac{\mathcal{R}T_L^2}{\Delta H_F^A(1 - X_L^B)} \tag{6.29b}$$

Considering both sides of the A − B diagram, the eutectic temperature, T_E, and the eutectic concentration, X_E^B, are given, respectively, from the following two equations

$$1 = \exp\left[\frac{\Delta H_F^A}{\mathcal{R}T_M^A}\left(\frac{T_M^A}{T_E} - 1\right)\right] + \exp\left[-\frac{\Delta H_F^B}{\mathcal{R}T_M^B}\left(\frac{T_M^B}{T_E} - 1\right)\right] \tag{6.30a}$$

and

$$\frac{1}{T_M^A} - \frac{\mathcal{R}}{\Delta H_F^A}\ell n(1 - X_E^B) = \frac{1}{T_M^B} - \frac{\mathcal{R}}{\Delta H_F^B}\ell n X_E^B \tag{6.30b}$$

(iii) General case For equilibrium to exist between a liquid and a solid phase, Eq. (6.26) is equivalent to equating the partial molar heats and entropies of each component in the following way:

$$\bar{H}_L - \bar{H}_S = T(\bar{S}_L - \bar{S}_S) \tag{6.31}$$

In general, each of these quantities is a function of both temperature and composition. We shall define an excess partial molar entropy for j in each phase by the equation

$$S_e = S + \mathcal{R}\ell nX \tag{6.32a}$$

and the partial molar heats and entropies of mixing by equations of the form

$$\Delta \bar{H}_\beta^j = \bar{H}_\beta^j - H_\beta^j \quad \beta = S, L \tag{6.32b}$$

$$\Delta \bar{S}_\beta^j = \bar{S}_{\beta,e}^j - S_\beta^j \quad \beta = S, L \tag{6.32c}$$

Utilizing this approach, we obtain

$$\ell n \tilde{k}_0^j = \frac{\Delta H_F^j + \Delta \bar{H}_L^j - \Delta \bar{H}_S^j}{RT} - \frac{\Delta S_F^j + \Delta \bar{S}_{L,e}^j - \Delta \bar{S}_{S,e}^j}{R} \tag{6.33}$$

When the liquid and solid solutions are ideal, $\Delta \bar{H}_L^j, \Delta \bar{H}_S^j, \Delta \bar{S}_{L,e}^j$ and $\Delta \bar{S}_{S,e}^j$ are each zero and Eq. (6.27b) is obtained. When the solid solutions are very dilute, $\Delta \bar{H}_S^j$ and $\Delta \bar{S}_{S,e}^j$ will not be significant functions of composition. If the dilute solid solutions are particularly simple (i.e., regular), $\Delta \bar{S}_{S,e}^j$ will be zero and, if the liquid solutions are ideal, Eq. (6.28b) will be obtained. In general, however, both $\Delta \bar{H}_S^j$ and $\Delta \bar{S}_{S,e}^j$ are expected to be large for impurities which have a solidus curve lying at very low compositions ($\tilde{k}_0^j << 1$). Under these conditions, ΔH_S^j and $\Delta S_{S,e}^j$ will dominate in Eq. (6.33). When these two quantities are independent of temperature, the logarithm of $\tilde{k}_0^j$ will be a linear function of $1/T$. Such a relationship has been observed for certain elements in Ge and Si. In some systems, where $\tilde{k}_0^j$ is not very small, it has been found that $\ell n \tilde{k}_0^j$ is not a linear function of $1/T$ and this can probably be accounted for in terms of the unexpected departures from ideality which occur in liquid solutions.

Returning to Eq. (6.26) and writing μ^j as

$$\mu_\beta^j = \mu_{0\beta}^j + RT\ell n \hat{\gamma}^j + RT\ell n X_\beta^j \tag{6.34}$$

we can gain other useful expressions for understanding correlations of $\tilde{k}_0^j$ with the physical properties of j. From Eqs. (6.26) and (6.34), we have for an n-component system,

$$\tilde{k}_0^j = \left(\frac{\hat{\gamma}_L^j}{\hat{\gamma}_S^j} \right) \exp \left(\frac{\Delta \mu_0^j}{RT_L} \right) \quad j = 1, 2, \ldots \tag{6.35a}$$

$$\left\{ \frac{1 - \sum_{j=1}^{n-1} \tilde{k}_0^j X_L^j}{1 - \sum_{j=1}^{n-1} X_L^j} \right\} = \frac{\hat{\gamma}_L^A}{\hat{\gamma}_S^A} \exp \left(\frac{\Delta \mu_0^A}{RT_L} \right) \tag{6.35b}$$

where

$$\Delta \mu_0^j = \mu_{0L}^j - \mu_{0S}^j \quad j = 1, 2, \ldots \tag{6.35c}$$

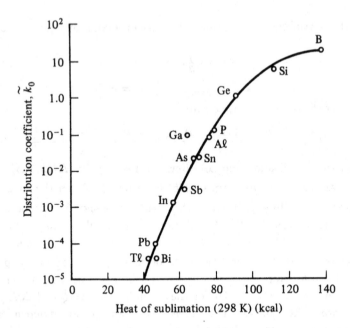

Fig. 6.30. Solute distribution coefficient, k_0^j, as a function of heat of sublimation of j, for most of the Group III, IV and V elements in Ge.

$$= \Delta S_F^j (T_M^j - T_L) + \int_{T_M^j}^{T_L} \Delta c_P^j dT - T_L \int_{T_M^j}^{T_L} \frac{\Delta c_P^j}{T} dT \ (6.35d)$$

Thus, Eqs. (6.35) give us all the $\tilde{k}_0^j$ and the liquidus surface temperature as a function of the X_L^j, provided we have the thermodynamic data in the form of $\hat{\gamma}$s and Δc_Ps (specific heat difference).

If we look at Eqs. (6.34) and (6.35) from a physical point of view, the μ_0 term can be seen to arise as a consequence of the intermolecular potential interactions of a j atom with neighbors either in the liquid or solid. Thus, the stronger is the binding, the more stable is the condensed phase and the smaller is μ_0 (relative to the gaseous state). Thus, neglecting entropy effects, we see that

$$\mu_{0S} \approx \mu_{0gas} - E_S^B \tag{6.36a}$$

$$\mu_{0L} \approx \mu_{0gas} - E_L^B \tag{6.36b}$$

and

$$\Delta\mu_0 \approx E_S^B - E_L^B = N_S \epsilon - N_L \epsilon' \tag{6.36c}$$

$$\approx (N_S - N_L)\Delta H'_{sub} \tag{6.36d}$$

where E^B refers to binding energy, N refers to number of bonds per atom, ϵ refers to an effective bond energy and $\Delta H'_{sub}$ is the heat of sublimation per bond. In practice, a solid and liquid differ only slightly in coordination so that $N_S - N_L \sim 1$ and the liquid has only a slightly different atom–atom distance compared to the solid so that $\epsilon' \approx \epsilon$. Thus, $\Delta\mu_0^j$ will be positive and of the order of one-tenth the heat of sublimation of j. We thus expect $\ell n k_0^j$ to increase with ΔH_{s*}^j as has been observed by Trumbore[6] and illustrated in Fig. 6.30.

The term $\mathcal{R}T\ell n\hat{\gamma}^j$ in Eq. (6.34) incorporates all of the non-idealities associated with placing the j solute in the solvent environment. Thus, in order to match Fig. 6.30 in a quantitative way, $\hat{\gamma}_L^j/\hat{\gamma}_S^j$ must be much less than unity. Since we expect $\hat{\gamma}_L^j$ to be close to ideal ($\hat{\gamma}_L^j \sim 1$), the solid must exhibit a large Henry's law constant for $j(\hat{\gamma}_S^j \gg 1)$ which is the kind of positive deviation from ideality that we might expect by incorporating solutes of very different size and valence into Ge or Si. To appreciate the size and valence difference effects, we need only write Eq. (6.34) in a different form; i.e.,

$$\mu^j = \bar{G}^j = \bar{E}^j + P\bar{v}^j - T\bar{S}^j \tag{6.37}$$

from Eqs. (6.14) and (6.15). From this formulation, neglecting non-ideal entropy effects, we can see that the valence difference is going to affect the $\bar{E}$ term so that $\bar{E}$ will increase as the valence difference increases (less effective binding). In addition, the size difference is going to affect the $\bar{v}$ term so that $P\bar{v}$ increases with the size difference. A rough correlation has been shown to exist between $\tilde{k}_0^j$, for various j in Ge and Si at T_M^{Ge} and T_M^{Si}, respectively, and the tetrahedral radius of j. As is expected, the larger is the tetrahedral radius, the smaller is $\tilde{k}_0^j$.

For donor and acceptor solutes that are completely ionized at the growth temperature and for solid solutions that are not degenerate, an activity coefficient due to a "common ion" effect has been calculated. This is given by

$$\hat{\gamma}_S^j = \frac{X_S^j + [(X_S^j) + 4X_{i*}^2]^{1/2}}{2X_{i*}} \tag{6.38}$$

for an otherwise ideal solid solution, where X_{i*} refers to the atom fraction of holes or electrons in the intrinsic semiconductor ($X_{i*} = (X_e X_h)^{1/2}$). Because of this effect, the change from $\tilde{k}_0^j$ (for $\hat{\gamma}_S^j = 1$) to $(\tilde{k}_0^j)'$ occurs where

$$\frac{(\tilde{k}_0^j)'}{\tilde{k}_0^j} = \frac{2X_{i*}}{X_S^j + [(X_S^j)^2 + 4X_{i*}^2]^{1/2}} \tag{6.39}$$

From Eq. (6.39), we note that, as long as $X_S^j > X_{i*}$ at any growth tem-

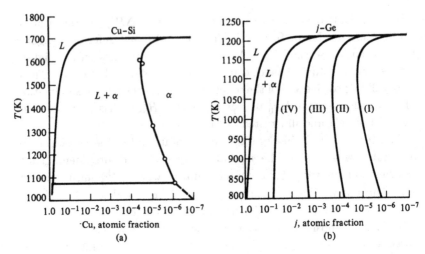

Fig. 6.31. (a) Si–Cu phase diagram showing retrograde solubility and (b) solidus lines for various values of $\Delta \bar{H}_S^j$ in kcal mole^{-1} ($\Delta \bar{H}_S^j = +22, +16.6, +11$ and $+5.5$ kcal mole^{-1} for (I), (II), (III) and (IV), respectively).

perature T, the distribution coefficient for j will be smaller than it would have been if no donor or acceptor ionization had occurred. This condition of $X_S^j > X_{i*}$ is met in a variety of heavily doped semiconductors and may account for the rapid drop in $\tilde{k}_0^j$ with temperature observed in such systems.

(iv) Retrograde solubility For most systems, the maximum solubility of a solute in a given phase occurs at the eutectic temperature; however, in a system showing retrograde solubility, it occurs at a considerably higher temperature. This effect may be very pronounced, as illustrated in Fig. 6.31, where the solubility of Cu in Si is ~ 25 times higher at ~ 1600 K than it is at the eutectic temperature of ~ 1075 K. In order to examine the thermodynamic causes of this effect, we shall follow Swalin[7] and assume that the solid α-phase behaves as a regular solution while the liquid phase in equilibrium with α above the eutectic temperature behaves as an ideal solution (simplifying rather than necessary assumptions). For this solid, the entropy of mixing is ideal (Eqs. (6.22)) but the heat of mixing is non-zero, thus, the equilibria between the two phases is described by Eq. (6.26). From Eq. (6.28a), the equation for

the liquidus line in equilibrium with α is given by

$$\ell n(1 - X_L^j) = \frac{\Delta H_F^A}{\mathcal{R}} \left(\frac{1}{T_M^A} - \frac{1}{T_L} \right) \tag{6.40}$$

If X_L^j from Eq. (6.40) is substituted into Eq. (6.28b) for $\tilde{k}_0^j = X_S^j / X_L^j$; then the solidus line, X_S^j, is obtained explicitly as a function of temperature, if the thermodynamic quantities $\Delta \bar{H}_S^j, \Delta H_F^j$ and H_F^A are known. For calculation purposes, let us assume that ΔH_F^A and $\Delta H_F^j / T_M^j$ are 8000 cal g^{-1} atom^{-1} and 3 entropy units respectively. Thus, from Eq. (6.28b), X_S^j may be plotted versus T for various values of $\Delta \bar{H}_S^j$ and T_M^j. Several solidus lines obtained in this manner are shown in Fig. 6.31(b) for Ge as solvent. There, the liquidus line is calculated from Eq. (6.40) while the solidus curves labeled (I), (II), (III) and (IV) were calculated from Eq. (6.28b) using values of $\Delta \bar{H}_S^j$ equal to 22 000, 16 600, 11 000 and 5500 cal g^{-1} atom^{-1}, respectively. It is observed that the tendency towards retrograde solubility is directly related to the relative partial molar enthalpy of solute in solid solvent. If the solute has a low heat of fusion, retrograde solubility will occur for much lower values of $\Delta \bar{H}_S^j$. The reason why retrograde solubility is not observed in most metallic systems is that $\Delta \bar{H}_S^j$ is too low for such systems. Extension of the analysis to include real, non-ideal and non-regular solutions does not invalidate the general conclusion that retrograde solubility is associated with a large value of $\Delta \bar{H}_S^j$. Figs. 6.32 and 6.33 present solid solubility data for Ge and Si respectively that illustrate this retrograde character.

6.3 The vapor–liquid–solid case

The relevant P–T–X information is usually presented in the two-dimensional T–X and P–T projections. These phase equilibrium plots are of extreme importance to heat treatment as well as to crystal preparation experiments. The compositional defect state of the material is specified by such considerations. For film deposition via the MBE technique, phase information applies to the source where thermal equilibrium is maintained. The T–X diagram is of interest mainly because it continuously specifies the state of the source. However, it is the P–T diagram that predicts the vapor pressure and thus the effusion rates of the various species. For a variety of compound semiconductors, vaporization

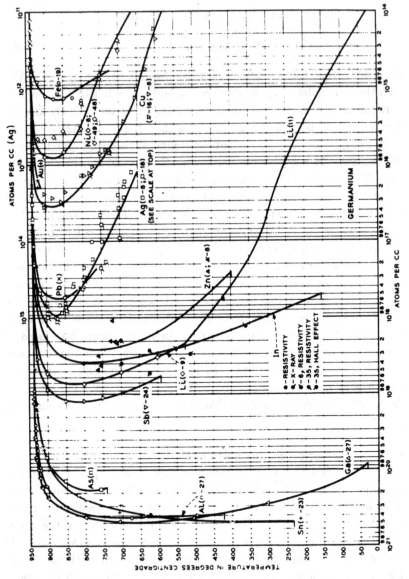

Fig. 6.32. Solid solubility for elements in Ge (after Thurmond)[5].

upon heating of the materials proceeds according to

$$MX(s) \to \alpha_* MX(g) + (1 - \alpha_*)\left[M(g) + \frac{(1 - \beta_*)}{2}X_2(g) + \frac{\beta_*}{4}X_4(g)\right]$$

$$(6.41)$$

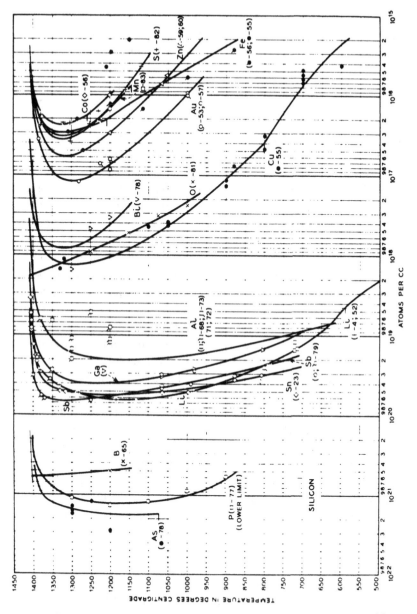

Fig. 6.33. Solid solubility for elements in Si (after Thurmond)[5].

where α_* is the fraction of molecular vaporization and β_* the fraction of dissociation into tetramer species. The constant for each reaction at a given temperature can be obtained and thus the vapor pressure of each species determined. Different kinds of compound semiconductors have quite different vaporization behavior and we shall illustrate the main features using GaAs as our example.

The T–X and P–T diagrams of the Ga–As system are plotted in Fig. 6.34. The liquidus curve of Fig. 6.34(a) gives the temperature at which a liquid of a given composition is in equilibrium with a solid and a vapor, which is not explicitly shown in this type of diagram. The dotted curve near $X = 0.5$ represents the solidus bounding the region of nonstoichiometry (exaggerated to display its existence). The region is generally not symmetrical with respect to the $X = 0.5$ line and depends upon the statistics of vacancy formation for the two sublattices. In fact, in many compounds, the solidus region lies completely off the stoichiometric composition.

The P–T diagram for the Ga–As system is shown in Fig. 6.34(b). These loops are the three phase lines since solid, liquid and vapor coexist along these lines. Taking the species As_4 as an example, the upper and lower legs of the loop give the As_4 pressure for GaAs in equilibrium with an As-rich liquid and with a Ga-rich liquid, respectively. As the temperature is decreased, the As_4 pressure represented by the upper leg approaches that in equilibrium with pure As. That represented by the lower leg approaches the hypothetical pressure of As as if it were in equilibrium with pure Ga. We see from the figure that the Ga, As_2 and As_4 species coexist in the vapor phase and that the pressure of As (As_2 and As_4) lies above the pressure for Ga over most of the temperature range. Upon heating GaAs, only the Ga-rich leg is of interest since the volatile As always vaporizes first leaving a Ga-rich solution. The contribution to the total As pressure from As_4 is small compared to that from As_2 except at very high temperatures. At very low temperatures ($T < 637$ °C) where the pressure of Ga equals that of $2As_2 + As_4$, congruent vaporization or sublimation occurs.

6.3.1   Point defects

In elemental lattices, we are concerned only with the formation of vacancies or interstitials or their combination as a Frenkel defect. In compound lattices, of the $A_m B_n$-type, we are concerned with vacancies on both the A and B sublattices as well as interstitials of the A or B-type and Frenkel defects of the A or B-type. Here, we are also interested in the substitutional anti-site defects of A on B sites or B on A sites. As is well known, although there is an enthalpy increase involved with

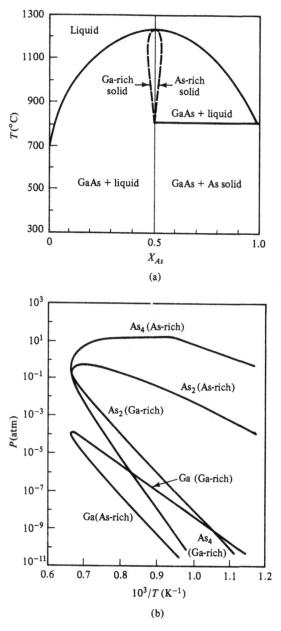

Fig. 6.34. (a) Phase diagram for the Ga–As system and (b) equilibrium partial pressures for Ga, As$_2$ and As$_4$ in the Ga–As system.

the formation of any one of these defects, there is also a configurational entropy increase and, at sufficiently small defect concentrations, the entropic part of the free energy change is always larger than the enthalpic part so that the free energy of the system is lowered by the introduction of such defects. For the formation of a substitutional defect in an AB compound the following reaction can be formulated:

$$A_A + B_B \rightleftharpoons A_B + B_A; \quad \Delta G_1^0 \qquad (6.42a)$$

At any temperature, this reaction will proceed as a homogeneous reaction until the crystal has achieved a new minimum in its total free energy as given by the temperature dependent equilibrium constant of Eq. (6.42a):

$$K_1(T) = \exp(-\Delta G_1^0/RT) \qquad (6.42b)$$

For point defects in ionic systems, the condition of electroneutrality for the entire crystal must be taken into account when determining the defect concentrations. This leads to the disorder type being characterized by two oppositely charged majority defect centers. For example, in AgBr and other Ag halides we find Frenkel-type disorder; i.e., nearly equal concentrations of cation vacancies and Ag ions exist in the interstices of the lattice. In alkali halides, we find Schottky-type disorder with nearly equal numbers of vacancies on the cationic and anionic sublattices.

As a simple example, let the neutral interstitial enthalpy of formation for Si_I in Si, SiO_2 or other phase be given by ΔH_α^I for any phase α. In each phase, the entropy of the I species, S_p, permuted amongst the mn_s available interstitial sites for m unit cells, is given by

$$S_p = \kappa \ln[(mn_s)!/(mn^I)!(mn_s - mn^I)!] \qquad (6.43)$$

Here, $n_s = 5$ for diamond cubic with one atom per site. Thus, the equilibrium number of interstitials per unit cell, n_α^{I*}, is given by

$$\frac{n_\alpha^{I*}}{n_{s\alpha} - n_\alpha^{I*}} = \exp\left(\Delta S_\alpha^v/\kappa\right)\exp\left(-\Delta H_\alpha^I/\kappa T\right) \qquad (6.44)$$

where ΔS^v is the vibrational entropy contribution to the I defects. In general, $n_\alpha^{I*} \ll n_{s\alpha}$ and is located at the minimum in the excess free energy plot of Fig. 6.35. For the conditions $n_{I\alpha} \gg n_{I\alpha}^*$, the excess free energy due to the presence of the interstitial defects, $\Delta G_{E\alpha}$, is given by

$$\Delta G_{E\alpha} = m\kappa T \left[n_\alpha^I \ln\left(\frac{n_\alpha^I(n_{s\alpha} - n_\alpha^{I*})}{n_\alpha^{I*}n_{s\alpha}}\right) \right.$$
$$\left. + (n_{s\alpha} - n_\alpha^I)\ln\left(\frac{n_{s\alpha} - n_\alpha^I}{n_\alpha^I}\right) \right] \qquad (6.45)$$

for the m unit cells. The chemical potential of the interstitials at these

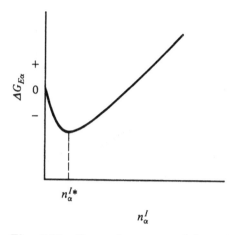

Fig. 6.35. Excess free energy of formation, $\Delta G_{E\alpha}$, for interstitial point defects in the α-phase as a function of defect content n_α^I.

excess concentrations is

$$\mu_\alpha^I = \frac{\partial G_{E\alpha}}{\partial n_\alpha^I} = \kappa T \ln \left[\frac{n_\alpha^I (n_{s\alpha} - n_\alpha^{I*})}{n_\alpha^{I*}(n_{s\alpha} - n_\alpha^I)} \right]. \tag{6.46}$$

An identical set of formulae holds for vacancies by substituting $\hat{v}$ for I in Eqs. (6.43)–(6.46).

A slightly more complex example deals with neutral and charged point defects in Si. Here, the equilibrium concentration of the charged defects depends upon the position of the Fermi energy, E_F. Although three entropically unique monomer neutral defects exist, vacancy, interstitial and interstitialcy (where two Si atoms share a substitutional position), we shall consider all the charged states to be associated with the vacancy for simplicity. This approximation illustrates all the principles involved.

In addition to the $\hat{v}^0$ state, $\hat{v}$ has been considered as a deep donor, $\hat{v}^+$, and a singly or doubly charged acceptor, $\hat{v}^-$ and $\hat{v}^{2-}$. Figure 6.36(a) shows the energy levels for these vacancies in Si[8] as a function of T where $E_C - E_{\hat{v}^-} = 0.111\,\text{eV}$, $E_C - E_{\hat{v}^-} = 0.44\,\text{eV}$ and $E_{\hat{v}^+} - E_V = 0.35\,\text{eV}$. Now let us consider the reaction equations for forming the different charged species as a function of the doping level; i.e.,

$$\hat{v}^0 \underset{\leftarrow}{\rightarrow} \hat{v}^- + e^+ \tag{6.47a}$$

and

$$np = n_i^2 = K_i \tag{6.47b}$$

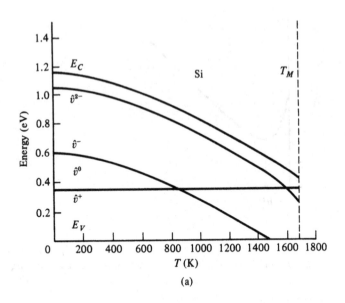

(a)

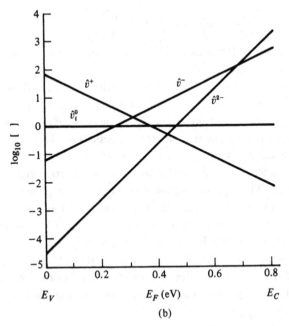

(b)

Fig. 6.36. Fig. 6.36(a) levels for vacancies in Si as a function of temperature and (b) relative concentration of charged vacancies in Si as a function of Fermi level, E_F, at $T = 750\ ^\circ$C. E_V and E_C denote the valence and conduction band free energies, respectively.

so that

$$K_1(T) = \frac{[\hat{v}^-][p]}{[\hat{v}^0]} \approx \frac{X_{\hat{v}_i^-} n_i}{X_{\hat{v}_i^0}} \tag{6.47c}$$

since $p = n_i$ at the intrinsic state. Thus, for extrinsic doping,

$$\frac{X_{\hat{v}^-}}{X_{\hat{v}_i^-}} = \frac{n_i}{p} = \frac{n}{n_i} \approx \exp[(E_F - E_{F_i})/\kappa T \tag{6.47d}$$

and the addition of donor atoms increases the charged vacancy density leading to the interesting effect of self-compensation; i.e., adding more donors to the crystal causes it to produce more acceptors.

In a similar fashion, we find that

$$\frac{[\hat{v}^{2-}]}{[\hat{v}_i^{2-}]} = \left(\frac{n}{n_i}\right)^2 ; \quad \frac{[\hat{v}^+]}{[\hat{v}_i^+]} = \frac{n_i}{n} \tag{6.48}$$

where $n_i = 10^{17}$, 5×10^{18} and 10^{19} at $T = 580\,°C$, $900\,°C$ and $1100\,°C$, respectively. In general, n/n_i is related to the doping concentrations N_D (donors) and N_A (acceptors) via

$$\frac{n}{n_i} = \frac{N_D - N_A}{2n_i} + \left[1 + \left(\frac{N_D - N_A}{2n_i}\right)^2\right]^{1/2} \tag{6.49a}$$

and

$$E_{F_i} \approx \frac{E_g}{2} + \frac{\kappa T}{4} \tag{6.49b}$$

where

$$E_g = 1.34 - 4.5 \times 10^{-4}T\,eV \tag{6.49c}$$

near $T \sim 1000\,°C$. Figure 6.36(b) shows the relative concentrations of charged monovacancies as a function of Fermi level, E_F, in Si at $750\,°C$. Here, we may note that, even for the intrinsic case, $[\hat{v}_i^-]\,/\,[\hat{v}^0] \sim 10$.

The equilibrium concentration of neutral vacancies is, of course, independent of the Fermi level position and is simply given by the mass action formula

$$C_{\hat{v}^0} = N_s \exp(\Delta S_f^0/\kappa) \exp(-\Delta H_f^0/\kappa T) \tag{6.50a}$$

where $N_S = 5 \times 10^{22}$ cm^{-3} is the concentration of Si lattice sites. Thus, the equilibrium concentrations of vacancies in the donor and acceptor ionization states are just

$$C_{\hat{v}^-} = C_{\hat{v}^0}\, g_A \exp[(E_F - E_{\hat{v}^-})/\kappa T] , \tag{6.50b}$$

$$C_{\hat{v}^{2-}} = C_{\hat{v}^0} g_A \exp[(2E_F - E_{\hat{v}^-} - E_{\hat{v}^{2-}})/\kappa T] \tag{6.50c}$$

and

$$C_{\hat{v}^+} = C_{\hat{v}^0}\, g_D \exp[(E_{\hat{v}^+} - E_F)/\kappa T] \tag{6.50d}$$

where $g_A = g_D = 2$ are degeneracy factors for acceptor and donor states, respectively. The entropy of ionization is thought to be about equal to the entropy of the band gap and this is assumed to be the same for ΔS_f^0; i.e., $\Delta S_f^0 = \Delta S_f^+ = \Delta S_f^- = \Delta S_f^{2-} \sim 5.1\kappa$. It should be clearly understood that the creation of ionized vacancies under equilibrium conditions does not diminish the equilibrium concentration of neutral vacancies. More neutrals are simply created to maintain their concentration at the level given by Eq.(6.50a) and the total concentration of vacancies increases. For dislocation-free Si crystals grown from the melt at T_M, since $\Delta H_f^0 > E_F - E_{F_i} < E_g/2$, the total equilibrium vacancy concentration decreases as T decreases below T_M regardless of the degree of doping; thus, there is always a sufficient potential supply of vacancies in the crystal to achieve the equilibrium concentrations at all temperatures. However, for sufficiently large band gap dislocation-free crystals where $E_g/2 > \Delta H_f^0$, sufficient doping can produce a condition where the potential source of vacancies is insufficient to provide the equilibrium concentrations at low temperatures where strongly extrinsic conduction conditions prevail.

The foregoing applies to the case of uniform electrostatic potential, ϕ_∞. If we have a varying potential, as in an interface region, then we can expect to find, for no change in E_F (electronic equilibrium),

$$\frac{C_{\hat{v}^j}(x)}{C_{\hat{v}^j}(\infty)} = \exp[-z^j e(\phi(x) - \phi(\infty))/\kappa T] \qquad (6.51)$$

under equilibrium conditions.

As a final comment, just as the foregoing holds for neutral and charged point defects in dislocation-free Si, the same concepts and equations apply to neutral and charged jogs on any dislocations present in Si crystals. Thus, the total jog population at any temperature will increase as the Fermi level moves away from the intrinsic condition. This means that the dislocation formation energy will decrease and its mobility under unit driving force will increase under such extrinsic conditions. This has important consequences for line defect formation in crystals and for the atomic roughness of internal interfaces in crystals, as well as for solid/solid phase transformations.

7

Thermodynamics of interfaces

7.1   Classical and homogeneous systems approach

If two homogeneous bulk phases, α and β, are in contact, the atoms or molecules at the interface are situated in an environment different from that in the interiors of the two phases. Here, both the density and local pressure vary and this difference may extend some distance from the interface. Nevertheless, we can think of the interface as being associated with energy, entropy and other thermodynamic quantities. Because the actual interfacial region has no sharply defined boundaries, it is convenient to invent a mathematical dividing surface – one that handles the extensive properties of the system $(F, E, S,$ etc.) by assigning to the bulk phases the values of these properties that would pertain if the bulk phases continued uniformly up to the dividing surface. The actual values for the system as a whole would then differ from the sum of the values for the two bulk phases by an excess or deficiency which is assigned to the surface region; i.e.,

$$E = E_\alpha + E_\beta + E_s \qquad (7.1a)$$

$$S = S_\alpha + S_\beta + S_s \qquad (7.1b)$$

$$v = v_\alpha + v_\beta + v_s \qquad (7.1c)$$

$$N^j = N_\alpha^j + N_\beta^j + N_s^j \qquad (7.1d)$$

where the subscript s relates to the surface excess quantity. The surface densities of the excess energy, entropy and component j will be symbolized here by e^j, s^j and Γ^j respectively.

For a system at thermal and chemical equilibrium, Gibbs[1] showed that the respective temperatures are equal and that the respective chemical potentials of the jth component occurs in both the α and the β phases

$$\mu_\alpha^j = \mu_\beta^j = \mu_s^j = \mu^j \tag{7.2}$$

For entropy reasons, there is no conceivable case wherein a small amount of j will not enter a phase if a contacting source of j is available.

The surface Helmholtz free energy density, f, is given by

$$f = e - Ts \tag{7.3}$$

while the surface Gibbs free energy density, g, is given by

$$g = \sum_j \mu^j \Gamma^j \tag{7.4}$$

From these, Gibbs defined the surface tension of a solid, γ, as

$$\gamma = f - g \tag{7.5}$$

The surface tension, γ, which is the work spent in *forming* a unit area of surface, needs to be clearly distinguished from the surface stress, $\tilde{f}_{mn} = \sigma_{mn}$, which is the work spent in *stretching* a unit area of surface. Surface stresses are compensated by stresses in the interior of the crystal. Thus, negative (stretching) stresses should tend to dilate the volume of a very small crystal or thin whisker and may be noticeable as an increase in lattice spacing.

The numerical values of e, s, f, g and Γ will all depend on the choice of a convention for locating the dividing surface between α and β. For solids, Gibbs recognized the convenience of locating the dividing surface so that the surface excess concentration of the solvent is zero. With this convention, $\gamma = f$ when we are dealing with a single component system (Eqs. (7.4) and (7.5) with $\Gamma^j = 0$). However, the existence of free electrons in materials suggests that we never have a truly one component system. For a polycomponent system, the surface excess is given by the Gibbs adsorption isotherm

$$\Gamma^j = -\left(\frac{\partial \gamma}{\partial \mu^j}\right)_{T,P,\mu^n} \tag{7.6}$$

and the surface entropy is given by

$$s = -\left(\frac{\partial \gamma}{\partial T}\right)_{P,\mu^j} \tag{7.7}$$

From Eq. (7.6), it follows that, with the introduction into a system of a component j that preferentially adsorbs on the surface ($\Gamma^j > 0$), by raising the corresponding μ^j; this causes a reduction in the work γ required to form unit area of interface. To express γ as a function of μ^j, it is necessary to know the dependence of Γ^j on μ^j and such information is not available theoretically using this purely classical approach.

The general equation relating surface stress and surface tension is

$$\tilde{f}_{mn} = \delta_{mn}\gamma + \left(\frac{\partial\gamma}{\partial\varepsilon_{mn}}\right); \quad m, n = x, y \tag{7.8}$$

where $\delta_{mn} = 1$ or 0 when $m = n$ and when $m \neq n$, respectively, and where all strains except ε_{mn} are held constant.[2] Here, the x and y coordinates lie in the tangent plane of the surface while the z coordinate lies perpendicular to this plane. Eq. (7.8) applies perfectly well to a liquid cube; however, in this case, $\partial\gamma/\partial\varepsilon_{mn}$ is ordinarily zero since the state of the surface of separation is not changed by the constrained extensions and the shear strains. New atoms are freely supplied to the expanding surface and dimensional changes along the z-axis can occur freely and prevent any volume strains in the liquid. Therefore, for a liquid

$$\tilde{f}_{mn} = \delta_{mn}\gamma \tag{7.9}$$

In this case, the surface stress reduces to a tensile force directed normal to a surface line, L, and having a constant magnitude γ per unit length of L.

For a solid, the key point that determines if $\tilde{f}_{mn}$ of Eq. (7.8) reduces to Eq. (7.9) is whether the surface is incrementally changed by the stretching process. In practice, this question hinges on the duration of the stretch compared to the relaxation time, τ, required for the surface (instantaneously stretched) to regain its original state by atomic migration. In glasses, both solid-like behavior ($\partial\gamma/\partial\varepsilon_{mn} \neq 0$) and liquid-like behavior ($\partial\gamma/\partial\varepsilon_{mn} = 0$) should be attainable by varying the temperature (and hence τ) and the rate of strain. Even in ordinary liquids, sufficiently high strain rates can give rise to non-vanishing values of $\partial\gamma/\partial\varepsilon_{mn}$.

In real solids, $\partial\gamma/\partial\varepsilon_{mn}$ and $\tilde{f}_{mn}$ may have either sign so that compressive stresses as well as tensile or zero stresses are possible in the interface (surface). Since the interior elastic stresses in a solid that is in equilibrium with its surface are determined by the requirement that they may compensate or balance the surface stresses, $\tilde{f}_{mn}$, which may even vanish, it should be evident that the capillary pressure (deduced from the application to a solid of the classical Gibbs–Thomson formula) which is valid for liquids is a fictitious quantity as a pressure for solids.

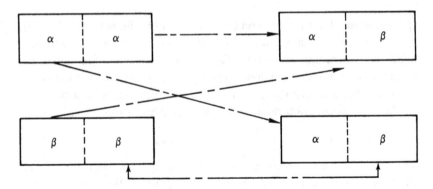

Fig. 7.1. Conceptual formation of two unit areas of α/β interface by the virtual separation of unit cross section single crystals of the α and β phases.

For a single crystal, perfect except for point defects, τ should be infinite since, as Herring has pointed out, the long-range order in the atomic positions prevents the relief of an imposed strain by atomic migrations.[2] The presence of incoherent grain boundaries, on the other hand, interrupts the long-range positional order and provides sources or sinks of atoms for the solid. Thus, external surface may be produced or swallowed along the line of emergence of a grain boundary by the migration of atoms to or away from the grain boundary surface. As a consequence, τ is finite for the surface of the specimen and $\tilde{f}_{mn} \approx \delta_{mn}\gamma$ for a strain rate slow compared to τ^{-1}.

The foregoing thermodynamic description of interfaces (surfaces) is completely capable of giving one a reliable procedure for categorizing experimental behavior; however, it gives little insight into the detailed mechanisms which lead to this behavior. To illuminate this situation somewhat, let us consider the various distinguishable contributions involved in forming the interfacial energy.

7.1.1   *The components of* γ

Let us consider the formation of an α/β interface from an α crystal and a β crystal (each of unit cross-sectional area) by the procedure illustrated in Fig. 7.1. We make imaginary cuts through the cross section of each crystal and separate the halves of each crystal without allowing either atomic position change or electron transfer over more than an atom distance. This requires work $W_{\alpha\alpha} + W_{\beta\beta}$ to be performed. We next bring the α and β half-crystals together along the cut surfaces gaining work $2W_{\alpha\beta}$. Next we let the free electrons relax their positions which does work $2W_e$. Finally, we let the atom positions relax which does work $2W_a$. Since we end up with two unit areas of α/β interface of specific interfacial free energy, $\gamma_{\alpha/\beta}$, the energy balance in the process requires that

$$\gamma_{\alpha\beta} = \left(\frac{W_{\alpha\alpha} + W_{\beta\beta}}{2}\right) - W_{\alpha\beta} - W_a - W_e \qquad (7.10a)$$

In Eq. (7.10a), all of the work terms are positive and we can associate $\gamma_{\alpha\beta}$ with five distinctly different sources; i.e.,

$$\gamma_{\alpha\beta} = (\gamma_c + \gamma_e + \gamma_t + \gamma_d + \gamma_a)_{\alpha\beta} \qquad (7.10b)$$

In Eq. (7.10b), γ_c is the short-range quasi chemical contribution due to the typical intermolecular potential interactions at the interface, γ_e is a longer-range electrostatic term associated with the equalization of the electrochemical potential of the free electrons, γ_t is a short-range transition structure contribution associated with the atomic diffuseness of the interface, γ_d is a long-range strain energy term associated with lateral atomic displacements (dislocations) at the interface or it is a surface reconstruction form associated with the rearrangement of dangling bonds at the interface and γ_a is a chemical adsorption contribution associated with the redistribution of different chemical species (or point defects) in the interface region. Of these interfacial energy contributions, only γ_c is a positive quantity. All of the other terms are relaxation effects and thus lower the free energy of the system. In reality, these contributions are not completely independent of each other; however, we can consider them to be so under somewhat idealized conditions and can evaluate the magnitude of each contribution.

7.1.2   *Quasi chemical contribution, γ_c*

All molecular and atomic forces ultimately find their root in the mutual behavior of the constituent parts of the atoms. To deal fully with this topic area, it is necessary to survey such forces as the exchange forces leading to covalent bonds, the coulomb forces between ions and dipoles, the electrostatic polarization of atoms or molecules by ions, the non-polar van der Waals (dispersion) forces, the repulsion forces and the energy states of the free electrons. A detailed understanding of these forces is not only important for a quantitative assessment of γ_c but is vital for the understanding of molecular adsorption to interfaces, crystal habit modification, layer kinematics modification, etc. In this book we will not explore the details of these forces as that information is available in many other places. Rather, we will utilize a generalized intermolecular potential function at this point to illustrate the general points to be made and will introduce specific functions for various examples or problems as they arise.

For an interfacial system of N spherical molecules in the α and β phases, the intermolecular potential energy, $\hat{U}$, depends only on the positions of the molecular mass centers, $\bar{r}_i$; i.e.,

$$\hat{U} = U(\bar{r}_1, \bar{r}_2, \ldots, \bar{r}_N) \tag{7.11}$$

This intermolecular potential is usually written as a series of terms which individually account for 2-body, 3-body, ..., N-body interactions:

$$\hat{U} = \sum_{i<j}\sum u(\bar{r}_i, \bar{r}_j) + \sum_{i<j<k}\sum\sum u(\bar{r}_i, \bar{r}_j, \bar{r}_k)$$
$$+ \sum_{i<j<k<\ell}\sum\sum\sum u(\bar{r}_i, \bar{r}_j, \bar{r}_k, \bar{r}_\ell) + \cdots \tag{7.12}$$

Much progress in recent years has been made towards understanding the gaseous, solid and liquid states via a "pair theory" wherein the series of Eq. (7.12) is truncated after the first term; i.e.,

$$\hat{U} = \sum_{i<j}\sum u(\bar{r}_i, \bar{r}_j) \tag{7.13}$$

This type of truncation has been extensively used and has been useful for determining a variety of bulk phase properties; however, even the simplest real fluids (the inert gases) do not obey the pairwise additive assumption of Eq. (7.13) at moderate to high densities. Even more important to us here is that the strictly pairwise approximation fails to account for the contraction that is generally found at a free surface, does not predict the proper atomic arrangement for small clusters of atoms, often fails to predict the proper order of stability of the different structural polymorphs of a pure material and does not predict the proper surface reconstruction features for covalent materials like Si. Thus, for surface and interfacial studies, one must really consider the use of both the two-body and the three-body potential functions as the simplest meaningful descriptions of surfaces.

Ab initio calculation involving electron orbitals is the most sophisticated and exact procedure for obtaining information about surfaces; however, with this approach one is presently limited to treating a very small number of atoms. Since one must deal with $\gtrsim 10^3$ atoms to obtain a reasonable representation of a relaxed surface, semiempirical potentials must be used. In this category, there are those involving many parameters (one using 13 parameters for Si gives good match to the elastic constants and the (7×7) reconstruction pattern) or those involving few

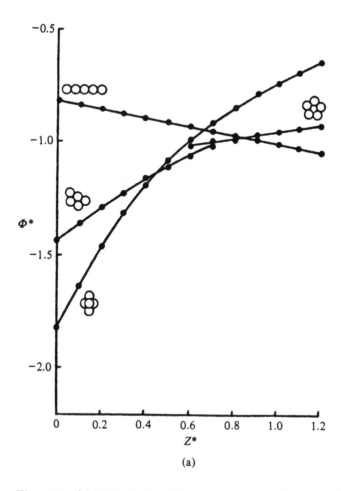

Fig. 7.2. (a) Plot of reduced interaction energy, Φ^*, versus the reduced three-body intensity parameter, Z^* for small clusters.

parameters. The minimum number of parameters needed for a unary system like Si is 3, 2 for a Lennard–Jones (L–J) pair potential and 1 for an Axilrod–Teller (A–T) triple dipole potential.

The L–J potential is given by

$$\hat{U}_{LJ}(r_{ij}) = 4\varepsilon \left[(r_0/r_{ij})^{12} - (r_0/r_{ij})^6 \right] \qquad (7.14a)$$

while the A–T potential is given by

$$\hat{U}_{AT}(r_{ij}, r_{ik}, r_{jk}) = \frac{Z_3 [1 + 3\cos\theta_1 \cos\theta_2 \cos\theta_3]}{(r_{ij} r_{ik} r_{jk})^3} \qquad (7.14b)$$

where r_{ij} is the scalar distance between atoms i and j, etc., θ_i is the

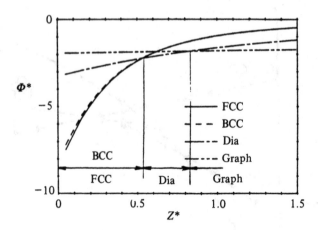

Fig. 7.2. (b) The same for crystalline phases of different structure.

angle formed at atom i by r_{ij} and r_{ik}, etc., while Z_3, r_0 and ε_0 are species-dependent constants. The parameters are obtained by matching to a data base of structural, thermochemical, surface energy, surface reconstruction pattern and phase diagram information. A best-fit parameter set is found. For a binary system like Ga–As, the minimum number of parameters is 10 (6 L–J-type potentials and 4 A–T-type potentials). For a ternary system like GaAs on Si, the minimum number of parameters is 22 (12 L–J-type for the six different pair interactions and 10 A-T type for the ten different triplet interactions). Because of this "best-fitting" procedure, the angular aspects of the PEF are not localized to the A–T parts of the PEF although it may appear so mathematically, but are present in all the parameters. This simplest possible many-body PEF does a reasonably good job of fitting a wide variety of material parameters; however, it doesn't do a very good job of second derivative quantities like elastic coefficients or bulk modulus. To also match these factors, additional three-body and perhaps even four-body terms need to be added.

Plots of reduced interaction energy, $\Phi^* = \phi/NE_0$ (where ϕ is the sum of the two and three-body interactions), versus reduced three-body intensity parameter, $Z^* = Z/\varepsilon_0 r_0^9$, can be made for any unary system and examples are shown in Figs. 7.2(a) and (b) for gas-phase clusters and crystalline phases, respectively. We see that, in both cases, as Z^* increases, open rather than close-packed structures are stabilized. Our very familiar pair potential favors close-packed structures while our many-body potential favors open structures so, including both in our PEF, we can properly treat a broad range of structures. It is also important to recognize that the degree of "openness" versus "close-packedness" increases as one moves from considering the bulk of a phase to the surface of a phase to a cluster of atoms or a point or line defect in a phase. In actual practice the proportion of three-body to two-body contribution may change from $\sim$ 5–10% for the bulk cohesive energy to $\sim$ 40–50% for the surface energy to $\sim$ 75–80% for the cluster energy. This means that the modulus of elasticity for the bulk will be determined largely by the two-body potential while that for the surface may be strongly influenced by the three-body potential. The surface may exhibit a softer modulus than the bulk because the curvature of the three-body contribution is less than that in the well of the two-body contribution. It is this property that allows for surface relaxation and the development of the surface stress tensor, $\sigma_{\alpha\beta}$, where

$$\sigma_{\alpha\beta} = \tau_{\alpha\beta} = \frac{1}{v}\left(\frac{\partial\phi}{\partial\varepsilon_{\alpha\beta}}\right) \qquad (7.14c)$$

and, for our simplest PEF,

$$\phi = \phi_2 + \phi_3 \qquad (7.14d)$$

while $\varepsilon_{\alpha\beta}$ is the Lagrangian strain parameter in the $\alpha\beta$ direction and v is the appropriate volume. Using this approach for Si(111), a plot of ϕ versus distance from the surface as seen in Fig. 7.3(a) doesn't show anything particularly noteworthy except for the excess energy present at the surface. However, plotting τ parallel to the surface, τ_{11}, we see in Fig. 7.3(b) that it is strongly compressive (because the surface relaxes inwards $\sim$ 25%) with an average stress in the first two layers being $\sim$ 0.75 Mpsi. It is the underlying crystal that holds the surface layers in compression (they want to expand laterally but cannot because of the substrate constraint). It is interesting to note that a tensile strain of 0.5% on the Si(111) gives a calculated reduction in the surface energy, γ of $\sim$ 5%, a very large effect indeed. Similar compressive surface stress behavior is found for GaAs(111) and ($\bar{1}\bar{1}\bar{1}$) as can be seen in Figs. 7.3(c) and (d). The PEF parameters used in the calculations leading to these results are given in Table 7.1.

There are really four levels of sophistication that one may take in an approximate theoretical description of the configurational energy, E_c, with pair potentials providing the simplest treatment. Next come cluster potentials such as we have just discussed; this is followed by pair functionals and then by cluster functionals.[3] A pictorial representation

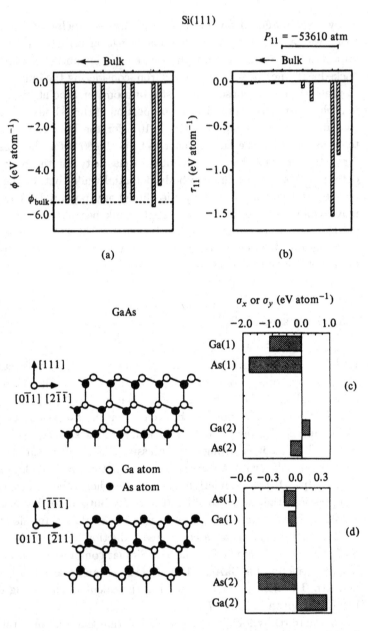

Fig. 7.3. (a) Plot of calculated potential energy, ϕ, for the Si(111) sheets of atoms from the surface inwards towards the bulk; (b) plot of calculated in-plane surface stress tensor, τ_{11}, for Si(111); (c) plot of calculated surface parallel stress, σ_x or σ_y, for GaAs(111); and (d) the same for GaAs($\bar{1}\bar{1}\bar{1}$). In all cases, the units are eV atom^{-1}.

Table 7.1.*Minimum parameter set for the Ga–As–Si system*

Two-body (L–J)		Three-body (A–T)
ε_0(eV)	r_0(Å)	Z (eV Å^9)
Ga–Ga	Ga–Ga	Ga–Ga–Ga
1.004	2.46070	1826.4
As–As	As–As	As–As–As
1.164	2.49132	2151.9
Si–Si	Si–Si	Si–Si–Si
2.817	2.29505	3484.0
Ga–As	Ga–As	Ga–Ga–As
1.738	2.44810	1900.0
Ga–Si	Ga–Si	Ga–Ga–Si
1.700	2.37800	2265.0
As–Si	As–Si	Ga–As–As
2.500	2.39300	4600.0
		Ga–As–Si
		5000.0
		Ga–Si–Si
		2809.0
		As–As–Si
		5000.0
		As–Si–Si
		4500.0

of the accuracy with which the various types of functions treat E_c is given schematically in Fig. 7.4. Here, in each frame, estimates of the relative energies of various atomic configurations obtained by one type of energy functional (solid curve) are compared with the supposed exact energies (dashed curve). Here, the "configuration axis" is meant to symbolize a variety of possible changes in configuration, such as de-

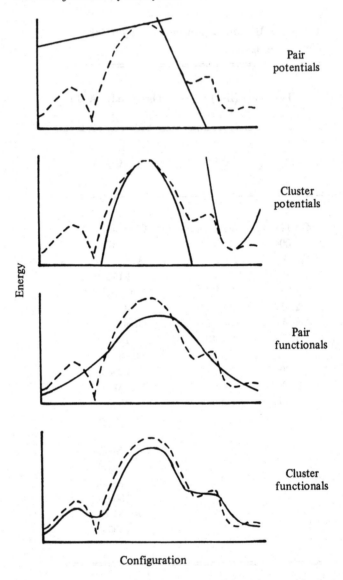

Pair
potentials

Cluster
potentials

Pair
functionals

Cluster
functionals

Energy

Configuration

Fig. 7.4. Schematic illustration of the relative accuracy with which configurational energies are obtained by simplified energy functionals. Solid curves denote calculated energies obtained by the indicated degree of approximation while the dashed curves denote the exact energy.

fect formation, crystal structure change, symmetry-breaking distortions, etc., which depend on the local environment. This local environment is usually associated with a single physical parameter such as background electron density or local electron bandwidth.

Because our present understanding of the many-body effect is so minimal, most of the following discussion will restrict itself to the pair-potential-only approximation. However, it is important for us to realize that a long-range interaction potential exists between groups of atoms and that this must be included in our PEF. Eqs. (7.18)–(7.23) illustrate the nature of one such long-range force called the electrodynamic force which arises due to an integration of the London dispersion forces.

7.1.3 Pair-potential-only approximation

Using only a two-body force approximation, it is convenient to say that each atom is linked to each other atom by a "bond" and to associate an energy with such a bond. When an infinite crystal is divided into two parts, as in Fig. 7.1, a number of these bonds will be broken (they will be increased to infinite length), and the energy of each new surface will be computed as one-half of the energy associated with breaking these bonds. If a particular bond is specified by the interatomic vector $\bar{r}_i$ of magnitude r_i, then the breaking of this bond will contribute to each surface an energy

$$E_i = \frac{1}{2}[\hat{U}(\infty) - \hat{U}(r_i)] \tag{7.15}$$

The division is assumed to occur along a mathematical plane of a particular orientation specified by the normal vector $\bar{h}$ of magnitude h and the energy computed is then treated as the surface energy $\gamma_c(\bar{h})$ for this orientation. Physically, this means first summing over all the bonds linking an atom to the atoms in a particular shell of neighbors and then summing over all shells which leads to

$$\gamma_c(\bar{h}) = \sum_j \hat{u}_j(\bar{h}) E_j \tag{7.16a}$$

where

$$\hat{u}_j(\bar{h}) = \frac{1}{\Omega h} \bar{h} \cdot \sum_k \bar{r}_k \tag{7.16b}$$

is a multiplicity factor for the surface term associated with a broken bond from the jth shell, the origin of $\hat{U}$ is chosen such that $\hat{U}(\infty) = 0$ and Ω is the volume of crystal per atom.

Nickolas[4] has applied this approach to the formation of free surfaces with the FCC and BCC crystal lattices using up to 300 shells of neighbors for a variety of Morse potentials and up to 500 shells of neighbors (22–23 lattice distances) for a variety of Mie potentials to obtain the requisite accuracy. He plotted energy contour maps on stereographic triangles for these two systems showing (a) that the FCC system exhibited an anisotropy in γ_c of less than 15% and exhibited a plateau in the middle portion of the diagram and (b) that the BCC system tends to one or two plateaus with a ridge joined to the (110) and exhibits a similar degree of anisotropy in γ. He then proceeded to calculate the situation wherein (1) only first NN bonds are considered and (2) both first and second NN bonds are considered (with E_2/E_1 of arbitrary magnitude). A most interesting result, illustrated in Fig. 7.5, was forthcoming. Case (b) differs considerably from the exact calculations (case (a)) and produces a much greater anisotropy (30–40%); however, case (c) bears a close resemblance to the exact calculations but with a slightly higher anisotropy for the BCC case. Thus, although it is common procedure in bond-type calculations to take account of NNs only, a significantly more reliable result is obtained by the inclusion of second NNs with an adjustable bonding parameter, E_2/E_1, rather than calculating the second NN bond strength from the potential function. A good estimate is obtained by taking $\frac{1}{3} < E_2/E_1 < \frac{1}{2}$ for FCC and $\frac{3}{4} < E_2/E_1 < 1$ for BCC

One reason why a high degree of accuracy is needed in this type of calculation is that small changes in γ cause large changes in the equilibrium shape of the crystal. This may be illustrated by considering the variation of the anisotropy parameter, $\hat{\lambda} = \gamma_{max}/\gamma_{min}$, over the entire stereographic triangle and the resulting features of the equilibrium form:

$\hat{\lambda} \sim$ 1.01–1.02 ;   no flat surfaces (nearly spherical form)

$\hat{\lambda} \sim$ 1.04–1.1 ;   flat areas surrounded by curved regions

$\hat{\lambda} \sim$ 1.15–1.25 ;   polyhedral form with rounded corners

$\hat{\lambda} > 1.3$;   polyhedron

We can utilize the understanding of this section to illustrate that the value of γ_c for a layer edge on the interface may be appreciably reduced below that for the same area of surface located in an infinite plane of the same orientation. Consider Fig. 7.5(d) where the hatched region is the crystal phase and let both the interface plane and the ledge plane be the (100) orientation in a cubic system. Regions 4, 5 and 6 have been removed by the separation process. Unit area of interface 1–6 will experience a decreasing γ_c as the 123 contact point is approached because of the interaction between regions 1 and 2. Likewise, unit area of interface 2–4 will experience an increased γ_c as the 245 contact point is approached. The lost interactions between regions 2 and 3 with regions 4 and 5 constitute the γ_c of the 2–3 interface. The lost interaction between

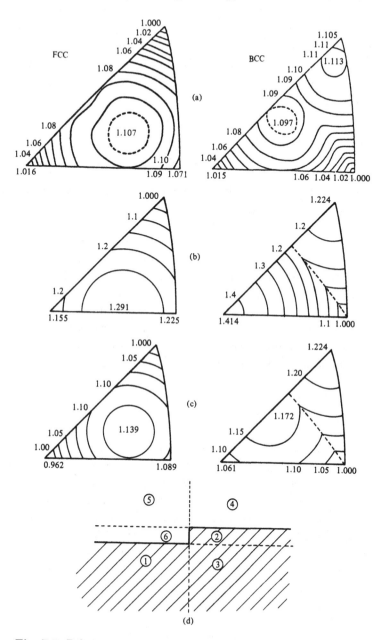

Fig. 7.5. Relative pair potential energy contour maps in stereographic triangles for FCC and BCC systems: (a) many neighbor bond calculations; (b) first NN bond only; and (c) first and second NN bonds only calculation; (d) construction used in evaluating various contributions to the ledge energy relative to the face energy.

regions 2 and 3 with region 6 and the gained interaction between region 1 and 2 constitute the γ_c of the 2–6 edge. Since the 3–6 and the 1–2 interactions are equal in magnitude but opposite in sign, γ_c for the 2–6 edge comes solely from the lost interactions between regions 2 and 6. Thus, we have

$$\frac{\gamma_c(\text{edge})}{\gamma_c(\text{plane})} = \frac{\left(\sum_j u_j E_j\right)_{2-6}}{\left(\sum_j u_j E_j\right)_{2-6} + \left(\sum_j u_j E_j\right)_{2-1} + \left(\sum_j u_j E_j\right)_{2-5}}.$$ (7.17)

For a material system with relatively long-range forces, we might find $\gamma_c(\text{edge}) < 10\%\gamma_c(\text{plane})$.

To gain an appreciation of the effective range of these van der Waals forces, we can consider the interaction of two separated macroscopic objects as occurring via the medium of the fluctuating electromagnetic field which is always present in the interior of any absorbing medium and always extends beyond its boundaries. As a result of statistical fluctuations in the position and motion of charges in a body, spontaneous local electric and magnetic moments occur in it. This electromagnetic field is present partially in the form of travelling waves radiated by the body and partially in the form of standing waves that are damped exponentially as they move away from the surface of the body. This field does not vanish even at absolute zero, at which point it is associated with the zero point vibration of the radiation field.

Lifshitz[5] developed a formula for this electrodynamic force that can be simply developed at separations small or large compared to the wavelengths important in the optical spectra of the materials involved. The force is significant in the range $50 \text{ Å} \lesssim z < 1000 \text{ Å}$ and experiments carried out in the region of 0.1–1.0 μm separation completely confirm the theoretical analysis. For two half-spaces of materials, 1 and 2 respectively separated by a small distance z which is filled with material 3, the force, $\hat{F}$, between the half-spaces can be either attractive or repulsive and is given by

$$\hat{F} = \frac{\hbar\bar{\omega}}{8\pi^2 z^3}$$ (7.18a)

where

$$\bar{\omega} = \int_0^\infty \left\{ \frac{[\varepsilon_{*1}(i\xi) - \varepsilon_{*3}(i\xi)][\varepsilon_{*2}(i\xi) - \varepsilon_{*3}(i\xi)]}{[\varepsilon_{*1}(i\xi) + \varepsilon_{*3}(i\xi)][\varepsilon_{*2}(i\xi) + \varepsilon_{*3}(i\xi)]} \right\} d\xi$$ (7.18b)

and where the ε_{*j} are the dielectric permeabilities of the three phases as a function of field frequency $\hat{\nu}$. Here, $\varepsilon_*(\hat{\nu})$ is a complex quantity ($\varepsilon_* = \varepsilon_*' + i\varepsilon_*''$) and its imaginary part is always positive so that it determines

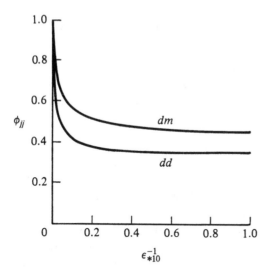

Fig. 7.6. Plot of the functions, ϕ_{dd} and ϕ_{dm} from Eq. (7.20) for two similar half-spaces separated by distance z.

the dissipation of energy in an electromagnetic wave propagated in the medium. In addition, $\hat{\nu}$ is a complex variable and iξ is the imaginary part of the argument of ε_*; thus, $\varepsilon_*(\text{i}\xi)$ is a real quantity which decreases monotonically from ε_{*0} (the electrostatic dielectric constant at $\xi = 0$). If $\varepsilon_{*1} = \varepsilon_{*2}$, then ω in Eq. (7.18) is always positive.

At large distances, when both bodies are metals, $\varepsilon_{*10} \to \varepsilon_{*20} \to$ very large values (generally approximated as ∞) and we obtain

$$\hat{F} = \pi^2 \hbar c / 240 (\varepsilon_{*30})^{1/2} z^4 \tag{7.19}$$

where c is the velocity of light. This force is independent of the nature of the metals and leads to an interaction energy proportional to z^{-3} at large distances. For z large in comparison with the principal wavelengths in the absorption spectrum of the bodies, $\varepsilon_{*i}(\text{i}\xi)$ can be replaced by the static dielectric constant and, when both half-spaces are the same material ($\varepsilon_{*10} = \varepsilon_{*20}$), we have

$$\hat{F} = \frac{\pi^2}{240} \frac{hc}{z^4} \frac{1}{(\varepsilon_{*30})^{1/2}} \left[\frac{\varepsilon_{*10} - \varepsilon_{*30}}{\varepsilon_{*10} + \varepsilon_{*30}}\right]^2 \phi_{dd} \tag{7.20}$$

where $\phi_{dd}(\varepsilon_{*10}/\varepsilon_{*30})$ is a function whose numerical value is given in Fig. 7.6. The same diagram shows the curve (dm) for the analogous function describing the attraction between a dielectric and a metal ($\varepsilon_{*20} \to \infty$). For this case, Eq. (7.20) is also used but with ϕ_{dd} replaced by ϕ_{dm}.

For the case when $\varepsilon_{*30} \to \infty$, these expressions tend to zero so that, if the gap is filled with liquid metal, the interaction force at large distances drops off as

$$\hat{F} = 0.0034\hbar c^2 / \sigma_{*3} z^5 \tag{7.21a}$$

where σ_{*3} is the electrical conductivity of the metallic fluid.

Using Eqs. (7.18) with $\varepsilon_{*3} = 1$ and metal–metal separation at the small separation limit, we find that

$$\hat{F} \approx \frac{\hbar e}{12\sqrt{2\pi}} \frac{1}{z^3} \left[\frac{1}{(m_2/N_2)^{\frac{1}{2}} + (m_1/N_1)^{\frac{1}{2}}} \right] \tag{7.21b}$$

Here, e is the electronic charge, m_j is the effective mass of the electron in metal j and N_j is the free electron density. According to Eq. (7.21b), two metals of high conductivity should adhere, at a given z, better than two metals of low conductivity. For the case of two metals of decidedly different conductivity, the adhesive force is governed by the one with the poorer conductivity. Even when two very well-polished metals (free of oxide) are placed in contact, they meet only at a few surface asperities in the size range 50–1000 Å so that Eq. (7.21b) can be used to calculate the average adhesive force or the interaction energy at an average separation in this range. Obviously, such equations apply to the case of a metal film interacting with a substrate through an oxide interlayer or to the case of crystal–crystal interaction in a solution. For other geometries, the separation dependences of the force interaction are given in Table 7.2.

The Lifshitz–van der Waals' constant, $\hbar\bar{\omega}$ in Eq. (7.18a), may be evaluated from the dielectric constant of the phase taken along the imaginary frequency axis $i\xi$, which can be obtained most conveniently from the imaginary part $\varepsilon''_{*j}(\hat{\nu})$ by using the Kronig–Kramers relationship

$$\varepsilon_{*j}(i\xi) - 1 = \frac{2}{\pi} \int_0^\infty \frac{\hat{\nu}\varepsilon''_{*j}(\nu)d\hat{\nu}}{\hat{\nu}^2 + \xi^2}; \quad j = 1, 2, 3 \tag{7.22}$$

In order of increasing photon energies, $\hbar\hat{\nu}$, the various peaks in a typical $\varepsilon''_*(\hat{\nu})$ spectrum result from the absorption of free electrons, interband transitions and collective electron (plasma) oscillations. Using these ε''_* values and numerical evaluation of Eq. (7.22) one finds, for $\varepsilon_{*3} = 1$ and $\varepsilon_{*1} = \varepsilon_{*2}$, that $\hbar\bar{\omega}$ varies between ε_* for KBr and ε_* for Ag. All other known spectra of ionic crystals, valence crystals, and metals lead to values roughly within these same limits. For molecular crystals, such as polymers, we may expect the minimum to be reduced to ~ 0.5 eV. In Table 7.3, a list of $\hbar\bar{\omega}$ values has been tabulated from some spectra. Values of $\hbar\bar{\omega}$ for the case of $\varepsilon_{*1} \neq \varepsilon_{*2}$ can be obtained from Table 7.3 by using the geometric mean relationship $(\bar{\omega}_{ij} \approx (\bar{\omega}_{ii}\bar{\omega}_{jj})^{1/2})$.

Table 7.2. *Non-retarded and retarded forces between macroscopic bodies in vacuum, calculated on the basis of pairwise additivity*

System	Non-retarded	Retarded
Atom–atom (separation d)	$U = -C/d^6$ $\hat{F} = 6C/d^7$	$U = -K/d^7$ $\hat{F} = 7K/d^8$
Atom–flat (separation D)	$\hat{F} = \dfrac{N_2 \pi C}{2D^4}$	$\hat{F} = \dfrac{4N_2 \pi K}{10D^5}$
Sphere–sphere (R_1 and R_2 at separation $D \ll R$)	$\hat{F} = \dfrac{N_1 N_2 \pi^2 C}{6D^2} \dfrac{R_1 R_2}{(R_1 + R_2)}$ $= \dfrac{A}{6D^2} \dfrac{R_1 R_2}{(R_1 + R_2)}$	$\hat{F} = \dfrac{N_1 N_2 \pi^2 K}{15D^3} \dfrac{R_1 R_2}{(R_1 + R_2)}$ $= \dfrac{2\pi B}{3D^3} \dfrac{R_1 R_2}{(R_1 + R_2)}$
Sphere–flat (separation $D \ll R$)	$\hat{F} = \dfrac{N_1 N_2 \pi^2 C R}{6D^2}$ $= \dfrac{AR}{6D^2}$	$\hat{F} = \dfrac{N_1 N_2 \pi^2 K R}{15D^3}$ $= \dfrac{2\pi B R}{3D^3}$
Perpendicular cylinder–cylinder (separation $D \ll R$)	$\hat{F} = \dfrac{N_1 N_2 \pi^2 C R}{6D^2}$ $= \dfrac{AR}{6D^2}$	$\hat{F} = \dfrac{N_1 N_2 \pi^2 K R}{15D^3}$ $= \dfrac{2\pi B R}{3D^3}$
Flat–flat (separation D)	$\hat{F} = \dfrac{N_1 N_2 \pi C}{6D^3}$ $= \dfrac{A}{6\pi D^3}$	$\hat{F} = \dfrac{N_1 N_2 \pi K}{10D^4}$ $= \dfrac{B}{D^4}$

Recently,[6] it has been shown that the metallic binding energy–distance curves can be approximately scaled into a single universal relationship that can be applied to physisorption and chemisorption on metal surfaces, metallic and bimetallic adhesion and cohesion of bulk materials. The expression is

$$E(a) = \Delta E \, E^*(a^*) \qquad (7.23a)$$

where $E(a)$ is the total energy as a function of separation distance, a, ΔE is a scaling parameter defined as the equilibrium binding energy and $E^*(a^*)$ is the approximately universal function which describes the

Table 7.3(a).*Lifshitz–van der Waals'
constants $h\bar{\omega}_{131}$ for a homogeneous
combination of two materials in vacuo or
in water*

Combination	$h\bar{\omega}_{131}$ (eV)	
	Vacuum	Water
Au–Au	14.3	9.85
Ag–Ag	11.7	7.76
Cu–Cu	8.03	4.68
Diamond–diamond	8.60	3.95
Si–Si	6.76	3.49
Ge–Ge	8.36	4.66
MgO–MgO	3.03	0.47
KCℓ–KCℓ	1.75	0.12
KBr–KBr	1.87	0.18
KI–KI	1.76	0.20
Aℓ_2O$_3$–Aℓ_2O$_3$	4.68	1.16
CdS–CdS	4.38	1.37
H$_2$O–H$_2$O	1.43	
Polystyrene–polystyrene	1.91	0.11

shape of the binding energy curve. The coordinate, a^*, is a scaled length defined by

$$a^* = (a - a_m)/\ell \tag{7.23b}$$

where a_m is the equilibrium separation and ℓ is a scaling length to be determined. The parameter ℓ is reasonably well determined by the Thomas–Fermi screening length. The universal function itself is given by

$$E^*(a^*) = -(1 + a^*)\exp(-a^*) \tag{7.23c}$$

while Fig. 7.7 illustrates its form for a variety of bimetals. Thus, in this representation, ΔE and ℓ are the two defining empirical parameters.

From the foregoing discussion, we can conclude that, during crystal growth, this type of electrodynamic force from the crystal will project into the adjacent nutrient phase an effective distance $\sim$ 50–1000 Å and

Table 7.3(b). *Lifschitz–van der Waals'*
constants $h\bar{\omega}_{132}$ *for a heterogeneous*
combination of two materials with
water and polystyrene as material 3

Combination	$h\bar{\omega}_{132}$ (eV)	
	Water	Polystyrene
Au–Ag		8.27
Au–Cu	6.41	5.93
Au–diamond	6.11	5.45
Au–Si	5.32	4.70
Au–Ge	6.50	5.93
Au–MgO	1.99	1.25
Au–KBr	0.73	0.00
Au–Aℓ_2O$_3$		2.60
Au–polystyrene	0.72	
Au–CdS		2.65

will influence the relative transport of both solute and solvent species as
well as the migration of small colloids in the vicinity of the interface.

7.1.4 *Electronic contribution, γ_e*

The macropotential When we think of the electrical potential
inside a solid, we generally think of the periodic potential field associated
with the molecules and have somewhat lost sight of the more primitive
viewpoint of a macroscopic continuum. To regain this perspective let
us consider an insulator as a dielectric between the plates of a charged
condenser as in Fig. 7.8. Then, by way of example, we have high potential values on the right side of this insulator and low values on the left.
Hence, because of its negative charge, an electron has a low electrostatic
energy on the right side and a high one on the left. To obtain the total
energy of the electron, E_T, in the insulator, we must add an electric
energy term, $-e\underline{V}$, due to the macropotential $\underline{V}$, to the lattice energy
term, E_L, exerted on the electron by the binding forces of the atomic
cores in the crystal lattice, i.e.,

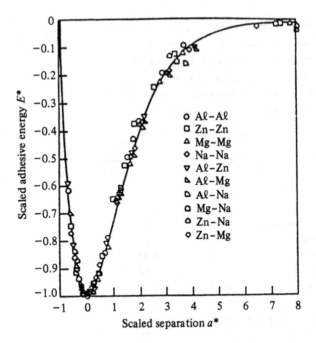

Fig. 7.7. Plot of the universal adhesive function, E^*, as a function of the scaled separation, a^* given by Eq. (7.23b).

$$E_T = E_L - e\underline{V} \tag{7.24}$$

Even if we remove the condenser in Fig. 7.8, we will find that a macropotential, ϕ, still exists in the insulator arising from the presence of the macroscopic surfaces. The origin of this macropotential and interface potential can be thought to reside in three simultaneously acting contributions (not totally independent) called the distribution, polarization and adsorption potentials.

The distribution potential arises as a result of the redistribution of charged particles, that are soluble in both phases, at the interface between the phases. In metals or semiconductors, prime examples are the redistribution of free electrons and the immersion into an electrolyte containing ions of the solid. For the former case, the interface is an ion-blocking electrode while, for the latter, the interface is an electron-blocking electrode. When a metal dips into an electrolyte as illustrated in Fig. 7.9, positive metallic ions cross the phase boundary and set up an electrical potential difference between the metal (e.g. Cu) and the

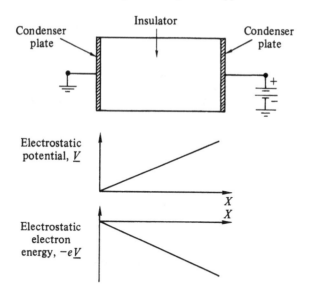

Fig. 7.8. Spatial distribution of electrostatic potential and electron energy in a charged condenser.

ionic solution (e.g. Cu^+ ions). If, initially, the rate of deposition of ions from the solution onto the metal is less than the rate of passage from the metal into the solution, the metal acquires a negative charge while the solution acquires a positive charge. The process of dissolution continues until a charged layer of such strength is set up that the energy level of the hydrated ions in the solution is raised and that of the positive ions in the metal lowered. When these two levels are equalized, net ion transfer ceases; i.e., transfer occurs until the electrochemical potentials of the ion in the metal and in the solution become equal.

The electrical double layer consists of a net accumulation of positive ions in the solution near the interface with the metal and of electrons near the surface inside the metal (see Fig. 7.9(c)). The surface potential is ϕ_s and the electrostatic potential decreases almost exponentially with distance to the bulk solution potential over a distance λ_D, called the Debye screening length in the solution where

$$\lambda_D \sim \left(\frac{\varepsilon_* kT}{4\pi n \hat{z}^2 e^2} \right)^{1/2} \tag{7.25}$$

Here, ε_* is the dielectric constant of the solution, n is the electrolyte content of the bulk solution and $\hat{z}$ the valence of the dominant ion. Thus, this non-zero electric field region can extend from ~ 50 Å in

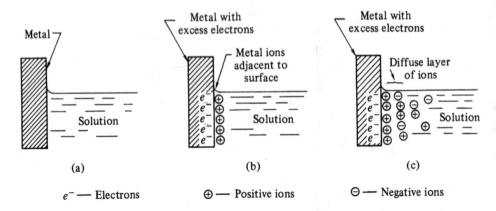

e^- — Electrons   ⊕ — Positive ions   ⊖ — Negative ions

Fig. 7.9. Representation of electrostatic double layer at a metal-solution interface: (a) at the instant of immersion; (b) at the first stages of double-layer formation; and (c) after complete double-layer formation.

strong electrolytes to ~ 1 μm in dilute electrolytes. In the metal the Debye length for the electrons is much smaller ($\lambda_{De} \sim 10$–30 Å) so that both the charge, q, and potential, ϕ, distributions will be as illustrated in Fig. 7.10.

The polarization potential occurs as a result of the orientation of neutral molecules having a dipole moment, permanent or induced, at the surface. Most molecules contain such dipoles and their presence is one of the principal reasons for the orientation of such molecules at surfaces; the oriented row of dipoles is a double layer which is not diffuse. At the water–air interface ~ 5 layers of H_2O molecules line up to lower the free energy of the system and, as a consequence, lead to a surface potential. If we consider a metal–vacuum interface, several sources of polarization potential exist. (1) Within the metal the electrons are in vigorous random thermal motion and those moving towards the surface overshoot the lattice of fixed positive ions before they turn back into the metal so that a thin negative skin is found beyond the positive skin of the ions (a double layer). (2) At the metal–vacuum boundary, the ions of the uppermost atomic layer are slightly displaced relative to their configuration in the bulk due to the anisotropy of intermolecular forces at the surface location and this contributes to the dipole double layer at the surface. (3) Rather than ion displacement occurring at the surface, ion deformation (polarization) may occur to produce surface dipoles. (4) Fi-

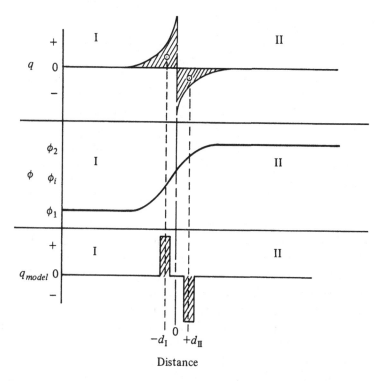

Fig. 7.10. Illustration of charge, q, and potential, ϕ, distribution across an interface between phases of similar electrical characteristics (I and II) plus the condenser approximation for the charge.

nally, adsorption of monatomic impurity layers leads to dipole formation at the surface.

The adsorption potential arises when neither the positively nor the negatively charged particles can leave either of the phases in appreciable amounts but where the particles of one sign of charge tend to be adsorbed at the interface to a greater degree than those of the other sign. For example, at the water–air interface, either the cations or the anions may be adsorbed in the greater quantity. In this case, there is negligible transference of charged particles between the air and the water but the electrical double layer due to the H_2O dipole orientation causes adsorption to occur. Only because of defects in the oriented dipole layer can a field gradient exist for any distance adjacent to the layer and produce long-range ion attraction or repulsion.

In general, for each phase one may distinguish an inner or Galvani potential, ϕ, and an outer or Volta potential, $\tilde{V}$. The inner potential,

ϕ, is measured relative to an infinitely distant point in charge-free vacuum and an electrical energy $q\phi$ is required to transfer a charge q from infinity to the interior of the phase. This quantity, $-e\phi$, is called the electron affinity by some people. An energy $q\tilde{V}$ is required to transfer a charge from infinity to the outside surface of a phase at a distance of 10^{-4} cm from the surface. The Galvani potential (inner) is different from the Volta potential (outer) only when a potential difference exists between the interior and the outer surface regions of the phase. This potential difference is called the surface potential, $\tilde{\chi}_s$. These three electrical potentials are related by the following equation

$$\phi = \tilde{V} + \tilde{\chi}_s \qquad (7.26)$$

In Fig. 7.11(a), the energy band picture of an electron in two solids that are isolated from each other is shown relative to the zero level of the electron energy. Here, ϕ is the macropotential so that $-e\phi < 0$ is the electrostatic binding energy (electron affinity in the isolated case). E_c is the chemical binding energy of the electron in the lattice at the lower edge of the conduction band; E_F is the chemical binding at the Fermi level (electrochemical potential level) and $\zeta = E_F - E_c > 0$ is called the Fermi energy. From Fig. 7.11(a), we note that the Fermi energy (electrochemical potential) is different in metal II than in metal I. Let us now allow the two metals to come into electronic equilibrium (by thermionic emission, say). The electron transfer creates an additional distribution potential between the two phases and continues until the electrostatic potential shift brings the two Fermi levels into alignment. Then there is no further driving force for electron transfer. From Fig. 7.11(b), we note that the surface potential $\tilde{\chi}_s$ differs in sign for the two phases and that the Volta potential difference will be different from the Galvani potential difference ($\Delta\phi_G = \phi_{II} - \phi_I$). We note also that the electron work function, $\tilde{\Psi}$, is the difference in electron energy between a point just outside the surface and the Fermi energy inside so that $\tilde{\Psi}$ is a function of crystallographic orientation. Some people define the work function as $\tilde{\Psi} + \tilde{V}$ which is orientation independent.

In the metal–metal example, the space charge layers at the interfaces have been neglected because they are so small, however, in the case of a semiconductor–semiconductor contact such layers cannot be neglected. Fig. 7.12(a) shows the energy band diagram for two isolated pieces of semiconductors of different band gaps, E_g, different permittivities, ε_*, different work functions, $\tilde{\Psi}$, and different electron affinities, χ. The work function and electron affinity are defined, respectively, as that energy

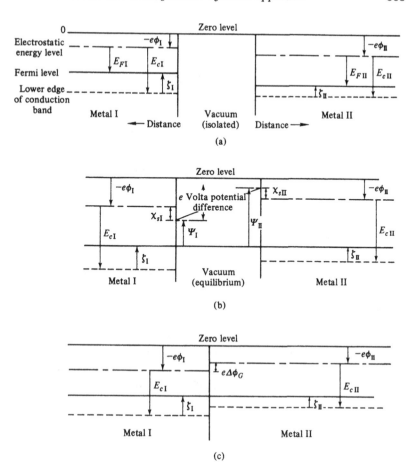

Fig. 7.11. Energy level diagram for two solids illustrating the macropotential condition in (a) the electrically isolated and separated condition; (b) the electronic equilibrium but separated condition; and (c) the intimate contact condition.

required to remove an electron from the Fermi level (E_F) and from the bottom of the conduction band (E_c) to the vacuum level. The difference in energy of the conduction band edges in the two semiconductors is represented by ΔE_c and that in the valence band edges is ΔE_V.

When a junction is formed between these semiconductors, the energy band profile is as shown in Fig. 7.12(b). This represents an n–p heterojunction between an n-type narrow gap semiconductor and a p-type wide gap semiconductor. For non-degenerate semiconductors, both E_g and X are invariant with doping level. The total built-in potential V_{bi}

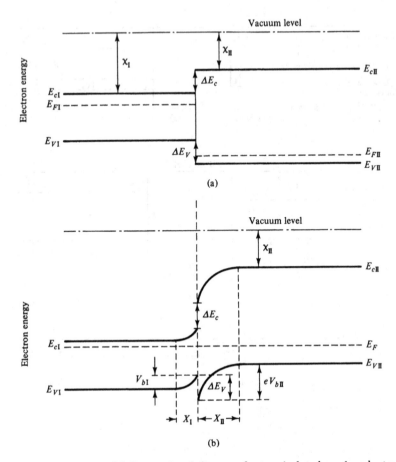

Fig. 7.12. (a) Energy band diagram for two isolated semiconductors in which space charge neutrality is assumed to exist in each region and (b) energy band diagram of an ideal n–p heterojunction at equilibrium.

is equal to the sum of the partial built-in voltages $(V_{bI} + V_{bII})$ where V_{bI} and V_{bII} are the electrostatic potentials supported at equilibrium by semiconductors I and II respectively. The case of an n–n heterojunction of the above two semiconductors is somewhat different. Since the work function of the wide-gap semiconductor is the smaller, the energy bands will be bent oppositely to the n–p case (see Fig. 7.13(a)). Also shown in Fig. 7.13 are the idealized band diagrams for p–n (narrow-gap p-type and wide-gap n-type) and p–p heterojunctions. If interface states are present, the above idealized conditions need to be modified. In this case, the energy band at the interface is free to move up or down with

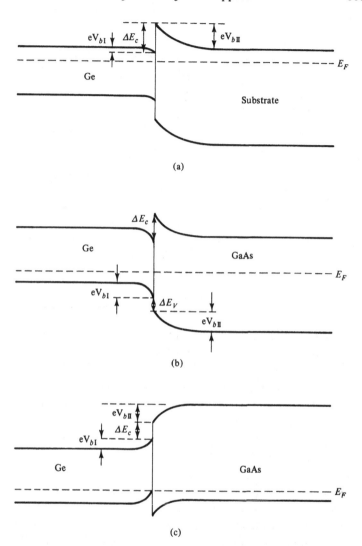

Fig. 7.13. Energy band diagrams for (a) an ideal abrupt n–n hetero-junction; (b) an ideal p–n junction; and (c) an ideal p–p heterojunction.

the necessary change being supplied by electrons (or their absence) in the interface states. The discontinuity in the conduction band is still equal to the difference in electron affinities; however, the height of the conduction band edge above the Fermi level at the interface is determined primarily by the interface states. Fig. 7.14(a) shows the energy band diagram of a Ge–Si, n–n heterojunction where both sides of the

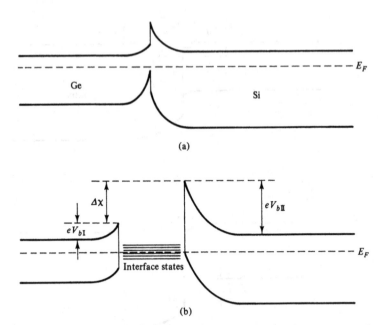

Fig. 7.14. (a) A Ge–Si n–n heterojunction with interface states; and (b) a magnified view of the interface region.

junction are depleted, a situation made possible by the acceptor nature of the interface states. A magnified energy band diagram of the n–n heterojunction with interface states is shown in Fig. 7.14(b). The interface states are assumed to be in a thin layer sandwiched between the two depletion regions.

One useful method for the evaluation of γ_e is to consider the electron redistribution that occurs on contact between two phases. If we take δn electrons from the phase with the higher Fermi level and place them in the phase with the lower Fermi level, this causes a decrease in free energy of the system by an amount $|\delta n \Delta E_F|$ where ΔE_F is the difference in Fermi level of the two materials and unit area of interface is assumed. Now let these electrons redistribute to build an equilibrium space charge at the interface with a potential change $\delta \phi$ which causes an increase in energy of $|\delta n e \delta \phi|$. Allow this process to continue until Δn electrons have been transferred and the potential difference $\Delta \phi$ generated which is sufficient to bring about Fermi level alignment. Thus, we have

$$\gamma_e = -|\Delta n \, \Delta E_F| + \int_0^\sigma \delta \phi \, dq \qquad (7.27a)$$

$$= -\int_0^{\Delta\phi} \bar{\sigma}_* \mathrm{d}\phi \tag{7.27b}$$

where $\bar{\sigma}_*$ is the charge per square centimeter of interface, $\Delta E_F = -e\Delta\phi$, partial integration of Eq. (7.27a) has been used to obtain Eq. (7.27b) and $\Delta\phi$ is the final Galvani potential difference between the two phases.

An alternate, and perhaps more general, procedure for developing a quantitative expression for γ_e is to consider the following form of the Gibbs adsorption equation

$$\mathrm{d}\gamma = -\sum_i \Gamma_i \mathrm{d}\mu_i - \bar{\sigma}_* \mathrm{d}\phi \tag{7.28}$$

Here, $\gamma_e = -\bar{\sigma}_*/e$ and $\mathrm{d}\mu_e = -e\mathrm{d}\phi$. For a pure system or a system where no atomic redistribution is allowed,

$$\gamma = \gamma_0 - \int_0^{\Delta\phi} \bar{\sigma}_* \mathrm{d}\phi \tag{7.29}$$

where γ_0 is the surface tension in the absence of any potential difference so that Eq. (7.27b) holds for γ_e. This equation is known as the Lippman equation, developed at the turn of the century for electrolyte systems.

If the electric surface potential is so small that $\bar{\sigma}_*$ is linearly related to $\Delta\phi$, the simple condenser model can be applied so that Eq. (7.27b) leads to

$$\gamma_e = -\frac{1}{2}\bar{\sigma}_*\Delta\phi \tag{7.30a}$$

Thus, in this case, the electric work to be done is exactly one-half of the chemical work gained in the process. For large values of potential change, the charge increases more rapidly with increasing potential (for a single double layer in an ionic system, it increases roughly according to $\exp(e\phi/kT)$) and, accordingly the magnitude of γ_e is generally greater than $(1/2)\bar{\sigma}_*\Delta\phi$; i.e.,

$$\gamma_e \lessgtr -\frac{1}{2}\bar{\sigma}_*\Delta\phi < 0 \tag{7.30b}$$

Referring back to Eq. (7.29), we note that γ shows a maximum for $\bar{\sigma}_* = 0$ and, on both sides of this point of zero charge, γ is lowered. For the range of systems to be dealt with during crystal growth, we shall find that the charge, $\bar{\sigma}_*$, needed to produce a given potential difference, $\Delta\phi$, is small for non-conductors and large for conductors so that $|\gamma_e|$ should be small for a non-conductor–non-conductor interface and large for a conductor–conductor interface. At a conductor–non-conductor interface, most of the $\Delta\phi$ will be developed in the non-conductor so that $\bar{\sigma}_*$ will be small and $|\gamma_e|$ will be small. For this range of cases, the Debye length, λ_D,

Table 7.4. *Calculated values of $\Delta\phi_{SL}$ and $|\gamma_e|$*

Element	$e\Delta\phi_{SL}$ (eV)	$|\gamma_e|$ (erg cm^{-2})	γ_{SL} (erg cm^{-2})
Li	0.047	10.1	
Aℓ	0.258	323.0	95
Cu	0.191	154.0	181
Ag	0.161	164.0	128
Au	0.194	142.0	135
Fe	0.186	17.6	208
Co	0.165	23.0	239
Zn	0.256	185.0	

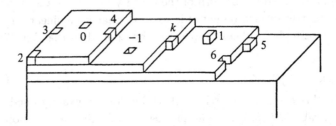

Fig. 7.15. Representation of the various, energetically unique, surface or interface sites for molecules.

will increase from ~ 25 Å for metallic systems to $\sim 10^{-2}$ cm for some insulators. Finally, calculated values of γ_e for the solid–liquid interface are given in Table 7.4 for a number of metallic systems.

7.1.5 Atomic diffuseness contribution, γ_t

The surface of a solid is heterogeneous on an atomic scale and Fig. 7.15 depicts schematically the various surface sites that are identified by experiments. There are atoms in terraces which are surrounded by the largest number of NNs. Atoms in steps have fewer NNs while atoms in kink sites have even fewer. Kink, step and terrace atoms have large equilibrium concentrations on any real surface while point defects such as adatoms and vacancies are also present and are important participants in atomic transport along the surface. Free surface studies of transition metals and oxides have yielded a great deal of experimental evidence that different types of surface sites have different chemistries. This is exhibited in the large heats of adsorption of molecules at the various sites and in the differing ability of the sites to break high energy bonds (H–H, C–H, N–N, N–O, etc.). Such effects have been explained theoretically in terms of large variations in the localized charge density distribution as a result of the structural differences and the appearance of large surface dipoles due to redistribution of the electron gas at these various sites in metals. Herein lies one of the important reasons for the complexity and the diversity of surface chemistry. Acting as a catalytic surface, the overall rate and product distribution in a surface reaction is the sum of the rates and products that form at each surface site. In Fig. 7.15, all the unique structural sites are important while the site labelled "k" has a special position in the theory of crystal growth; it is referred to as the "half-crystal" position. According to their positions, the atoms or molecules will have different energies.

The driving force for surface roughening is the creation of new unique states that provide configurational entropy to the system at the cost of an increase in enthalpy for each of the new states. Application of the minimum free energy condition yields the population of the various different states. We shall proceed to a description of the situation first by looking at the crystal–vapor case and then by proceeding to the crystal–liquid case.

Crystal–vapor case

(i) **Step roughness:**

At $T = 0$ K, steps tend to be as straight and as smooth as possible. At higher temperatures, thermal fluctuations produce roughness in the form of kinks. The kinks may be positive, forward jumps of the edge or negative backward drops and they may be single or multiple units of the lattice distance (see Fig. 7.16(a)). The excess energy needed to form a kink, of amount m, in a step on the (100) surface of a SC crystal is $m[(\varepsilon_1/2) + \varepsilon_2] - \varepsilon_2$ if we consider only first and second NNs where ε_1 and ε_2 are the first and second bond energies between atoms in the crystal. Despite our earlier criticism of the practice, we shall neglect the effects of neighbors higher than first. The roughness, $\hat{r}$, of the step is defined as the sum of all positive and negative jumps in a length of step divided

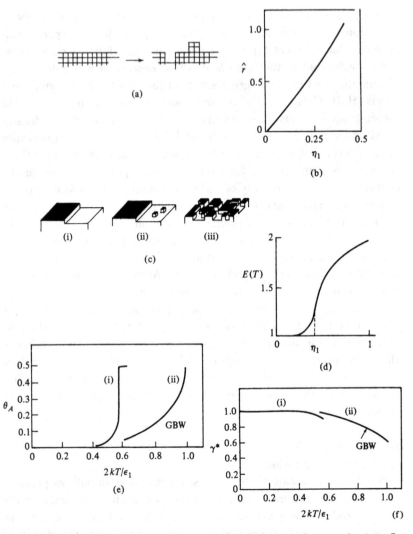

Fig. 7.16. Models and calculated results for surface roughening. In (d) $E(T)$ has units of $\epsilon_1/2a_0^2$ as does γ^* in (f).

by the number of sites in that length and is given by

$$\hat{r} = [\sinh(\varepsilon_1/2\kappa T)]^{-1} \tag{7.31}$$

which is plotted in Fig. 7.16(b) as $\hat{r}$ versus η_1 where $\eta_1 = \exp(-\varepsilon_1/2\kappa T)$. The partial excess Helmholtz free energy change associated with the roughening of unit length of step is given by

$$f_t = \frac{\kappa T}{a_0} \log \left[\tanh \left(\frac{\varepsilon_1}{4\kappa T} \right) \right] \tag{7.32}$$

where a_0 is the lattice spacing.

Example: development of Eq. (7.31)

If p_+, p_- and p_0 are the probabilities of there being, at a given site in Fig. 7.16(a), a positive, a negative or no kink of unit amount respectively, then

$$p_+ + p_- + p_0 = 1$$

For the [100] step, $p_+ = p_-$ and

$$\frac{(p_+ - p_-)}{2} = p_0 \exp(\varepsilon_1/2\kappa T).$$

If p_{+m} and p_{-m} are the probabilities for kinks of amounts $+m$ and $-m$, then

$$\sum_{m=1}^{\infty} p_{+m} + \sum_{m=1}^{\infty} p_{-m} + p_0 = 1$$

and, for a [100] step, $p_{+m} = p_{-m}$. The law of mass action applies to these entities in the form

$$p_{+1}p_{-1} = p_0^2 \eta_1^2,$$

$$p_{+m} = \left(\frac{p_+ \eta_2^2}{p_0} \right)^m \frac{p_0}{\eta_2^2}$$

and

$$\eta_j = \exp(\varepsilon_j/2\kappa T)$$

Using these relationships, it is found that

$$p_0 = \frac{1 - \eta_1 \eta_2^2}{1 + 2\eta_1 - \eta_1 \eta_2^2}$$

From its definition, the roughness $\hat{r}$ is given by

$$\hat{r} = \sum_{1}^{\infty} m \, p_{+m} + \sum_{1}^{\infty} m \, p_{-m}$$

$$= \frac{2\eta_1}{(1 - \eta_1^2 \eta_2^2)(1 + 2\eta_1 - \eta_1 \eta_2^2)}$$

When second NN interactions are neglected, $\varepsilon_2 = 0$ and $\eta_2 = 1$ so that

$$\hat{r} = \frac{1}{\sinh(\varepsilon_1/2\kappa T)}$$

(ii) Surface roughness

At low temperatures there will be small concentrations of self adsorbed molecules and of vacant sites on the surface (see Fig. 7.15). As the temperature is increased, more and more molecules jump out of the surface leaving more and more vacancies and, with increasing surface concentration, clusters of two, three or more molecules and vacancies appear on the surface. At still higher temperatures, if the crystal has not melted, more than three levels are involved. This progressive roughening can be viewed as a cooperative disordering of the surface.

Some time ago, Onsager found the exact solution of the eigenvalue problem for the "two-level" Ising model (with first NN-only bonding) and it is useful here to show his results. At $T = 0$ K, the flat (100) surface of a SC crystal is considered as being half covered by a monomolecular layer as illustrated in Fig. 7.16(c)(i). Neglecting the energy of the step, the potential energy of the surface at $T = 0$ K is $\varepsilon_1/2a_0^2$ per unit area on a first NN bond picture. At a slightly higher temperature (Fig. 7.16(c)(ii)), molecules in the half-layer on the left have evaporated onto the lower half-surface on the right forming self adsorbed molecules and leaving vacancies behind. The energy is increased by four "half-bonds" $(4\varepsilon_1/2)$ for each singlet vacancy or adsorbed molecule present. When the surface coverage of vacancies is θ_v and of adsorbed molecules is θ_A, the energy of the surface E becomes

$$E(0+) = \frac{\varepsilon_1}{2a_0^2} + \theta_A \left(\frac{4\varepsilon_1}{2a_0^2} \right) \tag{7.33}$$

on both the left and right since $\theta_A = \theta_v$.

At higher temperatures, the surface energy is given by

$$E(T) = \frac{\varepsilon_1}{2a_0^2} \left\{ 2 - \frac{1}{2} \left[1 + \frac{2}{\pi} (1 - k_1^2)^{1/2} K(k_1) \right] \cosh \left(\frac{\varepsilon_1}{2\kappa T} \right) \right\} \tag{7.34a}$$

where

$$k_1 = 2 \sinh(\varepsilon_1/2\kappa T) / \cosh^2(\varepsilon_1/2\kappa T), \tag{7.34b}$$

$$K(k_1) = \int_0^{\frac{\pi}{2}} (1 - k_1^2 \sin^2 \omega)^{-1/2} d\omega \tag{7.34c}$$

A plot of $E(T)$ versus η_1 is shown in Fig. 7.16(d) where we see that the curve has a vertical tangent at the transition temperature, T_c, given by $\sinh(\varepsilon_1/2\kappa T_c) = 1$ or $T_c \sim 0.57\varepsilon_1/\kappa$. The surface coverage, θ_A, is given by

$$\theta_A = \frac{1}{2} \left\{ 1 - \left[1 - \cosh^4(\varepsilon_1/2\kappa T) \right]^{1/8} \right\} \tag{7.35}$$

and its dependence on temperature is shown in Fig. 7.16(e), curve (i). At the transition temperature, $\theta_A = 0.5$ and the surface step disappears (Fig. 7.16(c)(iii)).

With increasing temperature, the partial surface tension, $\gamma^* = \gamma_c + \gamma_t$, begins to decrease slowly as the roughness increases (see Fig. 7.16(f), curve (i)). At the transition point, $\gamma^*(0) - \gamma^*(T_c) = 0.11\varepsilon_1/2a_0^2$, a decrease of $\sim 10\%$ from its 0 K value.

For inert gas crystals, it is found from experimental data that the melting point occurs at $T_M \sim 0.7\varepsilon_1/\kappa$ so that T_c occurs close to T_M for the (100) face. For the (111) face of an FCC crystal, theory yields $T_c \sim 0.91\varepsilon_1/\kappa$ so that T_c is probably above T_M for this face. For the (001) face, the coverage of self-adsorbed molecules is about 1% for $T = 0.75T_c \sim 0.6T_M$.

If we use the Gorsky–Bragg–Williams (GBW) approximation in the two-level problem (amounts to assuming a purely random arrangement of states on either level) the analysis is greatly simplified but the predictions of θ_A and γ^* are appreciably different as can be seen by curves (ii) in Figs. 7.16(e) and 7.16(f). In the GBW approximation, the $\theta_A \sim 0.5$ condition occurs at almost twice the temperature found by Onsager's exact calculation. This type of analysis is very sensitive to such simplifying approximations as the GBW approximation and the NN-only approximation. If one includes the effect of higher order bonds, each roughened state would cost more energy so that T_c would need to be increased. Thus, the GBW and the NN-only approximations produce offsetting errors so that the proper result may lie closer to curves (ii) in Figs. 7.16(e) and (f).

7.1.6 *Dislocation contribution, γ_d*

It is well known that the interface between any two crystals of the same material (the positions of whose atoms may be represented by a rotation of one crystal with respect to the other) can be described geometrically in terms of an array of dislocations. In the most general case, the description is a purely geometric one, since the elastic properties are non-linear in the vicinity of the interface and therefore the distortion associated with a closely spaced array of dislocations cannot be represented as a superposition of the distortions due to the individual components. This boundary energy, E_B, is a function of the orientation difference θ and is quite well represented by the formula

$$E_B = E_0\theta(A - \ln\theta) \qquad (7.36)$$

where the first term originates from the misfit energy at the boundary and is directly proportional to θ since it depends only on the number of rows of misfitting atoms; the second arises from the elastic distortion of

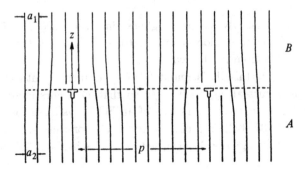

Fig. 7.17. Representation of the distortions associated with the mismatching interface of two epitaxial crystals, A and B, with a one-dimensional misfit (interpreted as misfit dislocations of spacing p).

the atoms far from the boundary and is the component of energy which dominates at small orientation differences.

A similar approach can be made for interphase boundaries. The main difference between the generalized dislocation model of interphase boundaries and the dislocation model of grain boundaries is the following: In the case of grain boundaries, the stable array of dislocations is always such that at large distances from the boundary the strain fields cancel, so that there is no resultant strain except at distances from the boundary which are comparable with the spacing of the dislocations. In the interphase model, the arrays of dislocations are such that, at long range, strain fields do not cancel in the reference lattice but are such as to produce a homogeneous strain of just the right amount required to represent the two crystal structures (see Fig. 7.17).

One way to picture this is to consider what happens when we try to fit together two crystals which differ from each other by a small distortion (assuming no rotation involved). The only way in which all the atoms could be made to fit along the boundary would be for the crystals on the two sides of the boundary each to be strained elastically and uniformly in such a way as to make the two lattices identical. Under these conditions there would be no energy localized at the boundary, but there would be a large volume strain, different in each crystal. The total energy could be substantially reduced by localizing the strain energy near the boundary by means of arrays of dislocations. These dislocations would be spaced so that their net effect at large distances would be to produce a uniform stress which just exactly cancelled the stress produced by the boundary tensions in the absence of the dislocations.

Now the total elastic energy of the system is that resulting from the superposition of the stresses due to the original boundary without dislocations upon the stress fields of the dislocations. Since the two stress fields cancel each other at large distances from the boundary, the total strain is localized within a distance from the boundary of the order of the spacing between the dislocations, just as in the grain-boundary case. Considering SC crystals with only a small difference in lattice parameter and joined at a cubic face for simplicity, let the lattice parameter of one crystal be represented by

$$a_1 = a\left(1 + \frac{\delta}{2}\right) \tag{7.37a}$$

and the other by

$$a_2 = a\left(1 - \frac{\delta}{2}\right) \tag{7.37b}$$

where δ is a small fraction and a may be described as the lattice parameter of the reference lattice. The dislocation array is of simple edge type with spacing $p = a/\delta$.

Brooks[7] shows that the specific surface energy for this interphase boundary is

$$E = \frac{1}{2}\tau_0 a\delta[A' - \ell n\delta] \tag{7.38a}$$

$$A' = \ell n\left(\frac{a}{2\pi r_0}\right) \tag{7.38b}$$

$$\frac{1}{\tau_0} = \left(\frac{1 - \sigma_1^*}{\tilde{G}_1} + \frac{1 - \sigma_2^*}{\tilde{G}_2}\right) \tag{7.38c}$$

where r_0 is a dislocation core radius chosen in such a way as to give the correct energy of atomic misfit at the center of a dislocation, σ^* is Poisson's ratio and $\tilde{G}$ is the shear modulus. It is interesting to note that the elastic models of the two contacting crystals are effectively in parallel so that the softer crystal predominantly determines τ_0.

In some situations (such as where one of the phases is a thin slice), the system may adopt a coherent interface rather than an incoherent one in order to minimize its free energy.

Consider a slab of phase α having unit cross-sectional area in the xy plane (a substrate). Let its lattice parameter in this plane be a_0. To this slab we wish to add a slice of thickness L which, in the unstressed state, has a lattice parameter a. If the slice is to be completely coherent with the slab after its addition, it must first be subjected to a strain δ

in the x and y directions given by

$$\delta = (a - a_0)/a \qquad (7.39a)$$

However, let us suppose that it is not completely coherent and that it adopts a lattice parameter a' and a strain δ' where

$$\delta' = (a' - a_0)/a' \qquad (7.39b)$$

For simplicity, we shall assume that the solid is elastically isotropic (which is never strictly true for a crystal); the stresses, σ_x and σ_y, required to produce this deformation in the absence of any stress in the z direction are given by the solution of

$$\delta' = \varepsilon_x = S_{11}\sigma_x + S_{12}\sigma_y \qquad (7.40a)$$

$$\delta' = \varepsilon_y = S_{12}\sigma_x + S_{11}\sigma_y \qquad (7.40b)$$

in which the Ss are the elastic compliances. Thus,

$$\sigma_x = \sigma_y = \delta'/(S_{11} + S_{12}) \qquad (7.41)$$

The reversible work, W, of deformation is given by

$$W = \frac{1}{2}(\varepsilon_x\sigma_x + \varepsilon_y\sigma_y) \qquad (7.42a)$$

$$= \delta'^2/(S_{11} + S_{12}) \qquad (7.42b)$$

Inasmuch as Young's modulus $E^* = 1/S_{11}$ and Poisson's ratio $\sigma^* = -S_{12}/S_{11}$, the above equation may be rewritten in the form

$$W = Y\delta'^2 \qquad (7.43)$$

in which $Y = E^*/(1 - \sigma^*)$. If we consider a thick slice of length L over which a transition layer occurs, it is apparent that the work done per unit volume, W_v, in achieving this partial coherency is given by

$$W_v = \frac{1}{L}\int_0^L Y\delta'^2 dz \qquad (7.44)$$

If this disregistry is due entirely to a compositional variation, we want to relate δ' to this compositional variation. Denoting the average composition by C_0 and the lattice parameter of the unstressed solid of this composition by a_0, we have (from a Taylor's expansion about C_0)

$$a' = a_0[1 + \eta(C - C_0) + \ldots] \qquad (7.45)$$

In which $\eta = (1/a_0)da'/dC$ evaluated at C_0. Thus

$$\delta' = (a' - a_0)/a' \approx \eta(C - C_0) \qquad (7.46)$$

which, on substitution into Eq. (7.44), yields

$$W_v = \frac{1}{L}\int_0^L \eta^2 Y(C - C_0)^2 dz \qquad (7.47)$$

In a real situation, the minimum free energy of the system will be a compromise between dislocation energy and strain energy so that

$$E_T = E_d + E_S \tag{7.48a}$$

where

$$E_d = \frac{1}{2}\tau_0 a(\delta - \delta')[A' - \ell n(\delta - \delta')] \tag{7.48b}$$

and

$$E_S = \int_0^L Y\delta'^2 dz \tag{7.48c}$$

and E_T is to be minimized with respect to variations of δ'.

From the foregoing, we see that the range of the interface forces for this contribution extend from the interface a distance $\sim$2–3 times the spacing between the dislocations; i.e., the forces are not likely to extend beyond $\sim$ 500 Å from the interface.

7.1.7 Surface reconstruction

For electron bonding, and thus free energy, reasons, the equilibrium symmetry of a crystal's surface structure may be quite different than that obtained by cleaving a perfect crystal. The symmetry of the surface structure is designated $(j \times k)$ where j and k are integers denoting the repeating distance on the surface in multiples of the solid lattice constant. While vacuum cleaving of Si and Ge crystals may produce (111) planes with (1×1) periodicity, heating this or any other planes usually produces larger unit cell structures. The close-packed (111) planes of Si and Ge exhibit (7×7) and (8×8) periodicities while the (100) planes both have (2×1) periodicities. The cause of the rearrangement of these surfaces from their bulk structures is the covalent bonding between atoms; the dangling bonds at the surface reconstruct to form special orbitals that lower the electron's energy and this requires large changes in the positions of the atoms in at least the first atomic layer.

For BCC metals (especially W, Mo, Ta and Nb), their clean surfaces give only (1×1) LEED patterns implying that no surface electron reconstruction occurs. For FCC metals, the (111) planes have (1×1) symmetry, however, reconstruction occurs on other surface orientations. On Ni, Pd and Cu the (100) planes also have (1×1) symmetry, however, clean (100) planes of Pt, Au and Ir exhibit a (5×1) symmetry. This appears to arise from a nearly hexagonal overlayer of metal atoms on the ideal (100) plane. The layer is slightly compressed from the ideal

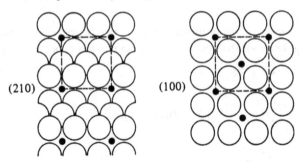

Fig. 7.18. Proposed adsorbate structures for N_2 on W(210) and W(100). N atoms occupy alternate sites of four-fold coordination on both planes.

(111) plane and all atoms are not coplanar. Registry with the underlying (100) plane occurs at five lattice spacings in one direction and one in the other to yield a (5×1) LEED pattern. The (110) plane is the plane of next closest packing and, while it may have a (1×1) symmetry on some metals, (110) Pt exhibits a (2×1) LEED pattern.

Specific adsorbates produce their own surface symmetry patterns and, in some cases, probably influence surface reconstruction of the substrate as well. N_2 on the (100) plane of W produces a (2×2) pattern as expected if N atoms occupy every other site on the surface (see Fig. 7.18). The (210) and (310) planes of W expose flat planes with four-fold sites identical to those on the (100) plane with steps consisting of substrate atom configurations identical to those on the (110) plane. When completely ordered, N atoms occupy alternate four-fold sites just as they do on the (100) plane. However, there is also a correlation of sites between rows. Above 900 K, the adsorbate orders both along and between rows while, below 900 K, ordering only occurs along rows. One remarkable feature of these results is the long-range correlation between rows, a distance of over 10 Å on the (210) plane.

The surface reconstruction and adsorbate ordering features occur because they lower the free energy of the system and thus may be expected to be occurring simultaneously with layer edge attachment during crystal growth. The "unreconstructing" of the surface will be a necessary precursor step for layer motion to proceed in some cases so that the growth kinetics may be appreciably slowed for such cases.

For simple dangling bond models, such as used in Fig. 7.16, the surface can only relax by roughening and cluster formation. *Real surfaces* relax by surface and ledge reconstruction as well as by roughening and cluster

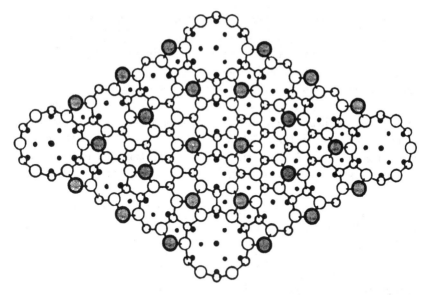

Fig. 7.19. Top view of Si(111) – 7 × 7 DAS model. Atoms on (111) layers of decreasing heights are indicated by circles of decreasing sizes. Heavily outlined dotted circles represent adatoms sitting at topsites. Larger open circles represent atoms in the stacking fault layer. Smaller open circles represent atoms in the dimer layer. Small circles and dots represent atoms in the unreconstructed layers beneath the reconstructed surface. J. E. Demuth, private communication (1989).

formation. The surface reconstruction process generates a surface stress tensor (see Fig. 7.3) which drives or influences all subsequent relaxation events on the surface. The presence of ledges alters the terrace reconstruction and sometimes so beneficially that $\gamma_\ell < 0$. This is what is calculated for $[\bar{1}10]$ ledges on GaAs(00$\bar{1}$). Likewise, surface point defect formation sometimes interacts with the reconstruction/stress tensor pattern such that $\Delta E_{PD} < 0$ which leads to spontaneous point defect formation up to some surface concentration. This is what is observed for the Si(111) – (7 × 7) reconstructed surface (see Fig. 7.19). Thus, any model of a surface that does not include surface reconstruction and surface stress tensor formation is likely to be in serious error for the modeling of real systems.

7.1.8   Chemical adsorption contribution, γ_a

From the foregoing, one finds that the chemical potential of a molecule can be quite different in an interface region compared to the bulk phase far from the interface. The change relative to the bulk can occur via both (1) the presence of an inhomogeneous field of stress, electrostatic potential, magnetic potential, etc. and (2) the gross structural changes in the molecular environment at the interface and thus in the intermolecular potential function for an atom situated there rather than in either bulk phase. Since thermodynamic equilibrium demands a constant electrochemical potential for each species throughout the entire system, a redistribution of solute atoms must occur in the interface region in response to these field effects. This gives rise to the common phenomenon of surface or interface segregation and involves the exchange of solute for solvent species between the bulk and the interface region.

Defining the total energy of interaction between the jth species ($j = 0$ is the solvent) and all of the interface fields as $\Delta \tilde{G}_0^j$, the extended electrochemical potential, $\tilde{\eta}^j$, for species j is given by

$$\tilde{\eta}_\beta^j = \mu_{0\beta}^j + \kappa T \ell n \hat{a}^j(z) + \delta \widetilde{\Delta G_0^j}(z); \quad \beta = S, L \qquad (7.49)$$

In general, we must expect that

$$\delta \widetilde{\Delta G_0^j}(z) = f[g^j(z), \phi^j(z), H^j(z), \sigma^j(z)] \qquad (7.50a)$$

where $g(z)$ is due to variations in local order, local coordination and local density, while $\phi(z), H(z)$ and $\sigma(z)$ are due to local variations in electrostatic potential, magnetic potential and stress, respectively. A little more explicitly, we have

$$\widetilde{\Delta G_0^j}(z) = z^j e\phi(z) + \frac{\partial}{\partial C^j} \left[\frac{\varepsilon_*}{2}(\nabla \phi)^2 + \frac{\mu^*}{2}H^2 + \frac{\sigma^2}{2E^*} \right] + \Delta \mu_B^j(z)$$

$$(7.50b)$$

In Eq. (7.50b), the first term is a very familiar one, the second involves the electrostatic, magnetic and strain energy storage and the third is due to the bonding energy changes. In the middle term of Eq. (7.50b), (i) ε_* is the permittivity of the medium relative to free space, $\varepsilon_* = \varepsilon_{*0}(1 + X_E^*)$ where X_E^* is the electric susceptibility of the medium, (ii) μ^* is the permeability of the medium relative to free space, $\mu^* = \mu_0^*(1 + X_M^*)$ where X_M^* is the magnetic susceptibility of the medium and (iii) E^* is the modulus of elasticity of the medium. The $\Delta \mu_B^j(z)$ term will always be non-zero at the interface because of the asymmetry of the interatomic potential fields at that location. Thus, the binding potential of the crystal for a molecule in the liquid adjacent to it will always be different than the binding potential for that molecule completely surrounded by n shells of liquid ($n \to \infty$).

Because of these interface fields, we must expect the equilibrium solute distribution in each phase, β, to be given by

$$\frac{X_\beta^j(z)}{X_\beta^j(\infty)} = \left[\frac{\hat{\gamma}'^j(\infty)}{\hat{\gamma}'^j(z)}\right] \exp\left[\frac{-\delta\widetilde{\Delta G}_{0\beta}^j(z)}{\kappa T}\right] \tag{7.51a}$$

where

$$\delta\widetilde{\Delta G}_{0\beta}^j = \widetilde{\Delta G}_{0\beta}^j - c_v\widetilde{\Delta G}_{0\beta}^0 \tag{7.51b}$$

since we are dealing with a process of exchange between the solute and solvent species; i.e., removing a solvent molecule from the interface region to the bulk and extracting a solute molecule from the bulk to place at the former solvent location in the interface region. Obviously size effects of these two species must be accounted for. In Eq. (7.51b), $c_v = v_0^j/v_0^0$ is the specific volume ratio for solute and solvent in their standard states. Depending upon the sign of $\delta\widetilde{\Delta G}_0^j$ in Eq. (7.51a), we have either an enhancement or a depletion of the jth species in the interface region. This interface adsorption increases as the temperature decreases and as $\delta\widetilde{\Delta G}_0^j$ becomes more negative. An orientation dependence is also expected since both the molecular density and the contact potential of the interface will be a function of crystallographic orientation.

As the closing segment to this section, we note that, for an ionic crystal where the values of the work to remove positive and negative ions from the crystal interior are unequal, the condition of zero space charge within the crystal can be satisfied provided a difference in electrostatic potential exists between the interior of the crystal and possible vacancy sources. Thus, for the process of formation of vacancies in the bulk from the crystal surface, the potential difference, $\Delta\phi_s$, between the bulk and surface is of such a magnitude and sign that the electrostatic energy difference of the two defects exactly cancels the difference in their separate works of formation. Corresponding to this potential variation between bulk and surface, a region of non-zero space charge will exist in which the concentration of vacancies of one type will be decreased while that of the other type will be increased.

For cation and anion vacancies, we have

$$\mu_+ = \mu_{0+} + \kappa T \ell n \hat{\gamma}_+ C_+ - e\phi \tag{7.52a}$$

$$\mu_- = \mu_{0-} + \kappa T \ell n \hat{\gamma}_- C_- + e\phi \tag{7.52b}$$

At equilibrium, $\mu_+ = \mu_- = 0$, so that

$$\hat{\gamma}_+ C_+ = \exp[-(\mu_{0+} - |e|\phi)/\kappa T] \tag{7.53a}$$

$$\hat{\gamma}_- C_- = \exp[-(\mu_{0-} + |e|\phi)/\kappa T] \tag{7.53b}$$

Since these equations are valid at any position in the crystal, the potential at large distance from the surface ($\phi_s = 0$ here by definition), $\Delta\phi_s$, is given from Eq. (7.53) by

$$\Delta\phi_s = \frac{1}{2|e|} \left\{ \mu_{0+}(\infty) - \mu_{0-}(\infty) + \kappa T \ln \left[\frac{\hat{\gamma}_+(\infty)C_+(\infty)}{\hat{\gamma}_-(\infty)C_-(\infty)} \right] \right\} \quad (7.54)$$

where the (∞) notation is used because these quantities will all be a function of position. We also note that

$$C_+(0) = C_+(\infty)\frac{\hat{\gamma}_+(\infty)}{\hat{\gamma}_+(0)} \exp\left(\frac{\mu_{0+}(\infty) - \mu_{0+}(0) - |e|\Delta\phi_s)}{\kappa T} \right) \quad (7.55)$$

For a pure crystal, neglecting defect–defect interactions so that $\hat{\gamma}_+ = \hat{\gamma}_- = 1$, we have $C_+(\infty) = C_-(\infty)$ and

$$\Delta\phi_s = \frac{1}{2|e|}[\mu_{0+}(\infty) - \mu_{0-}(\infty)] \approx \frac{1}{2|e|}[\Delta E_{0+}(\infty) - \Delta E_{0-}(\infty)] \quad (7.56)$$

when the entropies of formation of the positive and negative ion vacancies are equal and where the ΔEs are the corresponding increases in internal energy for forming the defects. For NaCl we find $\Delta\phi_s \approx -0.28$ V and the total excess negative charge, Q, within the space charge region per unit area of surface will be $\sim 1.6 \times 10^{11}$ at 600 K. In general, we have

$$Q = \frac{\varepsilon_* \kappa T}{\lambda_D 4\pi} \sqrt{8} \sinh \frac{|e|\Delta\phi_s}{2\kappa T} \quad (7.57)$$

where λ_D is the Debye length given by Eq. (7.25) ($\lambda_D \sim 2 \times 10^{-5}$ cm for NaCl at 600 K). Since this excess charge must be balanced by an equal and opposite charge on the surface layer, the surface layer must contain either a deficiency of negative ions or an excess of adsorbed positive ions. Using the above value for total charge, the excess concentration of negative ion vacancies or adsorbed positive ions is thus $\sim 10^{-4}$.

7.2   Thermodynamics of inhomogeneous systems approach

In the main body of this chapter, the method has been to use the thermodynamics of homogeneous systems approach for understanding and, with it, we have discriminated several field effects associated with the interface region but which influence the chemical potential of molecular species up to distances as large as a micrometer from the interface. These long-range effects seem to belie the assumption of homogeneity or required the use of a very thick interface region. A more acceptable approach was initiated some years ago by Cahn and Hilliard[8] (CH) who developed a method for expressing the free energy of a non-uniform system and applied it to obtaining the interfacial tension between two coexisting phases. In their method, the various spatial derivatives of density or composition were treated as independent variables and the free energy of the system was considered to depend upon them. One of the main advantages of this treatment is the splitting of the thermodynamic quantities into their corresponding values in the absence of a gradient with added terms due to the various gradients. The former are easily evaluated in terms of existing thermodynamic data for homogeneous systems.

To illustrate this method, we shall consider the equilibrium between two phases that differ in only one scalar property. Although the method may be presumed to be valid for any intensive scalar property of the system other than T and P, we shall restrict our consideration to a binary system with C as the non-uniform property. We expect the local Helmholtz free energy per unit volume, f, in the region of non-uniform concentration to depend upon both the local concentration C and that of the immediate environment. We therefore express f as the sum of the contributions that are, respectively, functions of C and the spatial derivatives of C.

Let us suppose that f in an infinitesimal volume element is a continuous function of a general potential $\hat{W}$ ($\hat{W} = C, \phi$, etc.) and its positional derivatives $\nabla \hat{W}, \nabla^2 \hat{W}$, etc., with respect to the space coordinates $x_i = 1, 2, 3$. Then, in general, the free energy density will be given by a Taylor's series expansion about f_0, the free energy density in the presence of a constant potential; i.e.,

$$f = f_0 + \sum_i \hat{L}_i \frac{\partial \hat{W}}{\partial x_i} + \sum_{ij} \hat{M}_{ij} \frac{\partial^2 \hat{W}}{\partial x_i \partial x_j} + \frac{1}{2} \sum_{ij} \hat{N}_{ij} \frac{\partial \hat{W}}{\partial x_i} \frac{\partial \hat{W}}{\partial x_j} + \cdots \quad (7.58)$$

For a cubic crystal or an isotropic medium where the free energy level of the crystal is raised by the presence of potential energy gradients fixed on external (Lagrangian) coordinates, the free energy of the system is invariant to the symmetry operations of reflection ($x_i \rightarrow -x_i$) and of rotation about a four-fold axis ($x_i \rightarrow x_j$). For this case with variable concentration, the integral free energy F in volume v is described by CH[8] as

$$F = \int_v [f_0(C) + \kappa_1 \nabla^2 C + \kappa_2 (\nabla C)^2 + \cdots] \mathrm{d}x \quad (7.59a)$$

In cases where the internal fields of the crystal are polarized by the external field or where the potential gradients in the crystal are fixed on internal (Eulerian) coordinates, the symmetry of the crystal is lowered. Some gradients, such as chemical potential gradients caused by concentration gradients within the crystal, are intrinsically fixed on internal coordinates and the symmetry operations do not hold. For such internally fixed coordinates, $\hat{L}_i \neq 0$ ($\hat{L}_i = \partial f/\partial(\partial C/\partial x_i)$) and, if both $\partial C/\partial x_1$ and $\partial C/\partial x_2$ gradients are present, rotation by π about the x_1-axis does not change the sign of the appropriate term in Eq. (7.58) containing $\hat{N}_{ij}$ or $\hat{M}_{ij}$ for the Eulerian coordinate system whereas they would change for the Lagrangian coordinates. For this kind of system, Eq. (7.59a) must be replaced by

$$F = \int_v \left\{ f_0(C) + \sum_{ij} \left[\alpha \left| \frac{\partial C}{\partial x_i} \right| + \kappa_1 \left(\frac{\partial^2 C}{\partial x_i^2} \right) + \kappa_2 \left(\frac{\partial C}{\partial x_i} \right)^2 \right. \right.$$
$$\left. \left. + \kappa_3 \left| \frac{\partial C}{\partial x_i} \right| \left| \frac{\partial C}{\partial x_j} \right| + \kappa_4 \left| \frac{\partial^2 C}{\partial x_i \partial x_j} \right| + \cdots \right] \right\} dx \qquad (7.59b)$$

where the coefficient κ_i in the free energy function, of a term arising from a positive curvature $(\partial^2 C/\partial x_i^2)_+$, can differ from that for a negative curvature $(\partial^2 C/\partial x_i^2)_-$ of the same magnitude. Thus, in general, $f(x)$ is not symmetrical about f_0 with respect to the curvature coordinate $(\partial^2 C/\partial x_i^2)$.

To determine the free energy of a flat interface, we shall consider an area A between two coexisting isotropic phases α and β of concentrations C_α and C_β. We shall assume that $f_0(C)$ is of the form shown in Fig. 7.20. If the intensive scalar property had been density rather than concentration (pure liquid and its vapor) one could substitute ρ_c and ρ_v for C_α and C_β so that the equations to be derived would apply specifically to an interface between the condensed phase and its vapor. In order to see how the α term in Eq. (7.59b) would compare with any of the second order terms, we shall make the simplifying assumption that the κ_1 term is zero so that the total free energy of the system is given by

$$F \approx AN_v \int_{-\infty}^{\infty} \left[f_0(C) + \alpha \left| \frac{dC}{dx} \right| + \kappa_2 \left(\frac{dC}{dx} \right)^2 \right] dx \qquad (7.60)$$

The specific interfacial free energy, γ, is, by definition, the difference per unit area of the interface between the actual free energy of the system and the value it would have if the properties of the phases were continuous throughout. Hence,

$$\gamma = N_v \int_{-\infty}^{\infty} \left\{ f_0(C) + \alpha \left| \frac{dC}{dx} \right| + \kappa_2 \left(\frac{dC}{dx} \right)^2 \right.$$

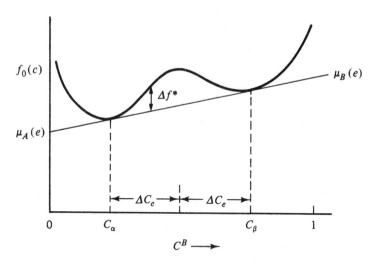

Fig. 7.20. Schematic plot of free energy versus composition for a binary solid solution illustrating the equilibrium concentrations C_α and C_β for this system that will decompose spinodally.

$$- [C\mu^B(e) + (1 - C)\mu^A(e)] \Big\} dx \qquad (7.61)$$

where $\mu^A(e)$ and $\mu^B(e)$ are the chemical potentials per molecule (referred to the same standard states as f_0) of the species A and B in the α or β phases which are in equilibrium with one another. Defining $\Delta f(C)$ by

$$\Delta f(C) = f_0(C) - [C\mu^B(e) + (1 - C)\mu^A(e)] \qquad (7.62a)$$
$$= C[\mu^B(C) - \mu^B(e)] + (1 - C)[\mu^A(c) - \mu^A(e)] \quad (7.62b)$$

Eq. (7.61) becomes

$$\gamma = N_v \int_{\infty}^{\infty} \left[\Delta f(C) + \alpha \left|\frac{dC}{dx}\right| + \kappa_2 \left(\frac{dC}{dx}\right)^2\right] dx \qquad (7.63)$$

Here, $\Delta f(C)$ may be regarded as a free energy referred to a standard state of an equilibrium mixture of α and β (Eq. (7.62a)) or as the free energy change associated with transferring material from an infinite reservoir of concentration C_α or C_β to material of concentration C (Eq. (7.62b)). According to Eq. (7.63), the more diffuse is the interface, the smaller will be the contribution of the gradient energy terms to γ. But this decrease in energy can only be achieved by introducing more material at the interface of non-equilibrium concentration and thus at the expense of increasing the integrated value of $\Delta f(C)$. At equilibrium, the concentration variation will be such that the integral in Eq. (7.63) is a minimum (equivalent to the requirement that the chemical potential be constant throughout the system). This leads to

$$\Delta f(C) = \kappa_2 (dC/dx)^2 \tag{7.64}$$

so that

$$\gamma = 2N_v \int_\infty^\infty [\Delta f(C)]\, dx \tag{7.65a}$$

or, by changing the variables of integration for x to C by means of Eq. (7.64),

$$\gamma = 2N_v \int_\infty^\infty [(\kappa_2 \Delta f/(C)]^{1/2} dC \tag{7.65b}$$

Utilizing this expression in the immediate vicinity of the critical temperature, T_c, we find that γ should be proportional to $(T_c - T)^{1/2}$ for non-zero values of α and proportional to $(T_c - T)^{3/2}$ if $\alpha = 0$. Experimental data for the cyclohexane–aniline system supports the $(T_c - T)^{3/2}$ dependence.

From this section, we see that surface adsorption and chemical segregation depend on both the local free energy and upon the spatial gradients of free energy. In the future, we will need to evaluate the magnitude of γ for the real case where spatial variations of $\gamma_c, \gamma_e, \gamma_t$ and γ_a for a binary system are all simultaneously taken into consideration.

8

General kinetics and the nucleation process

8.1 Elementary chemical reactions

While thermodynamics tells us what phases and what reactions are allowed for a given set of state variables of the system (T, C, P, Φ), kinetics tells us which phases and which reactions will actually occur at finite rates and which will dominate for a given set of state variables. The object of this chapter is to describe the frequency of occurrence of some important elementary processes, e.g., elementary chemical reactions and diffusion. Then, almost all other thermally activated rate processes can be analyzed in a straightforward way and, here, we shall apply these considerations to the nucleation process.

For the elementary chemical reaction

$$A + B \rightleftarrows C + D \qquad (8.1)$$

there exists in multidimensional space, a single potential energy surface on which the system will move to go from reactants, (A,B), to products, (C,D), and back. On this potential energy surface, there exists a path between reactants and products that will be the most economical in terms of the energy required for the reaction. This path is illustrated in Fig. 8.1 where position along the reaction path is determined by the value of the reaction coordinate. For this path, the reaction is exothermic from left to right and endothermic from right to left. At equilibrium the

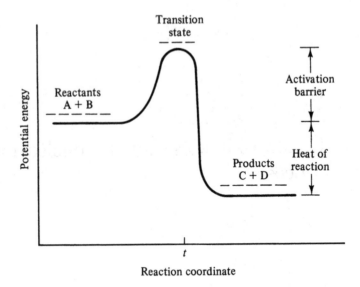

Fig. 8.1. Illustration of the most economical potential energy surface between reactants and products as a function of coordinate position along the reaction path.

following reaction holds

$$\bar{k}_+ = \bar{k}_- = \omega C_t \tag{8.2}$$

where $\bar{k}_+$ and $\bar{k}_-$ are the rates of the forward and reverse reactions respectively, C_t is the concentration of transition state species in equilibrium with the reactants A and B and ω is the frequency of transition. An observer situated at the top of the barrier would determine this frequency ω at which the transition state species cross the barrier going from reactants to products and vice versa.

In Fig. 8.1, the reaction coordinate expresses a measurement of progress along the reaction path. In solid state diffusion via vacancy diffusion, the reaction coordinate is a physical distance. In a phase transition involving only a change in the type of electronic binding, the reaction coordinate is a type of distance in configuration space where the different configurations are a series of electron states. In Fig. 8.1, the potential energy coordinate is a thermodynamic potential that measures the reversible work effect in a process subject to certain constraints, i.e., $\Delta G_{P,T}, \Delta F_{T,v}$, etc. Since we don't know the transition frequencies between these quantum states, we make an equilibrium assumption for such thermally activated rate processes. One assumes that equilibrium holds between the activated state species and the reactant.

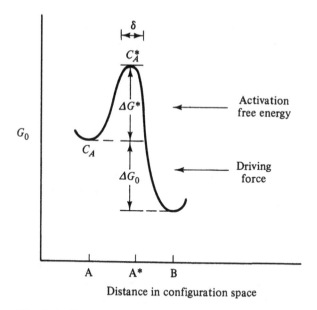

Fig. 8.2. Free energy surface connecting the transition states, A^*, to the reactant and product states, A and B, respectively.

8.1.1 Unidirectional process

For an essentially unidirectional elementary process of the following kind

$$A \rightleftarrows A^* \rightarrow B \tag{8.3}$$

where A^* is the activated state (transition state) illustrated in Fig. 8.2, equilibrium between A and A^* means

$$\hat{a}_{A^*}/\hat{a}_A = \exp(-\Delta G^*/\kappa T) \tag{8.4a}$$

where $\hat{a}$ refers to activity so that

$$C_{A^*} = C_A \frac{\hat{\gamma}_A}{\hat{\gamma}_{A^*}} \exp\left(\frac{-\Delta G^*}{\kappa T}\right) \tag{8.4b}$$

with C_{A^*} and C_A being the concentration of species in the activated state and reactant state, respectively. For any appreciable value of $\Delta G^*/\kappa T$, we expect a very dilute population of activated species; i.e., $C_{A^*}/C_A \ll 1$. We should note that, in Eq. (8.4b), ΔG^* is a *standard free energy* difference and is expressed per unit of reaction rather than per atom or per molecule. It is the free energy of formation of the activated species and, although in most elementary processes it refers to the activational free energy for net movement or change of state of only one or a few atoms or molecules, in complex processes like nucleation it refers to the net movement and change of state of hundreds of atoms.

The rate, $\hat{I}$, of the elementary process indicated by Eq. (8.3) in the forward direction is

$$\hat{I} = \omega C_{A^*} = \bar{k}_+ C_A \tag{8.5a}$$

where

$$\bar{k}_+ = \omega \frac{\hat{\gamma}_A}{\hat{\gamma}_{A^*}} \exp\left(\frac{-\Delta G^*}{\kappa T}\right) \tag{8.5b}$$

and where ω is the transmission frequency of the activated species across the top of the free energy barrier so that the units of $\hat{I}$ are # cm^{-3} s^{-1}. From statistical thermodynamics

$$\omega = \left(\frac{\kappa T}{2\pi m_r}\right)^{1/2} \frac{1}{\delta_*} \tag{8.6}$$

where m_r is the reduced mass in the separation coordinate reference frame ($m_r = m_1 m_2/(m_1 + m_2)$) and δ_* is a distance wherein the energy change is $\sim \kappa T$. This expression assumes a classical behavior of the activated species because the first term in Eq. (8.6) is the average velocity of the activated complex. It also assumes that the Heisenberg uncertainty principle is not violated in trying to describe the activated state so we must have a large de Broglie wavelength, $\lambda = h^*/2\pi m_r \kappa T$; i.e., we want $\lambda > \delta_*$.

Sometimes a transmission coefficient $K_* \sim 1$ is introduced into Eq. (8.5b) for $\bar{k}_+$. The case $K_* < 1$ describes a lack of thermal accommodation of the activated species where the species passes the barrier and then rebounds elastically back across the barrier without giving up its high kinetic energy to neighboring atoms. This arises in gaseous reactions but not generally in condensed phase reactions because there are always molecules there to absorb the excess energy. The case $K_* > 1$ describes quantum mechanical tunnelling through the potential energy barrier in cases where the barrier has a high curvature.

From statistical mechanics and Fig. 8.2 we have

$$\Delta G_0 = \Delta E_{0_0} - \kappa T \ell n \left(\frac{\hat{q}_B}{\hat{q}_A}\right)_0 \tag{8.7a}$$

and

$$\Delta G^* = \Delta E_0^* - \kappa T \ell n \left(\frac{\hat{q}_{A^*}}{\hat{q}_{A^0}}\right) \tag{8.7b}$$

neglecting terms $\Delta(pv)_0$ and $\Delta(pv)^*$, respectively, which are small with respect to κT. Here, ΔE_{0_0} is a potential energy and the $\hat{q}$s are canonical partition functions; i.e.,

$$\hat{q}_{A_0} = (\hat{q}_t \bullet \hat{q}_r \bullet \hat{q}_v \bullet \hat{q}_e \bullet \hat{q}_n)_{A_0} \tag{8.8a}$$

and

$$\hat{q}_{A^*} = (\hat{q}_{t^*} \bullet \hat{q}_t \bullet \hat{q}_r \bullet \hat{q}_{vib} \bullet \hat{q}_e \bullet \hat{q}_n)_{A^*} \tag{8.8b}$$

where

$$\hat{q}_{t^*} = (2\pi m_r \kappa T)^{1/2} \frac{\delta_*}{h^*} \tag{8.8c}$$

In Eqs. (8.8), the subscripts t, r, v, e and n refer to translation, rotation, vibration, electronic and nuclear degrees of freedom, respectively. The activated complex has no degree of freedom corresponding to vibration along the reaction path and, instead of this, we suppose the complex to have an additional degree of freedom, t^*, corresponding to translational motion along the reaction path. This motion along the reaction path is synonymous with disintegration of the complex. Chemists factored $\hat{q}_{t^*}$ out of $\hat{q}_{A^*}$ so that the unknown δ_* could be eliminated from Eqs. (8.5) and (8.6) and this has led to the Eyring formulation of transition state theory; i.e.,

$$\bar{k}_+ = \frac{\kappa T}{h^*} \exp\left(\frac{\Delta S'^*}{\kappa}\right) \exp\left(\frac{-\Delta H'^*}{\kappa T}\right) \tag{8.9}$$

where the prime on $\Delta H'$ denotes that one degree of translation is missing on the top of the barrier and the prime on $\Delta S'$ indicates that $\hat{q}_{t^*}$ is missing $\kappa T/2$. Wert and Zener[1] worried about the asymmetry in the degrees of freedom between the reactant and the activated species so they factored out a partition function for vibration, in the direction of the reaction coordinate, from the whole partition function for the reactant and obtained

$$\bar{k}_+ = \nu \exp(\Delta S''^*/\kappa) \exp(-\Delta H''^*/\kappa T) \tag{8.10}$$

Here, they assumed an independent simple harmonic oscillator form for the classical vibration partition function. This is entirely equivalent to the Eyring formulation,[2] but, in this case, two degree of freedom are missing, one from the reactant and one from the activated species. In terms of magnitudes, $\nu \sim 10^{12}-10^{13}$ s^{-1} and $\Delta S''^* \sim \pm 5\kappa$ at most for an elementary process so that the feature that limits the jump frequencies is the height of the potential barrier and, within $\sim \pm \kappa T/2$,

$$\Delta H''^* = \Delta E_0^* \tag{8.11}$$

8.1.2 Bidirectional process

Thus far we have considered only unidirectional processes for which $\Delta G_+^* \ll \Delta G_-^*$ where the subscripts $+$ and $-$ refer to the forward and the reverse directions respectively. We will now consider cases where $\Delta G_+^* \sim \Delta G_-^*$ which leads to a net rate, $\hat{I}$, given by

$$\hat{I} = \bar{k}_+ C_A - \bar{k}_- C_B \tag{8.12}$$

for the reaction

$$A \rightleftarrows A^* \rightleftarrows B \tag{8.13}$$

This does not imply a physical situation in which there is an equilibrium of A and B with A^*, for which the net rate would be zero. Rather, we assume that the forward and reverse currents act independently of each other because of the extreme dilution of the activated species; i.e.,

$$A \rightleftarrows A^* \rightarrow B \tag{8.14a}$$

and

$$A \leftarrow A^* \rightleftarrows B \tag{8.14b}$$

In analogy with Eq. (8.5), for this case we have

$$\hat{I} = \bar{k}_+ C_A \left[1 - \exp(\Delta\mu/\kappa T)\right] \tag{8.15a}$$

where

$$\Delta\mu = \mu_B - \mu_A \tag{8.15b}$$

At equilibrium, $\hat{I} = 0$, so we have from Eq. (8.12)

$$\frac{\bar{k}_+}{\bar{k}_-} = \frac{C_{Beq}^*}{C_{Aeq}^*} = \frac{\hat{\gamma}_A}{\hat{\gamma}_B} \hat{K} \tag{8.16}$$

where $\hat{K}$ is the equilibrium constant for the reaction. At small driving forces, $\Delta\mu/\kappa T \ll 1$, and we can expand Eqs. (8.15) to yield

$$\hat{I} = -\bar{k}_+ C_A \frac{\Delta\mu}{\kappa T} = \beta_* \Delta\mu \tag{8.17}$$

so we see that the current of reaction is proportional to the potential difference between the two states B and A. This linear form for the net rate is analogous to Ohm's law for electric current and to Fick's first law for the diffusion of chemical species given by

$$\hat{I} = -\frac{DC}{\kappa T} \nabla\mu \tag{8.18}$$

where D is the diffusion coefficient. It is probably for this reason that people have tried to equate $\bar{k}_+$ and D for the crystallization reaction and liquid diffusion respectively. However, there is no valid reason for equating these two quantities which could be different by many orders of magnitude depending upon the type of elementary processes involved. A clearer distinction between Eqs. (8.17) and (8.18) occurs as $\Delta\mu/\kappa T$

increases because additional terms need to be added to Eq. (8.17) until it reverts to Eq. (8.15a).

8.1.3 Diffusion

If we consider the elementary process of diffusion as an example for this section, we find that the Einstein condition for isotropic diffusion is given by

$$D = \bar{k}_+ \ell^2 \tag{8.19}$$

where $\bar{k}_+ = \Gamma/z_*$, Γ is the total mean jump frequency (n/t for n jumps), z_* is the number of equivalent jump directions, ℓ is the jump distance and $\bar{k}_+$ is given by Eq. (8.9) or (8.10) including the factor K_* which, in diffusion theory, is related to f_*, the correlation factor. The correlation factor is roughly the probability of an atom not jumping back to its original position on its next jump ($f_* = 0.866$ for self diffusion in pure metals). For volume diffusion, $z_* = 6$ and for surface diffusion $z_* = 4$. From a jump point of view, after an atom or molecule has made jumps of equal length, ℓ, the mean square displacement, $\bar{R}_n^2$, for an isotropic system with random jumps is given by

$$\bar{R}_n^2 = n\ell^2 \tag{8.20}$$

so we see that the actual linear distance moved, on a surface say, is given by $n^{1/2}\ell \ll n\ell$, the actual distance travelled by the particle.

For a crystal surface in contact with a vapor or a solution, a certain concentration of adsorbed, essentially mobile molecules, will be present on the surface. The mean displacement, $\bar{X}_s$, of an adsorbed molecule is given in terms of the Einstein formula

$$\bar{X}_s^2 = 4D_s \tau_s \tag{8.21}$$

where τ_s is the mean lifetime of an adsorbed molecule on the surface before moving back into the solution or vapor. For simple molecules, we have

$$D_s = a^2 \nu' \exp(\Delta G_S^*/\kappa T) \tag{8.22a}$$

$$\tau_s^{-1} = \nu'' \exp(-W_S'/\kappa T) \tag{8.22b}$$

where ΔG_S^* is the activation energy between two neighbouring equilibrium positions on the surface, a is the separation between two adjacent potential wells on the surface, W_S' is the energy of desorption from the surface into the vapor or solution and the factors ν', ν'' will be of the order of the atomic frequency of vibration. Combining Eqs. (8.22) with Eq. (8.21), we have

$$\bar{X}_s \approx 2a \exp[(W_S' - \Delta G_S^*)/2\kappa T] \tag{8.23}$$

For the growth of large perfect crystals, we wish to have $\bar{X}_s/a \gg 1$ so a need exists for $W'_S \gg \Delta G^*_S$. On the other hand for the formation of amorphous or fine grained polycrystalline layers, we need $\bar{X}_s/a \sim 1$ or $\Delta G^*_S \sim W'_S$. To make the magnitude of $\bar{X}_s/a$ tangible, let us consider migration on the (111) plane of a FCC crystal with first NN-only binding. Thus, the value of W'_S equals half the evaporation energy and ΔG^*_S, which is determined by the value of W'_S at an atom position half way between two equilibrium atom positions, is given by $\Delta G^*_S \sim W'_s/3$. For a typical vapor phase growth situation, $W'_s/\kappa T \sim 20$ so that $\bar{X}_s/a \sim 400$. For a typical solution growth situation, $W'_s/\kappa T \sim 10$ so that $\bar{X}_s/a \sim 2 \times 10^5$. In most real situations, the picture is not nearly so simple and the results depend upon whether or not reconstruction has occurred at the surface, how much free electron spill-over has occurred from the substrate to the adsorbed layer coordinates, the type of interface polarization that has occurred in the solution, the electrostatic potential of the surface, the charge nature of the surface species, etc. Thus, the above numbers should be considered only as very approximate guidelines at best.

8.1.4   *Multiple elementary step processes*

In a particular reaction step, like molecular attachment at the growing crystal interface with its nutrient phase, a multiple sequence of elementary rate steps may occur with one of them generally limiting the overall reaction. To illustrate with a real example, consider the growth rate of a Si film formed by CVD from one of a series of possible gases; i.e., $SiH_4, SiH_2Cl_2, SiHCl_3$ or $SiCl_4$ in H_2. The obvious sequence of reaction steps is illustrated in Fig. 8.3: (1) adsorption of the Si-containing compound on the surface; (2) surface chemical dissociation and reaction of the compound to give various molecular and atomic Si species in the adsorbed layer; (3) surface diffusion of the various Si species; (4) incorporation of the Si atom into the crystal at a kink site; and (5) desorption of the other chemical species from the surface. Any one of these steps could seriously hinder the film growth rate. Some experimental data for this situation is presented in Fig. 8.4 where we can see that, at intermediate temperatures, the activation energy is constant for the four different molecular species. Thus, it is unlikely that steps (1), (2) or (5) could be rate limiting since the data indicate that the limiting step involves the common denominator, Si, with the pre-exponential factor in the rate depending upon the presence and amount of H and/or Cl on the surface. Either step (3) or step (4) could be the rate limiting

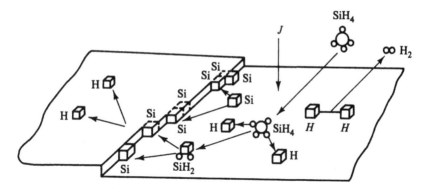

Fig. 8.3. Illustration of the sequence of steps involved in the heterogeneous decomposition of SiH4 on a Si surface during Si film growth from SiH4.

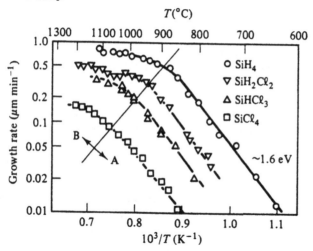

Fig. 8.4. The observed temperature dependence of Si film growth rate from various chlorosilane gases. J. Bloem and L. J. Giling in *VLSI Electronics: Microstructure Science*. Eds. N. G. Einspruch and H. Huff (Academic Press, New York, 1985). Vol. 12, Ch. 3.

step and a decision between one or the other cannot be made without more information.

8.1.5 Non-thermally activated processes

In many crystal growth situations, we may wish to enhance the rate of a critical elementary process by using optical, acoustical or other radiation, in addition to the thermal pool of fluctuations, to help an atom or molecule reach the activated state. In addition, we must be aware that

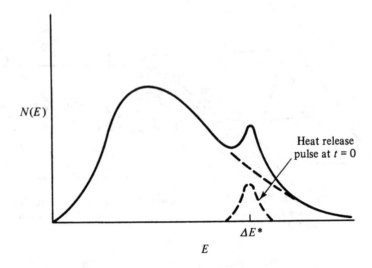

Fig. 8.5. Illustration of the superposition of a Maxwell–Boltzmann distribution of energies and a specific heat pulse of magnitude ΔE^* released at $t = 0$.

the thermal energy pool we are using to evaluate a fluctuation probability may have some non-equilibrium characteristics associated with other heat release reactions occurring in the near neighborhood. For example, one may be considering the formation of a film of an $A_x B_y$ compound via the sputtering process. The film crystallinity and perfection will be greater if the A and B species are separately deposited rather than via direct deposition from an $A_x B_y$ source. This is because the heat of formation for the compound is released at the substrate in the former case whereas it is absent in the latter case.

This heat release occurs in specific units of energy ΔE^* and these alter the Maxwell–Boltzmann distribution of energies in the fashion illustrated by Fig. 8.5. Such heat release at $t = 0$ produces a non-equilibrium phonon spectrum at the surface which will decay to the equilibrium spectrum in time τ. If new heat release pulses occur at repeat times much shorter than τ, then the surface phonon spectrum will continue to exhibit a non-equilibrium peak at the energy associated with ΔE^*. One consequence of this altered surface phonon spectrum is that surface diffusion becomes a more probable event so that $\bar{X}_s$ (see Eq. (8.21)) may increase greatly leading to larger crystallite sizes. Another mechanism which would lead to the same consequence is energetic impacts with the surface by other vapor atoms to "ring" the lattice.

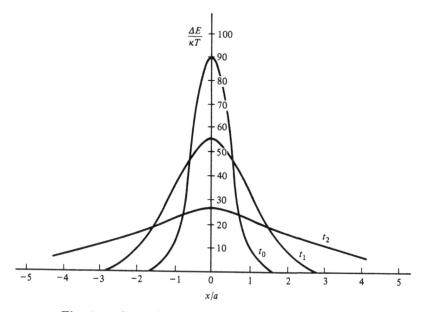

Fig. 8.6. Qualitative illustration of reduced energy profile in space and time for an energy pulse released at time t_0.

To illustrate the effect being discussed, consider a one-dimensional heat conduction situation with an energy release of $N'\kappa T$ by an atom at $x = 0$ at time t_0. This heat pulse can be considered as a δ-function of total energy content $\bar{Q}$ at $z/a = 0$ and $t = 0$ which decays to a peak value of $N\kappa T$ at $t = t_0$ as illustrated in Fig. 8.6 where a is the lattice spacing. The total energy content remains the same at any point in time but the distribution, $\bar{q}$, shifts according to the following formula

$$\bar{q}\left(\frac{z}{a}, \frac{t}{t_0}\right) = \frac{N\kappa T}{(t/t_0)^{\frac{1}{2}}} \exp^{-\beta \frac{(z/a)^2}{(t/t_0)}} \tag{8.24a}$$

where

$$\beta = a^2/4D_T t_0 \tag{8.24b}$$

and where D_T is the thermal diffusivity, $t_0 \sim a^2/4D_T \sim 10^{-14} - 10^{-13}$ s and N is a number. If $N \sim 10^2$, then it requires $\sim 10^{-9} - 10^{-10}$ s before the central atom is within κT of the background and the excess energy effect is spread over 50 atoms on either side of the initially excited atom. At any site within this zone, the excess energy of an atom, $\Delta E/\kappa T$, rises from 0 at time t' reaching a peak at $\Delta\varepsilon/\kappa T$ at time t'' and decays to zero again at time t''' as illustrated in Fig. 8.7. Because of this, the probability, p, of a fluctuation of energy ΔQ_A^* occurring at an atom in

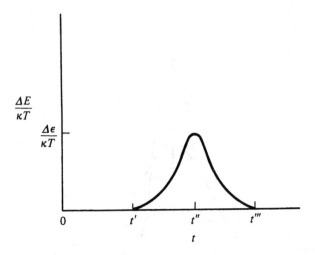

Fig. 8.7. Time variation of reduced energy, $\Delta E/\kappa T$, at any location due to the passage of a travelling pulse.

this region so that it may surmount an activation barrier, is given by

$$p = \exp\{-[\Delta Q_A^* - \Delta \varepsilon(t)]/\kappa T\} \qquad (8.25)$$

We thus note that the consequence of the initial energy pulse is effectively to lower the size of energy fluctuation needed for the atom to surmount its activation barrier. We can also conclude that, at extremely high reaction rates, for a specific reaction, the apparent activation barrier should be found to decrease.

In rapid thermal annealing (RTA) or other semiconductor processing using high intensity lamps, the photons not only produce a highly non-equilibrium surface condition due to the photoejected electron population but may also produce a very non-equilibrium phonon spectrum with a bump in the 0.3–3.0 eV range so that surface transport processes may be greatly enhanced in some cases.

8.2   Nucleation

Cooling or supersaturating a liquid to a state below its liquidus temperature decreases the stability of the liquid phase with respect to the solid. A similar situation occurs during the supersaturation of a vapor. At all conditions above or below the equilibrium transformation temperature, fluctuations in density, atomic configuration, heat content, etc., occur in the nutrient phase. These fluctuations always increase the

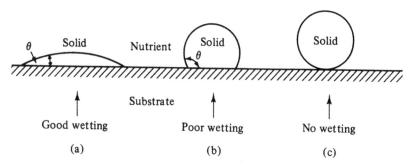

Fig. 8.8. Illustration of the wetting angle, θ, between a solid and a foreign substrate located in the nutrient phase.

free energy of the local environment when the nutrient is above its equilibrium temperature. However, at a condition below the equilibrium transformation temperature, these same fluctuations lead to the formation of stable nuclei of the solid phase.

Nuclei may form at random in homogeneous elements of volume of the nutrient phase or they may form preferentially at various interfaces in the system such as free surfaces, foreign particles, container walls, etc. The former type of nucleation is called "homogeneous" nucleation and generally a very large departure from equilibrium is required to initiate such a nucleation event. The latter type of nucleation is called "heterogeneous" nucleation and usually requires much less driving force for such nucleus formation; i.e., often a negligible departure from equilibrium. Heterogeneous nucleation is by far the most predominant and practically important mechanism of solid initiation and we may think of it as the prime operative mechanism with homogeneous nucleation as a limiting case when the wetting angle, θ, increases to $\theta = \pi$. The wetting angle θ is illustrated in Fig. 8.8 and is the angle between the embryo surface of the solid and a smooth flat substrate. When $\theta = \pi$, the embryo does not "wet" the substrate and homogeneous nucleation occurs.

8.2.1 *Classical model for a unary uncharged system*

To make the example specific, we shall consider nucleus formation from a supercooled liquid. There, the molecules of the liquid are colliding and continually forming short-lived clusters of solid-like particles, most of which redissolve because they are too small to be stable. A very few of these clusters may be struck by and combine with an additional molecule and, in turn, grow to a larger size. A nucleus is formed only as the end result of a whole sequence of probability events. Denoting α' as a molecule of liquid and β'_i as an embryo of solid containing i molecules, the sequence of bimolecular reactions leading to nucleus formation is

$$
\left.
\begin{aligned}
\alpha' \;\; &+ \alpha' \rightleftarrows \beta'_2 \\
\beta'_2 \;\; &+ \alpha' \rightleftarrows \beta'_3 \\
\beta'_3 \;\; &+ \alpha' \rightleftarrows \beta'_4 \\
\cdot \quad &\quad \cdot \quad\quad \cdot \\
\cdot \quad &\quad \cdot \quad\quad \cdot \\
\cdot \quad &\quad \cdot \quad\quad \cdot \\
\beta'_{i^*-1} &+ \alpha' \rightleftarrows \beta'_{i^*} \\
\beta'_{i^*} \;\; &+ \alpha' \rightleftarrows \beta'_{i^*+1}
\end{aligned}
\right\}
\tag{8.26}
$$

where a cluster containing i^* molecules is considered as a critical size nucleus which will continue to grow, on the average, with additional molecular collisions. Other possible mechanisms of cluster formation such as simultaneous collisions of i molecules and polymolecular collisions of lesser degree are highly improbable.

The outwardly homogeneous phase of supercooled liquid is considered to be a solution containing, as components, single atoms and embryos of various sizes. The entropy of mixing effect is what stabilizes the presence of such embryos (excess energy species). The equilibrium number of embryos containing i atoms can be obtained by minimizing the free energy of the system with respect to the number $n(i)$ of such embryos of size i. The free energy change of the system upon introducing $n(i)$ embryos is

$$
\Delta G = n(i)\Delta G_i - T\Delta S_{n(i)}
\tag{8.27}
$$

where $\Delta S_{n(i)}$ is the entropy developed by mixing $n(i)$ embryos with the remaining particles and ΔG_i is the free energy of formation for a single embryo of size i.

To compute the equilibrium distribution of embryos, it is assumed that an internal constraint is applied to the supercooled system such that particles larger than nuclei are not permitted to form. Under this condition, a dynamic equilibrium arises in which the number of embryos of a given size is fixed. The equilibrium distribution, $n^*(i)$, subject to the imposed constraint, is given by the operation

$$
[\partial \Delta G/\partial n(i)]_{n^*(i)} = 0
\tag{8.28}
$$

The entropy of mixing of $n(i)$ clusters of size i with $n(1)$ monomer species

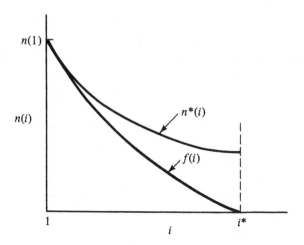

Fig. 8.9. Plot of the equilibrium, $n^*(i)$, and actual, $f(i)$, distribution of embryos as a function of embryo size, i.

is

$$\Delta S_{n(i)} = \kappa n(i)\{1 - \ell n[n(i)/n(1)]\} \tag{8.29a}$$

subject to

$$\sum_{i=2}^{\infty} n(i) \ll n(1) \tag{8.29b}$$

which is the dilute embryo solution approximation (the case where Eq. (8.29b) does not hold may also be considered). The equilibrium distribution $n^*(i)$ is plotted qualitatively in Fig. 8.9 and given quantitatively by

$$n^*(i) = n(1) \exp[-\Delta G_T(i)/\kappa T] \tag{8.30}$$

where $\Delta G_T(i)$ is the total free energy of formation of the embryo of size i. We shall return to the evaluation of $\Delta G_T(i)$ later.

The actual distribution of embryos in the liquid, $f(i,t)$ is different from $n^*(i)$. By solving the set of bimolecular reactions given by Eq. (8.26), and mentally withdrawing nuclei exceeding the critical size i^* from the system, the distribution $f(i,t)$ will attain a steady state value $f(i)$ which is less than $n^*(i)$ by some value as illustrated in Fig. 8.9. In this case, the constraint is that $f(i^*) = 0$.

The rate $I(i,t)$ at which embryos of size i become embryos of size $i+1$ at time t can be written

$$I(i,t) = \beta_*(i)\, A(i)\, f(i,t) - A(i+1)\, S(i+1)\, f(i+1,t) \tag{8.31}$$

where β_* is the rate at which monomer molecules join unit area of embryo surface, $A(i)$ is the surface area of size i, $f(i,t)$ is the number of embryos of size i existing at time t while $S(i+1)$ is the rate at which embryos of size $i+1$ lose molecules from a unit area of their surface. In Eq. (8.31), $I = I_{homo}$ is the number of embryos formed per unit volume per unit time while, for $I = I_{het}$, it is the number formed per unit area per unit time.

By replacing the quantities f in Eq. (8.31) with their equilibrium values, n^*, I may be set to zero and $S(i+1)$ determined in terms of the other quantities. Regarding i as a continuous variable, for $i \rightarrow i^*$, the transition occurs so slowly that $f(1,t)$ effectively represents an inexhaustible supply of monomer and a quasi steady state is set up in which f, for each value of i, is approximately independent of time so that the nucleation frequency $I^* = I(i^*)$ is given by

$$I^* = -\beta_* A n^* \frac{d}{di} \frac{f}{n^*} \tag{8.32}$$

The actual time dependence of $f(i,t)$ is governed by

$$\frac{\partial f}{\partial t} = \frac{\partial}{\partial i^+}\left[\beta_* A(i)\frac{\partial f}{\partial i}\right] + \frac{1}{\kappa T}\frac{\partial}{\partial i^+}\left[\beta_* A(i)f\frac{\partial \Delta G_T}{\partial i}\right] \tag{8.33}$$

This differential equation is mathematically similar to a transport equation with the first term relating to the purely diffusive portion with transport coefficient $\beta_* A(i)$. The second term is a corrective term representing the effect of some force field on the transport. Special solutions of Eq. (8.33) have shown that I^* may be given by

$$I^* = \omega f(i^*,t) \tag{8.34a}$$
$$= \omega \hat{\psi}(t)f(i^*) \tag{8.34b}$$
$$= \omega \hat{\psi}(t)\tilde{Z}\,n^*(i^*) \tag{8.34c}$$

where ω is the frequency at which a single atom from the unstable phase joins a critical sized embryo to promote it to a stable growing nucleus, $\omega(i) = \beta_* A(i)$, $\hat{\psi}(t)$ is the time lag factor for achieving steady state

$$\hat{\psi}(t) = \exp(-\tau/t) \tag{8.34d}$$

where τ is the time for an embryo to grow to critical size. Here, $\tilde{Z}$ is a correction factor (called the Zeldovich factor)

$$\tilde{Z} = f(i^*)/n^*(i^*) = \left[4\xi/9\pi\kappa T(i^*)^{4/3}\right]^{1/2} \tag{8.34e}$$

$$\xi = 4\pi\gamma_A(3\Omega/4\pi)^{2/3} \tag{8.34f}$$

where Ω is the molecular volume of the monomer and γ_A is the average surface energy of the embryo–liquid interface. All the excess energies due to any non-uniformity of the embryo and its environment must be incorporated into γ_A. We expect $\tilde{Z} \sim 10^{-2}$ and $\tau \sim$ microseconds for normal liquids.

8.2.2   Free energy change associated with an embryo of size i, $\Delta G_T(i)$

It has only been very recently appreciated that several important contributions to the free energy of formation of embryos had been heretofore neglected. These contributions are of a quantum statistical nature and arise from consideration of the absolute entropy of the embryos.

Since the embryos are treated as unique molecules in the evaluation of the equilibrium distribution, the absolute entropy changes in forming the species from the monomer must be considered. To evaluate $\Delta G_T(i)$, consider the following reversible operation for forming n spherical embryos of size i: (1) crystallize n_i molecules of liquid from a reservoir containing a very large number of molecules, N, to the bulk solid phase; (2) mechanically comminute this solid to produce n units of equal size i; (3) energize these n units to make unique "liquid" molecules (spheres) of size class $i^{(3)}$; and (4) crystallize these molecules to form solid embryos of size class $i^{(3)}$. The sum of the reversible work involved in these four steps is the total Gibbs free energy of formation of an embryo of size i, $\Delta G_T(i)$, in the liquid; i.e.,

$$\Delta G_T = \Delta G_0 + \Delta G_s + \Delta G_t + \Delta G_r + \Delta G_v \qquad (8.35)$$

Here, $\Delta G_v, \Delta G_r$ and ΔG_t arise because, as a "molecule", the embryos are not at rest and these are the vibrational, rotational and translational components, respectively, necessary to energize the embryos to the state of "liquid" molecules. The term ΔG_s, is the free energy change involved in separating a group of i molecules from a larger ensemble to an ensemble of the embryo size.[3] The term ΔG_0 is the familiar bulk free energy change associated with an embryo's formation involving the increase in free energy due to the surface formation and the decrease in free energy due to the phase change. These two contributions are illustrated in Fig. 8.10 and, we have for a spherical embryo,

$$\Delta G_0 = i\Delta\mu + \xi i^{2/3} \qquad (8.36)$$

Here, $\Delta\mu$ is the electrochemical potential difference of the molecules between the liquid and the solid and ξ is a parameter including both geometry and average surface tension γ_A and is given by Eq. (8.34f) ($i = \frac{4}{3}\pi r^3/\Omega$). This term is the only contribution considered in the original nucleation theory and $\tilde{i}$, which yields the maximum value of ΔG_0, becomes i^*, because ΔG_0 continues to decrease for $i > \tilde{i}$.

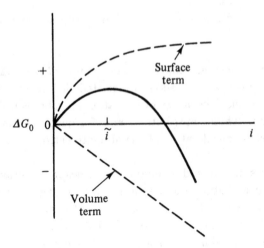

Fig. 8.10. Plot of the bulk free energy change, ΔG_0, associated with embryo formation as a function of embryo size, i.

Turning to the various contributions in Eq. (8.35), we expect that (i) $\Delta G_v \sim 0$ since the force of attraction between embryos will be small for dilute solutions of embryos, (ii) ΔG_t can be neglected due to the very small amount of free volume available for embryos in the monomer solution (the situation is different on a free surface) and (iii) ΔG_r will not be zero but the motion of the embryos will be hindered. For unhindered spherical embryo rotation

$$\Delta G_r^{uh} \simeq -\kappa T \ell n \left[(2\kappa T)^{3/2} (\pi \widetilde{I}^3)^{1/2} / \hbar^3 \right] \qquad (8.37a)$$

and

$$\widetilde{I} = 2mr^3/5 \qquad (8.37b)$$

where m is the mass of the sphere. For Cu, with $i \sim 10^2$–10^3, $\Delta G_r^{uh} \sim -26\kappa T$ which would increase $n^*(i)$ by $\sim 10^{11}$ so it would be a very large effect if totally unhindered. However, viscous hindrance increases strongly with r so the true value of ΔG_r is much less than this.

If there are q energy states available to the molecules of the small system associated with the embryo, the number of complexions of that system is

$$\widetilde{w}_i = q^i/i! \qquad (8.38a)$$

For the larger system containing ni molecules

$$\widetilde{w}_{ni} = (nq)^{ni}/(ni)! \qquad (8.38b)$$

Thus, the entropy of separation per droplet of i molecules is given by

$$\Delta S_{s'} = \frac{\kappa}{n} \ell n \left(\frac{\tilde{w}_i^n}{\tilde{w}_{ni}} \right) = -\frac{\kappa}{2} \ell n(2\pi i) \tag{8.39a}$$

A second contribution, $\Delta S_{s''}$, arises from the requirement of degrees of freedom conservation. Six degrees of freedom are needed for energizing the embryo for the homogeneous nucleation case; thus, six degrees of freedom in the molecules of the embryo must be deactivated. This entropy reduction (increase in $\Delta G_{s'}$) will be given approximately by the molecular entropy, S_s so that

$$S_{s''} \approx S_s \tag{8.39b}$$

and the total contribution is

$$\Delta G_{s'} \approx \frac{\kappa T}{2} \ell n(2\pi i) + T S_s \tag{8.40}$$

For many liquids, $S_L \sim 5\kappa$ and the entropy of fusion $\sim 2.5\kappa$ so that $S_s \sim 2.5\kappa$. Since we find that $i \sim 10^2$–10^3 in practice, $\Delta G_{s'} \sim 6\kappa T$ and the effect is such as to increase $n^*(i)$ by $\sim 10^{-2}$. Since $\Delta G_{s'}$ and ΔG_r are of opposite sign we may expect total cancellation when the hindrance to rotation is just right and a net increase in $n^*(i)$ when negligible hindrance occurs.

Returning to a consideration of Eq. (8.36), variations of $\Delta\mu$ occur with temperature, pressure, electric and magnetic fields, strain, bulk liquid concentration, etc., i.e.,

$$\Delta\mu = \Delta\mu_T + \Delta\mu_P + \Delta\mu_{\hat{E}} + \Delta\mu_{\hat{H}} + \Delta\mu_\sigma + \sum_j \Delta\mu_{C_j} + \cdots \tag{8.41}$$

In addition, variations of ξ (via γ_A) occur with embryo size, electronic equilibrium, interface diffuseness, bulk liquid concentration, shape, etc., i.e.,

$$\gamma_A = \gamma_0(r) + \Delta\gamma_e + \Delta\gamma_t + \sum_j \Delta\gamma_{C_j} + \cdots \tag{8.42}$$

For $\Delta\mu_T$ we have

$$\Delta\mu_T = -\Delta H_F' \left(1 - \frac{T}{T_M} \right) \tag{8.43}$$

For glass-forming systems, $\Delta H_F'$ must be multiplied by T/T_m to correct for the fact that the entropy difference between the supercooled liquid and the solid falls below $\Delta H_F'/T_M$ as T decreases below T_M. Here, $\Delta H_F'$ is the heat of fusion per molecule ($\Delta H_F' = \Omega\Delta H_F$). For $\Delta\mu_P$, application of pressure stabilizes the phase with the smaller specific volume. If the materials contract on freezing (metals), T_M increases with increase of P. If the material expands on freezing (most semiconductors, water, etc.), T_M decreases with increase of P. Thus, we have

$$\Delta \mu_P = (\Omega_S - \Omega_L)\Delta P \tag{8.44}$$

For many systems, over a range of several kilobars pressure change, the melting temperature, $T_M(P)$, changes linearly with pressure such that

$$2 \times 10^{-3} < \frac{[T_M(P) - T_M(1)]}{\Delta P} < 2 \times 10^{-2}\,^\circ\mathrm{C}\ \mathrm{atm}^{-1} \tag{8.45}$$

The use of Eq. (8.44) in nucleation studies would be to produce super-cooling by the rapid application of hydrostatic pressure. However, in volumes larger than $\sim 10\ \mathrm{cm}^3$, apparatus limitations lead to a limit on the practical $[T_M(P) - T_M(1)]$ attainable to less than 10–20 °C for metals and ~ 100 °C for certain semiconductors and polymers.

The magnitude of $\Delta \mu_{\hat{E}}$ due to embryo formation in an electric field is generally small for the usual dielectric constant difference between liquid and solid and leads to $\Delta T_M \lesssim 2$ °C. However, if we are interested in the effect of an ionized center upon the nucleation of a crystal, we need consider only the difference in electrostatic energy stored in a sphere of radius r about the ion when it is surrounded by liquid or solid. This energy difference, ΔE, is given by

$$\Delta E = \frac{Q^2}{8\pi\varepsilon_{*L}r_0}\left(\frac{\varepsilon_{*L}}{\varepsilon_{*S}} - 1\right)\left(1 - \frac{r_0}{r}\right) \tag{8.46}$$

where r_0 is the ion radius, Q is the charge of the ion and ε_* is the dielectric constant. If the ion is singly charged, $r_0 \sim 1.5$ Å and $\varepsilon_{*L}/\varepsilon_{*S} \sim 0.5$, $\Delta E \sim -8 \times 10^{-19}$ J for r/r_0 slightly in excess of 2. This exactly balances the surface creation term for an embryo radius ~ 8 Å or an $i \sim 30 - 50$ atoms. If the ion were doubly charged, the effect would be significantly enhanced. Thus, this contribution can have a significant effect upon nucleation.

Example: $\Delta \mu_P = \Delta \mu_{\hat{E}} = \Delta \mu_{\hat{H}} = \Delta \mu_\sigma = 0$, pure material, γ isotropic

In this case the embryos will be spherical and we will assume that $\Delta G_T \approx \Delta G_0$ so that we can use Eq. (8.36) to find $i = i^*$ which is given from

$$\left(\frac{\partial \Delta G_0}{\partial i}\right)_{i^*} = 0 \tag{8.47}$$

so that

$$i^* = -\left(\frac{2\xi}{3\Delta\mu_T}\right)^3 \tag{8.48a}$$

and

$$\Delta G_0^*(i^*) = \frac{4}{27} \frac{\xi^3}{\Delta \mu_T^2} = \frac{16 \pi \gamma^3}{3 \Delta G_v^2} \qquad (8.48b)$$

where ΔG_v is the volume free energy driving force given by melt:

$$\Delta G_v = -\Delta S_F \Delta T_\infty \qquad (8.48c)$$

solution:

$$\Delta G_v = -\frac{RT}{v_m} \ell n \left(\frac{C_\infty}{C^*} \right) \qquad (8.48d)$$

vapor:

$$\Delta G_v = -\frac{RT}{v_m} \ell n \left(\frac{P_\infty}{P^*} \right) \qquad (8.48e)$$

where the molar volume, v_m, is used to put the driving force on a per cubic centimeter basis and $C^* > C_\infty$ for the solute means supersaturation for the solvent. Thus, the critical size embryo, i^*, varies inversely as the third power of the supercooling ΔT_∞ while the free energy fluctuation to form the critical size embryo varies inversely as the square of ΔT_∞. Since $r \propto i^{1/3}$, we find that the radius of the critical size embryo varies inversely as ΔT_∞.

Defining a_0 as the effective molecular radius ($\Omega = \frac{4}{3}\pi a_0^3$), the interfacial free energy per molecule, γ', is given by $\gamma' = \pi a_0^2 \gamma_A = \xi/4$. Turnbull found from his nucleation studies[4] that, $\gamma'/\Delta H_F' \approx 0.46$ for metals and $\gamma'/\Delta H_F' \approx 0.30$ for semiconductors, water, etc. Thus, if we define in general that $b = 8\gamma'/3\Delta H_F'$, we have (treating ΔT_∞ as a positive quantity)

$$i^* = b^3 (T_M/\Delta T_\infty)^3 \qquad (8.49a)$$

$$\Delta G_0^*/\gamma' = \frac{4}{3} b^2 (T_M/\Delta T_\infty)^2 \qquad (8.49b)$$

and

$$r^*/a_0 = b(T_M/\Delta T_\infty) \qquad (8.49c)$$

For Cu, $b \approx 1.2$ and, for $\Delta T_\infty = 1\,°C$, $10\,°C$ and $10^2\,°C$, $r^*/a_0 \approx 1500$, 150 and 15, respectively, $\Delta G_0^*/\kappa T_M \approx 3 \times 10^6, 3 \times 10^4$ and 3×10^2, respectively, and $i^* \approx 4 \times 10^9, 4 \times 10^6$ and 4×10^3, respectively. These results are shown more fully in Fig. 8.11. We note that $r^* \to \Delta G_0^* \to i^* \to \infty$ as $\Delta T_\infty \to 0$. Another representation of this information is given in Fig. 8.12 for different bath supercoolings ΔT_∞.

Combining the foregoing leads to a critical nucleation frequency for

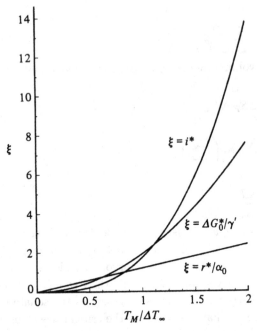

Fig. 8.11. The variation of critical embryo reduced radius, r^*/a, number of molecules in the critical embryo, i^*, and reduced bulk free energy change, $\Delta G_0^*/\gamma'$, for the critical embryo as a function of reduced supercooling, $T_M/\Delta T_\infty$, all for Cu nucleating homogeneously from a pure melt.

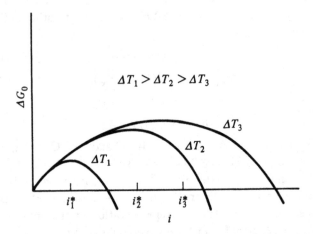

Fig. 8.12. Variation of bulk free energy change for embryo formation, ΔG_0, as a function of embryo size for the different melt undercoolings.

homogeneous nucleation, I^*_{homo}, given from Eq. (8.34) by

$$I^*_{homo} = \frac{n_s \nu}{6} 4\pi r^{*2} \hat{\psi}(t) \tilde{Z} \, n(1) \exp\left(-\frac{\Delta G_A}{\kappa T}\right)$$

$$\times \exp\left(-\frac{\Delta G^*_T}{\kappa T}\right) \tag{8.50a}$$

where n_s is the number of atoms per square centimeter at an embryo surface in the unstable phase. Using Eqs. (8.49), this becomes

$$I^*_{homo} \approx A'_{homo} \exp -\{[\Delta G_A + \frac{4}{3}b^2 \gamma'(T_M/\Delta T_\infty)^2]/\kappa T\} \tag{8.50b}$$

$$= A_{homo} \exp(-B/T\Delta T_\infty^2) \tag{8.50c}$$

where A' is essentially a constant with respect to variations of ΔT_∞ and $A'_{homo} \sim 10^{42}/\Delta T_\infty^2$ per cubic centimeter per second. The critical behavior of I^* enters through the agency of the exponential factor in Eqs. (8.50b). A hundred-fold error in the pre-exponential factor does not appreciably alter I^* in the important range of supercooling. In contrast, those factors entering ΔG^*_T are very important. Plots of I^*_{homo} versus ΔT_∞ are given in Fig. 8.13, where, from (a), we can see that the nucleation rate changes from essentially zero to extremely high values over a very small temperature range ($\Delta T_\infty \sim 5$–10 °C). In Fig. 8.13(c) the scale of ΔT_∞ is very different than in Fig. 8.13(a) and we can see that, as the supercooling becomes very large, the nucleation barrier is no longer limited by the available driving force $\Delta \mu_T$ but by individual molecule transport across the transition state at the embryo surface; i.e., by ΔG_A (leads to glassy or amorphous state). Fig. 8.13(b) illustrates the use of experimental data to obtain the two important grouped material parameters of Eq. (8.50c). When the cooling rate is extremely high, the peak of I^*_{homo} in Fig. 8.13(c) is passed without nucleation occurring so we do not get crystal formation.

8.2.3 Heterogeneous nucleation

A foreign surface will serve as an effective catalyst for nucleation only if it tends to be "wet" by the solid more readily than by the melt (see Fig. 8.8). In such cases, the average surface energy is reduced by solid formation. It is the contact angle, θ, between the spherical cap of a solid embryo and the substrate, as illustrated in Fig. 8.14, that is a measure of the potency of the catalyst substrate for nucleation. The thermodynamic/kinetic treatment of heterogeneous nucleation is exactly the same as for homogeneous nucleation except for a few differences of detail. For simplicity, one generally assumes that γ is isotropic and

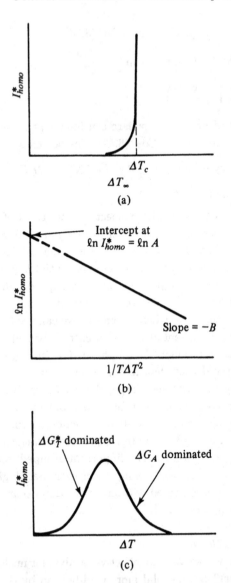

Fig. 8.13. Plots of homogeneous nucleation rate, I^*_{homo}, versus melt supercooling: (a) illustration of the critical undercooling, ΔT_c, for rapid nucleation; (b) proper plot for matching theory and experiment; and (c) illustration of the variation over a very large range of ΔT.

independent of embryo radius so that spherical symmetry develops. The wetting angle, θ, is determined by the condition of minimum free energy of the embryo with respect to constant mass and shape and this becomes

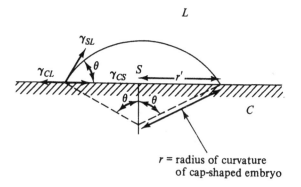

r = radius of curvature
of cap-shaped embryo

Fig. 8.14. Geometrical and surface energy balances for a cap-shaped embryo on a foreign substrate involved in heterogeneous nucleation.

translated into the condition of mechanical equilibrium at the locus of three phase contact; i.e.,

$$\gamma_{CL} = \gamma_{CS} + \gamma_{SL}\cos\theta \tag{8.51}$$

where the subscripts C, L and S refer to catalyst, liquid and solid, respectively. Eq. (8.51) is the well known Young's equation and assumes an inert substrate (non-dissolving). As illustrated in Fig. 8.8, as $\theta \to 0$ we have perfect wetting and as $\theta \to \pi$ we have no wetting (homogeneous nucleation).

In this case, no matter what the embryo geometry, the standard free energy of embryo formation, ΔG_0, is given by

$$\Delta G_0 = V_e \Delta G_v + \gamma_{SL}(A_{SL} - A_{SC}\cos\theta) \tag{8.52}$$

where V_e and A refer to the volume and surface area of the embryo, respectively. In terms of Eq. (8.35), the $\Delta G_{S'}$ contribution will be the same as for the homogeneous nucleation case, $\Delta G_v \sim 0$ but ΔG_r and ΔG_t will not be zero in all cases because, for some systems, the binding to the substrate will be small so that Brownian motion of the embryos can occur on the substrate. However, an additional correction factor, ΔG_d, may arise when considering the number of complexions, $\tilde{w}$, due to distributing the crystalline embryos amongst the p sites on the substrate surface. Taking an embryo concentration of C and invoking Fermi–Dirac statistics, we have

$$\tilde{w} = p!/C!(p-C)! \tag{8.53}$$

and the contribution to the free energy of formation is

$$\Delta G_d = \frac{-\kappa T}{C}\left[p\ell n\left(\frac{p-C}{p}\right) + C\ell n\left(\frac{C}{p-C}\right)\right] \tag{8.54a}$$

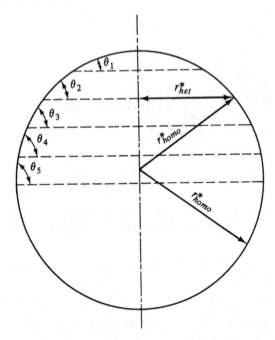

Fig. 8.15. Relationship between the cap-shaped embryo involved in heterogeneous nucleation at wetting angle θ and the spherical embryo involved in homogeneous nucleation ($\theta = \pi$).

In most cases we expect that $p \gg C$ so that

$$\Delta G_d \approx -\kappa T[\ell n(p/C) - 1]; \quad C/p \ll 1 \qquad (8.54b)$$

Since the formation of these different complexions requires translational movement on the substrate, the two contributions may cancel in some cases. For those cases where there is no surface mobility, we must expect $\Delta G_d = 0$.

Neglecting any volume accommodation strains between the embryo and the substrate, we find for the spherical cap-shaped embryo of Fig. 8.14 that

$$\frac{r'^*_{het}}{r^*_{homo}} = \sin \theta \qquad (8.55a)$$

$$\frac{\Delta G^*_{0\ het}}{\Delta G^*_{0\ homo}} = f(\theta) = \frac{(2 + \cos \theta)(1 - \cos \theta)^2}{4} \qquad (8.55b)$$

and

$$\frac{i^*_{het}}{i^*_{homo}} = f(\theta) \sin^3 \theta \qquad (8.55c)$$

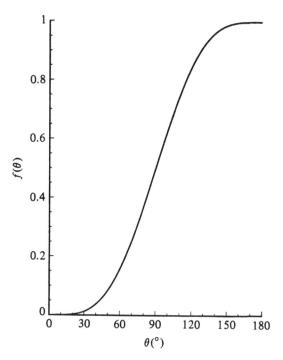

Fig. 8.16. Variation of $f(\theta) = \Delta G^*_{0_{het}}/\Delta G^*_{0_{homo}}$ as a function of wetting angle, θ, between the nucleus and the substrate.

In Fig. 8.15, starting with a sphere of radius r^*_{homo} for the particular undercooling ΔT, the critical embryo for a given θ is represented by the upper cap of the sphere obtained by making a horizontal slice through the sphere at angle θ with the surface tangent at that point. Thus, we readily see how r'^* decreases and i^* decreases as $\theta \to 0$. A plot of $f(\theta)$ is given in Fig. 8.16 and, with it, we can readily relate the enhanced nucleation frequency to θ. We can conclude that all values of $\theta < \pi$ reduce the nucleus size and ΔG^*_0 relative to the homogeneous nucleation limit. As $\theta \to 0$, complete wetting of the substrate occurs and $\Delta G^*_0 \to 0$. However, here the capillary approximation breaks down and we must use a monomolecular high circular disk as the embryo shape since this is the minimum possible thickness of solid. Obviously, as γ_{CS} decreases, θ decreases and heterogeneous nucleation is favored over homogeneous nucleation. In general we find that Eq. (8.50a) holds for I^*_{het} and, more simply,

$$I^*_{het} = A'_{het} \exp -\{[\Delta G_A + \frac{4}{3}b^2\gamma' f(\theta)(T_M/\Delta T_\infty)^2]/\kappa T\} \quad (8.56a)$$

$$= A_{het} \exp[-Bf(\theta)/T\Delta T_\infty^2] \qquad (8.56b)$$

For the freezing of liquids, homogeneous nucleation takes place at $\Delta T_\infty \sim 100\text{--}200$ °C and heterogeneous nucleation at $\Delta T_\infty \sim 10\text{--}20$ °C with the magnitude of $A'_{het} \sim 10^{34} f(\theta)/\Delta T_\infty^2$ per square centimeter per second. The same form for I^* occurs on superheating a phase for the nucleation of the higher temperature phase and again there is a "critical" superheating needed for initiation of the high temperature phase. For the supercooling case (or the superheating case), in a very short temperature range ($\delta T \sim 5\text{--}10$ °C) the nucleation rate changes dramatically. Above this degree of supercooling ($\Delta T_\infty \gtrsim \Delta T_c + \delta T/2$), I^* is too rapid to observe while below this range ($\Delta T_\infty \lesssim \Delta T_c - \delta T/2$), the process goes too slowly for convenient observation. For very small nuclei, such as needed during deposition from vapor beams at very high supersaturation, the capillary approximation cannot be used. Finally, if one is interested in superheating a perfect crystal, internally focused heat is required because, if the free surface temperature at an edge exceeds T_M, heterogeneous nucleation of a critical liquid embryo occurs very easily at the crystal edge.

8.3   Experimental results

When a bulk liquid is uniformly undercooled, the first nucleus to form generally does so on the most effective catalytic singularity in the system. In very fluid liquids, crystal growth is then so rapid that the latent heat of evolution from one center appreciably warms the entire liquid. Thus, subsequent formation of nuclei may not occur at the same supercooling. Because of the severe difficulty of specifying and controlling the presence of nucleation catalysts in a bulk liquid sample, little progress was made until experiments were designed to isolate the centers of heterogeneous nucleation from the majority of the supercooled liquid. Droplet techniques were developed and, since the probability of existence of an accidental singularity should decrease directly with sample volume or sample surface area, it should in principle be possible to subdivide a liquid into droplets so small that most are singularity-free. Secondly, the droplets must be physically separated from each other so that the freezing of one droplet does not initiate nucleation in adjacent droplets.

Due to such an isolation technique, the undercooling required to initiate freezing in the majority of the droplets should greatly increase with decrease in droplet size. Experimental support for this supposition was

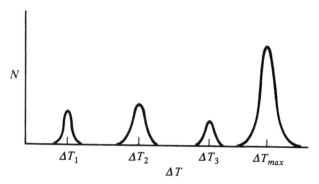

Fig. 8.17. Schematic illustration of number of liquid droplets, N, that were observed to solidify on a quartz wafer as a function of droplet supercooling, ΔT.[5]

found by a number of workers. Further, if the time of crystallization of a droplet is negligible compared to the nucleation incubation period, τ, meaningful data on I^* could be readily determined from counting the rate at which droplets freeze. Finally, the last droplets to crystallize will correspond to θ_{max} and these may represent an example of homogeneous nucleation.

The droplet-types of experiment fall into two classes. One utilizes direct visual observations of droplets placed on a substrate while the other is an indirect observation using either a dilatometric or an X-ray technique. In a direct observation technique, Turnbull and Cech[5] used separated droplets of pure metal, 10–50 μm in diameter, on quartz wafers and observed the supercooling at which the droplets solidified. They used a hot stage technique and determined when solidification had taken place by the sudden change in surface appearance of a droplet (optical reflectivity change). Some droplets solidified before a maximum supercooling was reached (see Fig. 8.17). However, when all the droplets were solid, this maximum supercooling could be reproduced. They felt that these results represented homogeneous nucleation at ΔT_{max} and heterogeneous nucleation at $\Delta T < \Delta T_{max}$. Using Eq. (8.50a), they calculated γ from their data with the results illustrated in Fig. 8.18. They found that the ratio of the molar interfacial tension and the molar heat of fusion was roughly constant for a given type of liquid ($\gamma_g/\Delta H_g \approx 0.5$ for metals and $\gamma_g/\Delta H_g \approx 0.3$ for semimetals like Bi and Sb, semiconductors like Ge and Si, certain organic compounds and water). From Table 8.1, $\Delta T_{max}/T_M$ was found to be greater than 0.15 for most liquid metals.

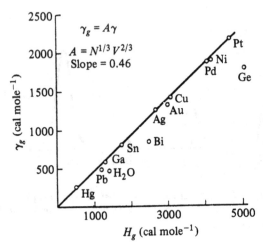

Fig. 8.18. Plot of molar interfacial tension, γ_g, versus the molar heat of fusion, Hg, generated from early nucleation experiments.[5]

A very powerful indirect observation technique separates the droplets from each other by a relatively inactive surface film and places a very large number of coated droplets in a dilatometer with the bulb placed in a constant temperature bath. The rate of freezing of the droplets was observed by the volume change of the fluid in the dilatometer. With this technique, the nucleation of crystals in the supercooled liquid droplets may be studied under isothermal conditions because the rapid crystal growth is arrested at the droplet boundary. Each droplet requires only one nucleus for its solidification and hence measurement of droplet size distribution and fraction frozen as a function of time provide a measure of the nucleation rate.

In the experiment, the temperature of the dilatometer is raised until the droplets are completely melted. The spheres do not coalesce because of the surface film. The dilatometer is then quickly immersed in a thermostat bath at a given temperature below the melting point. The course of the isothermal reaction is followed by observing the drop of the meniscus caused by the solidification shrinkage of the droplets. From the character of the time dependence of the shrinkage, the data can be analyzed to determine if the nucleation events were heterogeneous or homogeneous.

In the experiments of Pound and LaMer,[6] essentially monodisperse droplets of Sn coated with SnO were investigated using oil as the fluid in the dilatometer measurements. Their data is quantitatively represented

Table 8.1.*Relationship between maximum supercooling,
solid–liquid interfacial energy and heat of fusion*

Metal	Interfacial energy γ_{SL} (erg cm^{-2})	γ_g (cal mole^{-1})	$\gamma_g/\Delta H$	ΔT_{max} (°C)
Hg	24.4	296	0.53	77
Ga	55.9	581	0.44	76
Sn	54.5	720	0.42	118
Bi	54.4	825	0.33	90
Pb	33.3	479	0.39	80
Sb	101	1430	0.30	135
Ge	181	2120	0.35	227
Ag	126	1240	0.46	227
Au	132	1320	0.44	230
Cu	177	1360	0.44	236
Mn	206	1660	0.48	308
Ni	255	1860	0.44	319
Co	234	1800	0.49	330
Fe	204	1580	0.45	295
Pd	209	1850	0.45	332
Pt	240	2140	0.45	370

in Fig. 8.19. Here, Fig. 8.19(a) represents the raw isothermal transformation data as a function of time for a certain average droplet size
and various ΔT_∞ while Fig. 8.19(b) represents the change in isothermal
transformation data at a fixed temperature for various average droplet
sizes.

If the nucleation of the monodisperse droplets were of first order,
as would be expected for homogeneous nucleation or for heterogeneous
nucleation with a uniform distribution of a single species of impurity
among the droplets, the process would be described by

$$\frac{\mathrm{d}X_d}{\mathrm{d}t} = -\bar{k}_d X_d \tag{8.57}$$

where X_d is the fraction of diameter d remaining unfrozen at time t and
the isothermal data of Fig. 8.19(a) would be straight lines with a slope

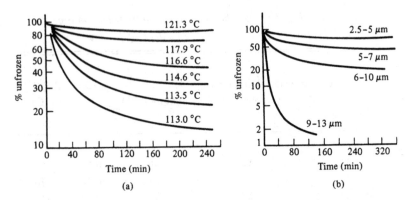

Fig. 8.19. Nucleation data for Sn droplets: (a) % unfrozen versus time for 2.5–5.0 μm size at different undercoolings; and (b) % unfrozen at $T = 117.9$ °C for different droplet sizes.

of $\bar{k}_d$. However, the data are seen to exhibit a marked curvature with decreasing slope. Such data can be fitted fairly well by hypothesizing that heterogeneous nucleation occurred as a result of randomly distributed impurity particles located amongst the droplets according to Poisson's distribution law so that

$$X_d = \exp\left\{-m\left[1 - \exp(-\bar{k}_d t)\right]\right\} \tag{8.58}$$

where m is the average number of foreign nuclei or centers per droplet that are active at the supercooling of interest. Applying Eq. (8.58) to Fig. 8.19(b), one finds that m is more nearly proportional to the droplet surface area than to the droplet volume. This indicates that the droplets were nucleated at their surface so that one should select

$$k_d = A_{het} \exp(-B'/T\Delta T_\infty^2) \tag{8.59a}$$

with

$$A_{het} = a_D \widetilde{A}_{het} \tag{8.59b}$$

where a_D is the droplet area. Evaluating the data according to Fig. 8.13(b) led to good agreement with heterogeneous nucleation theory.

In subsequent experiments of Turnbull,[7] a polydispersity of droplet sizes was used. He made the important assumption that no more than one catalyst type is present in a specimen and that this, if present, is distributed among the droplets in amounts uniformly proportional to droplet area or volume so that the radioactive decay law can be applied to each droplet size class. Once again, Eq. (8.57) applies so that X_d decreases exponentially with time. Defining $X_d^0 = v_d^0/v^0$ where v^0 is the initial volume unfrozen and v_d^0 the initial volume of droplets with diameter d unfrozen and considering all size classes in the specimen

$$X = \frac{1}{v^0} \int_0^\infty v_d^0 \exp(-\bar{k}_d t) \mathrm{d}d \tag{8.60}$$

If heterogeneous nucleation occurs and is due to a surface film on the droplet, the solidification rate is proportional to the droplet surface area and

$$\bar{k}_d = a_D I^* \tag{8.60a}$$

For homogeneous nucleation, the data would be best fitted by

$$\bar{k}_d = v_D I^* \tag{8.60b}$$

where v_D is the volume of the droplet of diameter d. For mercuric acetate coated droplets of Hg, Turnbull found that $\Delta T_{max} = 45\,°\text{C}$, $\bar{k}_d \propto a_D$ and $A'_{exp} \approx 10^{28}$ cm^{-2} s^{-1} ($A'_{theory} = 10^{28}$) lending good support for heterogeneous nucleation from a single type of catalyst. For mercury stearate coated Hg droplets, the data were more complicated and several nucleation catalysts of different potency were needed to fit the data ($\Delta T_{max} < 45\,°\text{C}$). In the case of Hg droplets coated with mercuric laurate, it was found that $\Delta T_{max} \approx 75\,°\text{C}$ and $A'_{exp} = 10^{42}$ cm^{-3} s^{-1} and the isothermal rate data was best fitted by Eq. (8.60b) (but $A'_{theory} \sim 10^{35}$). This was interpreted as an example of homogeneous nucleation. By 1970, the only other well-established example that might be homogeneous nucleation from a liquid was that of water at $\Delta T_{max} = -40\,°\text{C}$.

For some years it was thought that the data of Table 8.1, particularly for Hg and Ga, where dilatometric measurements found the nucleation rate proportional to droplet volume, represented homogeneous nucleation. However, recently, Perepezko and coworkers[8] have shown by the use of more refined techniques that the older limits of undercooling can be greatly extended. Their differential thermal analysis measurements have demonstrated that ΔT_{max} for Aℓ, Bi, Sn, Pb, In, Ga and Hg can be increased to 175, 230, 187, 150, 110, 158 and 90 °C, respectively. Large increases in liquid metal specific heat have also been observed to occur below the melting point.

On another front, Sharaf and Dobbins[9] recently succeeded in measuring the actual nucleation rates in homogeneous nucleation of water droplets from supersaturated vapor in an expansion cloud chamber. They stated Eq. (8.50a) in a more general form

$$I^* = A_0 \psi_*(P, T) \exp[-B_0 \eta_*(P, T)]$$

so that the various forms of Eq. (8.50a), depending upon the choice of replacement partition function, could each be tested by a best fit to

Table 8.2.*A comparison of five nucleation theories with experiments in an expansion cloud chamber*

Theory	A_0^{th}/A_0^{exp}	B_0^{th}/B_0^{exp}
Classical	$2.0 \times 10^{+7}$	1.54
Lothe–Pound	$5.6 \times 10^{+16}$	1.58
Reiss–Katz–Cohen	$1.0 \times 10^{+8}$	1.38
	$5.1 \times 10^{+11}$	1.84
Reiss	$2.3 \times 10^{+7}$	1.56

the data. Here, $\psi_*(P,T)$ is the pre-exponential appropriate for a given theory while $\eta_*(P,T)$ is the exponential characteristic of that theory. A_0 and B_0 represent the factors of deviation of theory from experiment in a test of each of the various theories by a best fit to the data. Their results are shown in Table 8.2 where it is seen that, even for the classical theory, the agreement with experiment appears to be fortuitous.[10] One sees that a theoretical pre-exponential quantity that is too large by a factor of 2×10^7 is almost exactly offset by the negative argument of the exponential that is too high by a factor of 1.54. In this assessment, the planar surface tension of water was used and very different results would be obtained if the surface tension, γ, for the nucleus were a function of embryo size. Accordingly, due to the apparently fortuitous nature of the agreement between measured and calculated critical supersaturations in the homogeneous nucleation of droplets from the vapor, such results cannot be used as strong substantiation of nucleation theory.

8.3.1 *Possible rationalizations for the theory/experiment mismatch*

Present computer simulation-type calculations, by this author and coworkers, of the ledge energy, γ_ℓ, at the free surfaces of Si and GaAs as a function of ledge spacing, λ_ℓ, show that (1) $\langle 211 \rangle$-type ledges on Si(111) interact with each other to spacings larger than 50 Å with $\partial\gamma_\ell/\partial\lambda_\ell < 0$ while (2) $\langle 110 \rangle$ and $\langle 100 \rangle$-type ledges on GaAs($00\bar{1}$) interact with each other to spacings larger than 30 Å with $\partial\gamma_\ell/\partial\lambda_\ell > 0$. In both cases, γ_ℓ changes by factors of 2–5 over this range of λ_ℓ. The quantitative aspects of these results are demonstrated in Fig. 5.7. In addition, a long-range interaction effect is also illustrated for calculated values of interfacial energy, $\gamma_{\alpha\beta}$, for a GaAs/AℓAs superlattice as a function the layer spacing, $\lambda_{\alpha\beta}$. Such changes in γ_ℓ are expected to have large consequences for the magnitude of I^*. The physical phenomenon occurring here is that the surface reconstruction of the crystal face lowers γ_f and produces a strong in-plane compressive stress tensor. The ledge can act like a dilatational stress contribution and alter both the local reconstruction state and the stress state. Since stress is a relatively long-range force, one obtains long-range ledge–ledge interactions. For such systems, one must also be aware of the additional kinetic barriers to surface cluster formation on an already reconstructed surface. In the critical size cluster formation, a patch of surface must unreconstruct while the embryo forms and the new top surface may reconstruct or remain unreconstructed or only partially reconstructed out to some larger size. In addition, at the expanding edge of the disk, unreconstruction of the old surface must occur while reconstruction of the new surface layer must also occur simultaneously.

To evaluate the simple two-dimensional pill-box model of embryo formation on a surface for materials where $\partial\bar{\gamma}_\ell/\partial r \neq 0$, it is perhaps best to assume a general functional form for $\bar{\gamma}_\ell(r)$ where r refers to the pill-box radius. Let it be

$$\bar{\gamma}_\ell(r) = \bar{\gamma}_0 + \Delta\bar{\gamma}[1 - \exp(-br)] \tag{8.61a}$$

which leads to

$$\partial\bar{\gamma}/\partial r = b\Delta\bar{\gamma}\exp(-br) \tag{8.61b}$$

Defining $x = br$, $j_1 = b\bar{\gamma}_0/\Delta G_v$ and $j_2 = \Delta\bar{\gamma}/\bar{\gamma}_0$, the critical nucleus formation condition is given by

$$j_1 = x^*/[(1 + j_2) - j_2(1 - x^*)\exp(-x^*)] \tag{8.61c}$$

and

$$\frac{\Delta G^*(j_2)}{\Delta G^*(0)} = (1 - j_2^2) + 2j_2^2(1 - x^*)\exp(-x^*) - j_2^2(1 - x^*)\exp(-2x^*) \tag{8.61d}$$

The first important thing to notice about this result is that $\Delta G^*(j_2)/\Delta G^*(0) < 1$ independent of the sign of j_2. Under conditions where $x^* \ll 1$, one finds that

$$\frac{\Delta G^*(j_2)}{\Delta G^*(0)} \approx 1 - 5j_2^2 x^* \tag{8.61e}$$

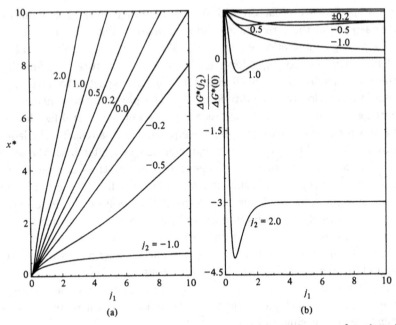

Fig. 8.20. Dimensionless nucleation parameter plots as a function of the dimensionless surface to volume energy parameter, $j_1 = b\hat{\gamma}_0/\Delta G_v$ for a range of the dimensionless ledge energy parameter, $j_2 = \Delta\hat{\gamma}/\hat{\gamma}_0$: (a) critical radius, $x^* = br^*$; and (b) critical fluctuation parameter, $\Delta G^*(j_2)/\Delta G^*(0)$.

and

$$x^* \approx j_1/(1 - 2j_1 j_2) \qquad (8.61f)$$

This condition occurs only for large values of $|j_2|$. From Eq. (8.61f), one sees that $x^*(j_2)/x^*(0) > 1$ for $j_2 > 0$ and < 1 for $j_2 < 0$. This general trend is expected at larger values of x^*. Thus, for Si(111), $j_2 < 0$ so both x^* and ΔG^* are decreased. For GaAs$(00\bar{1})$, $j_2 > 0$ so x^* is increased while ΔG^* is decreased. In Figs. 8.20(a) and (b), x^* and $\Delta G^*(j_2)/\Delta G^*(0)$ are plotted as a function of j_1 for a range of values for j_2. It is interesting to note that a triangle-shaped embryo on Si(111) with $[2\bar{1}\bar{1}]$ and $[\bar{2}11]$ ledges has $j_2 \approx -1$ and $b^{-1} \approx 50$ Å so that $\Delta G^* \to 0$ at $j_{1c} \approx 0.7$.

The importance of this result to nucleation and crystal growth theory is profound! Not only could $\Delta G^* \to 0$ for certain pure materials under some special conditions of surface reconstruction/stress tensor development, but heterogeneous nucleation of two-dimensional pill-boxes on foreign substrates is expected to exhibit similar behavior.

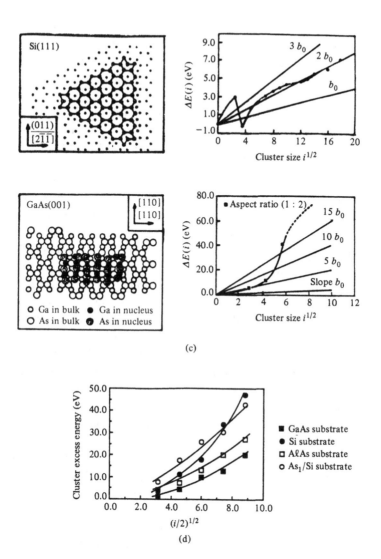

Fig. 8.20. (c) Calculated excess energy, $\Delta E(i)$ for clusters containing i atoms on the Si(111) and GaAs(001) planes as a function of cluster perimeter parameter $i^{1/2}$ and (d) $\Delta E(i)$ for GaAs clusters on four different substrates ((1:2) aspect ratio, monolayer high, pyramidal ledge).

Using semiempirical potential energy functions, cluster formation for Si on Si(111) and GaAs on GaAs(00$\bar{1}$) has been simulated and the excess energy associated with cluster formation, $\Delta E(i)$, as a function of number of atoms in the cluster, i, determined. These results for the homoepitaxy case are shown in Fig. 8.20(c) where $\Delta E(i)$ is plotted versus $i^{1/2}$. For the classical dangling bond model, one expects the excess energy to come only from the ledge which is of constant energy so that $\Delta E(i)$ versus $i^{1/2}$ should yield a straight line whose slope depends upon the geometry of the cluster and the ledge energy. The calculations, which allow surface and ledge reconstruction to occur, show a very non-linear behavior at small i although the results should asymptotically approach a straight line at sufficiently large i. For the GaAs case, $\Delta E(i)$ differs by at least 5 eV from this asymptotic slope so that, at crystal growth temperatures in the 500–600 °C range, one expects the cluster population to differ from the dangling bond model expectations by $\sim 10^{10}$. In Fig. 8.20(d), such calculations have been carried out for GaAs cluster formation (pyramidal shape with 1:2 aspect ratio) on a variety of different substrates and we see that the same type of non-linear behavior with $i^{1/2}$ occurs. However, this is not so surprising in this case since $\Delta E(i)$ contains both a ledge energy term and a face energy term and even from the classical dangling bond picture, $\Delta E(i)$ should have a term in $i^{1/2}$ plus a term in $(i^{1/2})^2$.

The origin of a second possible source of error in homogeneous nucleation theory comes from the neglect of cluster–cluster interactions. This arises because we have assumed a dilute solution theory. However, suppose we consider a relevant example; i.e., the case of colloid stability. From this area of understanding, we learn that the small particles are attracted to each other via their van der Waals dispersion forces acting through the intervening medium and repelled from each other by electrostatic forces due to the presence of an electrical double layer at their surfaces. This leads to a potential energy, $\hat{U}$, versus particle separation curve which is often like that shown in Fig. 8.21. In particular, we see that such colloids can be metastably bound in a secondary potential minimum that may be $\sim 5\kappa T$–$15\kappa T$ deep.

Applying the insights of Fig. 8.21 to our cluster formation situation we can see that, for a solution containing electrolyte or small dipoles, clusters may form cooperatively at some range of distance from each other such that they remain stably dispersed via an electrostatic repulsion force and stably attracted via their dispersion force such as to move in a potential minimum relative to each other. The loss of entropy due to this configurational cluster arrangement should not be too large so that, for such systems, this effect might increase the equilibrium cluster population by factors of 10^4–10^{10}.

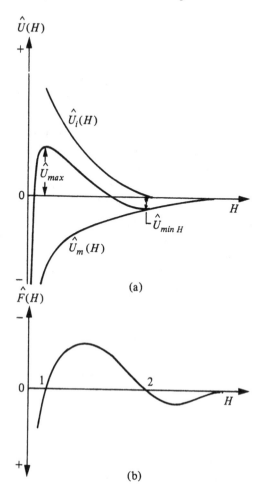

Fig. 8.21. (a) Interaction energy, $U(H)$; and (b) resultant force, $F(H)$, between two electrically charged particles as a function of separation distance, H, between the particles.

8.4 The γ_e contribution to heterogeneous nucleation

In addition to the interesting surface reconstruction/stress tensor related effects on heterogeneous nucleation discussed in the last section, the electrostatic contribution to the interfacial energy (see Chapter 7, Eqs. (7.27)–(7.30)) is expected to have a significant effect on decreasing the wetting angle θ for heterogeneous nucleation.

Turnbull and Vonnegut[11] developed a crystallographic theory of nucleation catalysis which predicted that the order of the catalytic potency between different catalysts should be identical with the order of the reciprocal of the disregistry between the catalyst and the forming crystal on low index planes of similar atomic arrangement. Experimental data for the nucleation of NH_4 on mica and of ice on AgI and quartz tended to support these theoretical expectations. However, additional experimental data have since appeared to show the incompleteness of the simple disregistry viewpoint. It was found that the compounds WC, ZrC, TiC and TiN were much more potent catalysts for the nucleation of Au droplets than a variety of oxides even though the disregistries were similar. Even among these intermetallic compounds a dominant disregistry effect did not appear. Additional studies showed that the nucleation of bulk Sn samples occurred more readily on metallic substrates than on non-metallic substrates with little relationship to substrate disregistry effects (γ_d effect from Chapter 7, Eqs. (7.44)–(7.55)).

Sundquist and Mondolfo[12] considered the ΔT_c needed for the nucleation of the second phase by the primary phase in 60 binary eutectic alloys with startling qualitative conclusions. The most significant finding of the study was that, in a system of α and β solid phases, α nucleates β at low undercoolings ($\Delta T_c \lesssim 10 \,°C$) whereas β does not nucleate α except at very high undercoolings ($\Delta T_c \sim 100 \,°C$). They also found that the metals investigated could be placed in a series of the form I(Tℓ, Pb):II(Ag{Au, Cu, Ni, Co, Fe}) :III(Ge, Sn, Zn{Bi, Sb}). In this series, any element of Group II can readily nucleate any element of Group I or I–II compounds and has no observable nucleating effect upon any element of Group III or II–III compounds. Within any group, the exact location of an element was not so well established.

From Eqs. (7.10), one recognizes that the interfacial energy, γ, has several component contributions which may be the dominant factor in heterogeneous catalysis in a particular instance and all must be considered in the general case. Tiller and Takahashi[13] took this viewpoint and attempted to evaluate the importance of the electrostatic contribution, γ_e, to the wetting angle. They wished to see how the γ_e values between catalyst substrate ($\bar{c}$) and liquid (L) and the catalyst and solid (S) might be arranged to favor embryo formation; i.e., $\theta \to 0$.

By minimizing the free energy of the embryo in Fig. 8.14(a), one is led to the following relationship

$$\cos\theta + \frac{\gamma_{\bar{c}L} - \gamma_{\bar{c}S}}{\gamma_{SL}} = \frac{\{(\gamma_c + \gamma_d)_{\bar{c}L} - (\gamma_c + \gamma_d)_{\bar{c}S}\}}{\gamma_{SL}} + \frac{\delta\gamma_e}{\gamma_{SL}} \quad (8.62)$$

where $\delta\gamma_e = \gamma_{e\bar{c}L} - \gamma_{e\bar{c}S}$. For a pure material, one might expect $\gamma_{c\bar{c}L} \approx \gamma_{c\bar{c}S}$ and $\gamma_{d\bar{c}L} \sim 0$ so that, for zero disregistry between catalyst substrate and solid, the first term on the RHS of Eq. (8.62) is close to zero. As the disregistry increases to about 15%, the energy of the dislocation array at the interface corresponds to that of a low angle grain boundary of about $10°$. Such a boundary has an energy about one-half that of a high angle grain boundary or about equal to γ_{SL}. Thus, as the disregistry increases from 0 to ~ 0.15, the first term on the RHS of Eq. (8.62) decreases from 0 to ~ -1 and, if $\delta\gamma_e \sim 0$, this leads to $\pi/2 \lesssim \theta \lesssim \pi$; i.e., to poor wetting. From this, one can see that $\delta\gamma_e/\gamma_{SL} \gtrsim 2$ is needed for a substrate to lead to a very low value of θ and thus become a good nucleation catalyst. It is also apparent that a necessary, but not sufficient, condition for the substrate to be an effective nucleation catalyst is that $\delta\gamma_e > 0$.

Defining the capacitance, $\widetilde{C}_{ij}$, of the heterogeneous capacitor at the ij interface (see Fig. 8.22(a)) as

$$\widetilde{C}_{ij} = \frac{1}{4\pi} \left(\frac{1}{d_i/\varepsilon_{*i} + d_j/\varepsilon_{*j}} \right) \tag{8.63}$$

an expression for $\delta\gamma_e$ can be written as

$$\delta\gamma_e = \frac{1}{2} \left(\widetilde{C}_{\bar{c}S} \Delta\Phi_{\bar{c}S}^2 - \widetilde{C}_{\bar{c}L} \Delta\Phi_{\bar{c}L}^2 \right) \tag{8.64a}$$

where $\Delta\Phi_{ij}$ is the macropotential difference between the bulk phases i and j and where all size effects of the substrate and embryo have been neglected. If one assumes further that $\widetilde{C}_{ij}$ is largely determined by the properties of the substrate so that $\widetilde{C}_{\bar{c}S} \approx \widetilde{C}_{\bar{c}L} \approx \widetilde{C}$, one has

$$\delta\gamma_e \approx \frac{1}{2}\widetilde{C}(\Delta\Phi_{\bar{c}S}^2 - \Delta\Phi_{\bar{c}L}^2) \tag{8.64b}$$

$$\approx \frac{1}{2}\widetilde{C}[2\Phi_{\bar{c}} - (\Phi_S + \Phi_L)](\Phi_L - \Phi_S) \tag{8.64c}$$

Using free electron theory, it can be shown that for metals[13]

$$e\Delta\Phi_{LS} \approx E_{\bar{F}}(S) - E_{\bar{F}}(L) \approx \frac{2}{3}\frac{\Delta v}{v} E_{\bar{F}}(S) \tag{8.64d}$$

where Δv is the volume change upon melting, v is the molar volume of the solid and $E_{\bar{F}}$ is the Fermi energy for the electrons. From this, one expects that $\Delta\Phi_{LS} = \Phi_L - \Phi_S > 0$ for close-packed or BCC metals which expand upon melting and < 0 for those open structures which contract upon melting; e.g., Bi and Ga. Thus, for the case of $\Delta\Phi_{LS} > 0$, a good nucleation catalyst must have a macropotential (in the contact situation) which is greater than the average of that for the solid and liquid. The reverse applies for $\Delta\Phi_{LS} < 0$. Further, in order that the magnitude of $\delta\gamma_e$ be large, $\widetilde{C}$ must be large; i.e., the substrate must approach metallic-like conduction characteristics with respect to d/ε_*. Finally, one notes that if a catalyst is a good nucleating agent for the solid ($\Phi_{\bar{c}} > \Phi_S$, say

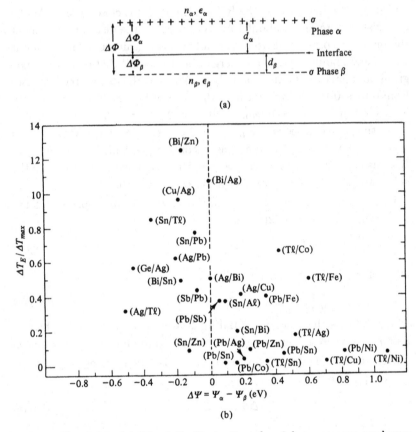

Fig. 8.22. (a) Schematic illustration of an inhomogeneous condenser model for a two-phase interface; and (b) plot of normalized supercooling, $\Delta T_E/\Delta T_{max}$, versus work function difference, $\Delta\Psi$, for several α/β eutectics. The parenthesis (β/α) means that the β phase is nucleated by the α phase.

and $\delta\gamma_e > 0$), then the solid is a poor nucleating agent for the catalyst since $\delta\gamma_e$ changes sign between these two situations.

One cannot actually measure Φ for a phase and there are few reliable measures of $\Delta\Phi_{ij}$ between phases; however, a crude approximation may be obtained by substituting work function differences, $\Delta\Psi_{ij}$ for $e\Delta\Phi_{ij}$. Work function data for the Sundquist and Mondolfo nucleation series are given in Table 8.3. In Fig. 8.22(b), the $\Delta T_E/\Delta T_{max}$ data from Sundquist and Mondolfo[12] have been plotted versus $\Delta\Psi = \Psi_\alpha - \Psi_\beta$ where ΔT_E is the supercooling below the eutectic temperature for nucleation, ΔT_{max} is the maximum supercooling of the major constituent in the nucleating

Table 8.3. *Sundquist and Mondolfo's nucleation series*

Group	Element	m.p. (K)	ΔS_f (cal mole^{-1} °C)	γ_{SL} (cal mole^{-1})	ψ (eV)
I	Tℓ	577	1.78	512	3.70
	Pb	600.6	1.90	479	4.00
II	Ag	1234	2.31	1240	4.30
	Au	1336	2.21	1320	4.30
	Cu	1357	2.30	1360	4.40
	Ni	1725	2.40	1860	4.50
	Co	1768	2.32	1800	4.41
	Fe	1812	2.02	1580	4.31
III	Ge	1210.4	6.28	2120	4.76
	Sn	505	3.40	720	4.38
	Zn	692.7	2.55	870	4.24
	Bi	544.5	4.78	825	4.40
	Sb	903	5.28	1430	4.08

solid, Ψ_α is the work function of the major constituent in the primary phase while Ψ_β is the work function of the major constituent in the nucleated solid. It should be noted that, when the α phase nucleates the β phase; i.e., $\Delta T_E / \Delta T_{max} < 1, \Delta \Psi > 0$. When the α phase does not nucleate the β phase, $\Delta \Psi < 0$. This striking contrast is exactly in agreement with the predictions of Eq. (8.64c).

To be perfectly rigorous, the work function comparison should be made with the respective alloy phases rather than with those of the major constituents of the phases. Unfortunately, there is almost a complete lack of information on the work function change with alloying. However, since the work function is roughly proportional to the number of valence electrons per atom, one should expect a linear variation in Ψ with alloying provided that no molecular complex formation occurs in the solid. In any event, the sign of $\Delta \Psi$ will not change with alloying and the foregoing conclusions still hold.

In this section, it has been shown that the contribution of γ_e to γ is negative and large when two good conductors are in contact, i.e., when

$|\gamma_e|$ (calculated) is large, γ_{SL} (experimental) is small.[13] In addition, it has been shown[13] that the adhesion between a non-conducting film and a conductor is small if both are thick relative to the Debye lengths but may be large if the non-conducting film thickness is a small fraction of the Debye length of the space charge in the non-conductor. Such a non-conducting film may be a good nucleation catalyst for solid metal formation provided it is sufficiently thin and is immersed in the metal. This same type of size effect is expected to enter homogeneous nucleation considerations making γ a function of embryo radius for small embryos.

For metals which contract upon freezing, $\Delta\Phi_{LS} > 0$ and a good nucleation catalyst must have a macropotential which is greater than that of the average for the solid and liquid values. For metals that expand upon freezing, $\Delta\Phi_{LS} < 0$ and the reverse situation applies for a good nucleation catalyst. It appears that the work function differences, $\Delta\Psi/e$, can be used as a reasonable approximation to macropotential, $\Delta\Phi$, differences. Finally, one finds that, when γ_e is a significant portion of γ, if α nucleates β ($\delta\gamma_e > 0$) then β will not nucleate α ($\delta\gamma_e < 0$). Clearly, the electrostatic contribution to γ is an important factor in the selection of an effective nucleation catalyst.

As a concluding segment to this section, although it is more γ_c-related than γ_e-related, it is useful to consider a simple computation performed to compare the effect of atomic disregistry at an interface with altered atomic potential interactions on stresses developed in the interface region. A detailed picture of film formation requires a knowledge of the map of stored strain energy in the substrate/film combination. Explicit analytical forms can be calculated for the strain derivatives near the interface for both two-body and three-body PEFs. Following this procedure,[14] it was found that the entire variation from one stress component or elastic constant to another, as one crosses the interface, is determined by the general symmetry of the two-particle and three-particle distribution functions so that the net result can be separated into a geometrical combination factor and a material combination factor. Thus, it seemed possible that a certain disregistry effect could be completely offset by the proper PEF effect or a certain PEF effect without disregistry could give the same net stress effect as obtained with a different PEF but with disregistry.

To test this general result, the adlayer–substrate interaction effects were investigated by using a system of 1000 atoms, using only two-body forces, for an Ag(111) substrate (~ 600 atoms) with an adlayer, X, of ~ 400 atoms where X = Cu(111) or Au(111). The bottom three layers of the six substrate layers were held fixed and all the quantities needed to evaluate the stress effects were calculated considering only the "core" region of the hexagonal (111) slab (in order to minimize edge effects).

For the Cu–Ag system, in spite of the 11.5% mismatch between the bulk phases, the interface is found to be coherent with increased compressive forces in the substrate parallel to the interface while the adlayer itself is in a tensile state. For σ_{xx} (lying in the interface), only the top Cu layer exhibits a stress dipole behavior while, for σ_{zz} (perpendicular to the interface), a much larger dipole effect exists in the first layer with a smaller dipole effect existing for the lower layers while, for σ_{xy} (in-plane shear stress), no stress dipole effect was present at all. For the Au–Ag system, although the atomic mismatch is only 0.1%, a very strong stress effect was observed probably because the Au–Au interaction energy is very different than the Au–Ag interaction energy. The net result was very similar to that found for the Cu–Ag system and illustrates that the cross-interaction parameters, ε_{X-Ag}, are an important quantity to consider when evaluating stress effects in the absence of atomic mismatch.

8.4.1   *Athermal nucleation*

The earlier theoretical considerations applied to the case where a sample of material is cooled from a temperature where the liquid phase is stable to a temperature where a solid phase is stable. If the temperature change is rapid, the initial distribution of embryos at the new temperature will not differ appreciably from that corresponding to the old temperature. There will not have been sufficient time for them to shrink or grow to their steady state concentration. To illustrate how the transient effects change with conditions, consider the following two cases: (1) The specimen is quenched from a temperature T_1 above the melting temperature, T_M, to a temperature T_{-1} slightly below T_M. If the cooling rate is sufficiently rapid, the high temperature equilibrium distribution of embryos will be retained during quenching. Fig. 8.23(a) indicates, however, that in this distribution there are no embryos exceeding the critical size. But when steady state conditions are reached at the lower temperature, a substantial flux of embryos will continue to grow to critical size. Fig. 8.23(b) illustrates the curve of nucleation frequency versus time for this heat treatment where we see that the time, τ, for the retained embryos to reach critical size by the chain of bimolecular reactions (Eq. (8.26)) appears as an incubation period. (2) For the second case, the specimen is quenched from T_1 to a temperature T_{-2} considerably below T_M. Again the high temperature distribution

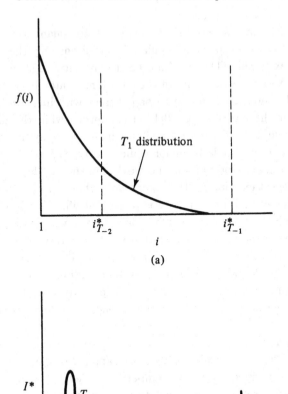

Fig. 8.23. (a) Embryo distribution, $f(i)$, versus size, i, at temperature T_1, with critical embryo size, i^*, for temperatures T_{-1} and T_{-2}; and (b) corresponding nucleation frequency, I^*, versus time, t, at temperature T_{-1} and T_{-2}.

of embryos is retained during quenching. Fig. 8.23(a) indicates that, for this case, there are a large number of embryos in the T_1 distribution which exceed the critical value of $i \approx i^*_{T_{-2}}$. These embryos are thus im-

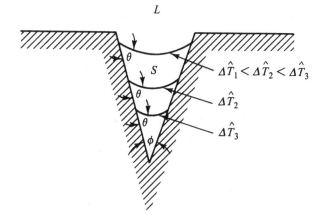

Fig. 8.24 Illustration of heterogeneous nucleation in a pore or crevice of a foreign surface.

mediately promoted to nuclei and grow as independent regions of solid. Similarly, embryos from the T_1 distribution of size i less than i^* for the T_{-2} distribution are present in smaller amount than their equilibrium concentration and grow in number towards the T_{-2} steady state distribution. The dependence of I^* for this case is also shown in Fig. 8.23(b) with no incubation period. A heterogeneous nucleation mechanism with a limited supply of catalyst substrate particles was assumed so that I^* reaches a maximum and decays as the supply is exhausted.

The nucleation mechanism whereby embryos grow and become nuclei through the action of thermal fluctuations has been termed *thermal nucleation*. The second nucleation mechanism, whereby embryos are automatically promoted to nuclei as the peak of the free energy of formation curve, ΔG_T^*, moves past them to smaller values of i, is called *athermal nucleation*.

8.4.2 Substrate geometry effect for heterogeneous nucleation

For a substrate material that is favorably wet by the solid, $0 < \theta < \pi/2$, the substrate prefers to be coated by solid than by liquid because the free energy of the system is lowered in this case if we are at or below the wetting temperature of the solid. Thus, if a groove, pore or crevice exists in the substrate such as illustrated in Fig. 8.24, a greater surface to volume ratio for the embryo exists inside the groove, pore or crevice than on the flat making the solid embryo more thermodynamically stable than on the flat. In fact, solid embryos are stable inside such favorable geometrical sites at temperatures above T_M and can exist at quite high superheating, $\Delta \hat{T}$. In this case it is the volume free energy that destabilizes the embryo while the surface free energy is the stabilizing factor. As we increase $\Delta \hat{T}$, the solid embryo retreats into the crevice because it requires an increasingly higher surface to volume ratio to remain thermodynamically stable. Above some high superheat level, $\Delta \hat{T}_c$ the embryo is completely melted out of the pore.

If a large melt is superheated an amount $\Delta \hat{T} < \Delta \hat{T}_c$ then, on cooling to the supercooled state, the bulk melt will always nucleate at some critical supercooling, ΔT_c. This occurs because, at this supercooling, the embryo can slowly expand out of its pore or crevice onto the flat catalyst surface and then expand rapidly into the melt. At supercoolings $\Delta T_\infty < \Delta T_c$, the embryo may expand out towards the mouth of the pore or crevice but it cannot expand beyond this. If the superheating is to a value of $\Delta \hat{T} > \Delta \hat{T}_c$, then the initial solid embryo seed is gone and its reappearance at smaller superheatings or even supercoolings is a small probability event (because the bulk melt contains so very few pores or crevices with these favorable characteristics). This type of behavior for bulk melts is well documented in the foundry literature.

8.4.3   Nucleation from binary solutions

The general nucleation frequency expression can always be written in the form

$$I^* = \tilde{Z}\,\beta_* A(i^*)n(1)\exp(-\Delta G_T^*/\kappa T)\exp(-\tau/t) \qquad (8.65)$$

in which β_* is the impingement flux of monomer, $A(i^*)$ is the surface area of the critical nucleus, $n(1)$ is the surface layer monomer concentration, $\tilde{Z}$ is the non-equilibrium factor while τ is the induction period. This general expression applies for both homogeneous and heterogeneous nucleation. As we turn from considerations of a unary melt to a binary solution, we need to consider the changes that will occur in the various parameters of Eq. (8.65).

The key ideas relative to the volume free energy change, ΔG_v, have been discussed in Chapter 6 (see Fig. 6.2 and Eq. (6.4)). If we consider a binary alloy melt at low solute content, then Fig 1.5 applies and Eq. (8.50a) still applies. On the other hand, for the solution growth case, Fig 1.8(b) is the appropriate diagram and the volume free energy driving force per cubic centimeter, ΔG_v, is

$$\Delta G_v = \frac{RT}{v_m}\ell n\left(\frac{C_\infty}{C^*}\right) \qquad (8.48d)$$

where C^* is the liquidus concentration for the temperature T_∞. In this case, we shall express ΔG_0^* and r^* by

$$\Delta G_0^* = 16\pi\gamma^3/3\Delta G_v^2 \qquad (8.66a)$$

and

$$r^* = -2\gamma/\Delta G_v = a_0 i^{*1/3} \qquad (8.66b)$$

while

$$\tau = 3\pi\kappa T i^{*2}/\beta_* A(i^*)\Delta G_0^* \qquad (8.66c)$$

and

$$\tilde{Z} = (\Delta G_0^*/3\pi\kappa T i^{*2})^{1/2} \qquad (8.66d)$$

Thus, for comparable driving force, surface energy, etc., no significant change occurs in these important parameters when we consider either a unary liquid or a binary solution at the same temperature. Since many solution-grown crystals form at very low temperatures, this is a factor to be considered. The remaining item in Eq. (8.65) to be considered is the surface monomer concentration, $n(1)/h_a = C_i$, which will be different than the bulk concentration, C_∞, because of the interface field effects at the embryo and substrate surfaces (h_a = adlayer thickness). These interface field effects have been discussed in Chapter 7 and Eqs. (7.18) give the ratio C_i/C_∞ for some general interface field. Thus, in Eq. (8.65) we must use

$$n(1) = C_i h_a \qquad (8.67)$$

Because of interface segregation effects at the catalyst or embryo surfaces, C_i/C_∞ may be very different from unity so that, not only will γ and ΔH_F be affected but, for the case where $C_i/C_\infty \ll 1$, $n(1)$ can be greatly depleted and A' reduced by many orders of magnitude.

8.4.4 Nucleation from the vapor phase

In this case, Eq. (8.65) still holds but with a few changes in the parameters. Instead of β_* being given by

$$\beta_* = \frac{\nu}{6a_0^2} \exp\left(-\frac{\Delta G_A}{\kappa T}\right) \qquad (8.68a)$$

it becomes

$$\beta_* = P/(2\pi m\kappa T)^{1/2} \qquad (8.68b)$$

for embryo growth via direct impingement from the vapor where P is the partial pressure of the concerned constituent. For embryo growth via atom diffusion from an absorbed layer, Eq. (8.68a) must be used. In general, one should include a sticking coefficient, S_*, multiplying P

in Eq. (8.68*b*) ($S_* < 1$). Provided the capillary approximation holds, Eqs. (8.66) hold for this case. To evaluate $n(1)$, we need to consider the mean residence time, τ_s, of an adsorbed molecule on the surface before desorption (see Eq. (8.22*b*)). If atoms are incident on the surface at rate β_* per unit area given by Eq. (8.68*b*) and desorption is the only process by which monomer is removed, the equilibrium or steady state monomer density is

$$n(1) = \beta_* \tau_s \qquad (8.69)$$

so that $n(1)$ can be very small in some cases. Provided Eq. (8.29*b*) holds, we need not be concerned about monomer removal via embryo formation depleting $n(1)$.

8.5   Bulk crystallization kinetics from a unary system

Before leaving this chapter, it is well to place the nucleation and growth events in a practical perspective insofar as they both contribute to the bulk transformation of a nutrient phase. For simplicity, we shall consider a unary melt.

If one crystal is growing in a supercooled melt, then the amount of converted material at any time is related to the dimensions of the crystal. If we consider several crystals growing from various nucleation centers, the extent of conversion to the crystallized state is a combined function of the crystal growth rate and the nucleation density. We wish to consider the fraction X transformed of the initial volume v_0 as a function of time.

Case 1 The assumed conditions will be (i) constant nucleation frequency, I^*, uniformly located as in homogeneous nucleation, (ii) constant linear growth rate V, (iii) spherical crystal symmetry and (iv) that the growing crystals don't impinge on each other. The number of nuclei formed between τ and $\tau + d\tau$ is $I^* v_0 d\tau$ and, at time t, the volume transformed due to these nuclei will be given by

$$\frac{dv}{v_0} = -\frac{4}{3}\pi V^3 (t - \tau)^3 I^* d\tau \qquad (8.70)$$

Since $dX/(1 - X) = dv/v$, we have

$$\int_0^X \frac{dX'}{1 - X'} = -\int_0^t \frac{4}{3}\pi V^3 (t - \tau)^3 I^* d\tau \qquad (8.71)$$

which leads to

$$X(t) = 1 - \exp\left(-\frac{\pi}{3} V^3 I^* t^4\right) \qquad (8.72)$$

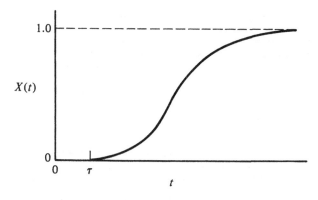

Fig. 8.25. Typical S-shaped curve giving the fraction, X, of the initial volume that is transformed at time t after quenching at $t = 0$.

which is a typical S-shaped curve as illustrated in Fig 8.25. For the general case of different crystal shapes, different nucleation conditions or different growth rate time dependences,

$$X(t) = 1 - \exp(-Bt^n) \qquad (8.73)$$

where B and n depend upon the particular system. This is called the Avrami (or Johnson–Mehl or Schmid–Boas) equation. In Table 8.4, the exponent, n, is listed for various types of nucleation or growth. Half-power exponents will occur when $V \propto t^{-1/2}$.

Case 2 Assume $V = $ constant but that we have $\bar{N}_0$ uniformly distributed heterogeneous nucleation sites. Then,

$$\frac{d\bar{N}}{d\tau} = -I^* A_{het} \bar{N} = -\nu^* \bar{N} \qquad (8.74)$$

where A_{het} is the surface area of this substrate material and $\bar{N}$ is the number of nucleation sites. This leads to

$$\bar{N} = \bar{N}_0 \exp(-\nu^* \tau) \qquad (8.75)$$

and we must replace I^* in Eq. (8.70) by $d\bar{N}/d\tau$. This leads to two limiting situations: (i) if ν^* is extremely small, then Eq. (8.73) holds but with $n = 4$ and $B = (\pi/3)V^3\nu^*\bar{N}_)$ and (ii) if ν^* is very large and all nuclei form very close to $t = 0$, $\bar{N}_{sites} \rightarrow \bar{N}_{nuclei}$ so $n = 3$ and $B = (4\pi/3)V^3\bar{N}_0$.

8.6   Spinodal decomposition

The assumption of homogeneous phases was removed by Cahn

Table 8.4. *Time exponent, n, for various nucleation and growth situations*

n	Conditions
$3 + 1 = 4$	Constant V, spherical growth from sporadic nuclei ($I^* \propto t'$)
$3 + 0 = 3$	Constant V, spherical growth from instantaneous nuclei
$2 + 1 = 3$	Constant V, disk-like growth from sporadic nuclei
$2 + 0 = 2$	Constant V, disk-like growth from instantaneous nuclei
$1 + 1 = 2$	Constant V, rod-like growth from sporadic nuclei
$1 + 0 = 1$	Constant V, rod-like growth from instantaneous nuclei

and Hilliard.[15] They obtained an expression for the work required to form a critical nucleus in an initially homogeneous solution at constant v and P of the form

$$W = \int_v [\Delta f + \kappa_2 (\nabla C)^2] dv \tag{8.76}$$

where Δf is the Helmholtz free energy density and κ_2 is a coefficient utilized in Eqs. (8.70)–(8.76). The gradient energy term can be made arbitrarily small by making the interface more diffuse and hence decreasing the gradient. However, this can only be accomplished by making more material of non-equilibrium composition which increases Δf. The critical nucleus characterized by the least work of formation for the given conditions reflects a balance of these two influences.

At low supersaturations (near C_α in Fig. 7.20), the critical nucleus

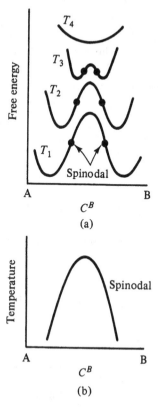

Fig. 8.26. (a) Schematic plots of free energy versus composition for a series of temperatures ($T_4 > T_3 > T_2 > T_1$) showing the location of the spinodal ($\partial^2 f/\partial C^2 = 0$); and (b) the corresponding T–C plot (phase diagram) for the spinodal.

resembles that for the abrupt interface treatment. The composition at the center of the nucleus approaches the equilibrium composition, C_β, while the specific surface free energy of the interface between nucleus and matrix approaches that for a planar abrupt interface. With increasing supersaturation, the interface between nucleus and matrix becomes progressively more diffuse, the composition at the nucleus center decreases from C_β and the thickness of the interface increases. As the inflection point on the free energy curve (see Figs. 7.20 and 8.26) called the spinodal is approached ($C_\infty \to C_{Sp}$), the work of forming the nucleus goes to zero as $(C_{Sp} - C_\infty)^{3/2}$, the composition of the center of the nucleus, C_n, approaches C_∞ as $(C_{Sp} - C_\infty)$ and the radius at which the composition is $\frac{1}{2}(C_n - C_\infty)$ increases as $(C_{Sp} - C_\infty)^{-1/2}$. In the immediate vicinity of the spinodal, as $C_\infty \to C_{Sp}$, an increasingly diffuse interface is favored by a reduction in the free energy required to form material of non-equilibrium composition. At and within the spinodal, Δf decreases monotonically with increasing composition difference and there is no barrier to the formation of an infinitely diffuse interface. Thus, γ and the work of forming the critical nucleus should vanish at the spinodal.

Near the spinodal, the critical nucleus no longer resembles a cluster of the new phase, but rather represents a fluctuation small in degree but large in its spatial extent. Since $\Delta f(C) = f(C) - f(C_\infty)$ in Eq. (8.76), this is positive when $C_\infty < C_{Sp}$ (outside the spinodal) so the system is stable against all such fluctuations. When $C_\infty > C_{Sp}$ (inside the spinodal), the bulk free energy term is negative and the system may be unstable against some fluctuations.

Suppose we consider an infinitesimal sinusoidal fluctuation of the form

$$C - C_\infty = A \cos wx \tag{8.77}$$

Any actual fluctuation can be Fourier analyzed into such components and treated independently. From Eq. (8.76), the difference in free energy becomes

$$W = \frac{A^2}{4} \left[\left(\frac{\partial^2 f}{\partial C^2} \right)_{C_\infty} + 2\kappa_2 w^2 \right] \tag{8.78}$$

When $W < 0$ for a particular w, the homogeneous solution is unstable with respect to sinusoidal fluctuations of that w. When $W > 0$, the homogeneous solution is stable against such fluctuations. Because $\kappa > 0$, when $(\partial^2 f/\partial C^2)_{C_\infty} > 0$ (which occurs outside the spinodal), the solution is stable against infinitesimal fluctuations of all wavelengths. When $(\partial^2 f/\partial C^2)_{C_\infty} < 0$ (inside the spinodal), the solution is unstable against fluctuations of wavelengths greater than

$$\lambda_c = \frac{\pi}{w_c} \left[\frac{-8\pi^2 \kappa_2}{(\partial^2 f/\partial C^2)_{C_\infty}} \right]^{1/2} \tag{8.79}$$

The presence of the gradient term in Eq. (8.76) prevents decomposition of the solution on too small a scale. As the spinodal is approached, $\lambda_c \to \infty$ but, for a finite system, this limit changes to L, the largest linear dimension in the system.

Allowing the molar volume to vary with composition introduces a strain energy term into the above formalism and, for an infinite isotropic binary solid solution

$$W = \int_v \left[\Delta f(C) + \frac{2\beta_c^{*2} E^*}{1 - \sigma^*}(C - C_\infty)^2 - \kappa_2(\nabla C)^2 \right] dv \tag{8.80}$$

where β_C^* is the linear expansion per unit composition change, σ^* is Poisson's ratio and E^* is Young's modulus for $C = C_\infty$. In this case, the solution is stable against infinitesimal fluctuations of all wavelengths, even within the spinodal limit until

$$\left(\frac{\partial^2 f}{\partial C^2}\right)_{C_\infty} < \frac{-2\beta_C^{*2} E^*}{(1-\sigma^*)} \tag{8.81a}$$

since, for the fluctuation of Eq. (8.77),

$$\Delta F_v = \frac{A^2}{4}\left[\left(\frac{\partial^2 f}{\partial C^2}\right)_{C_\infty} + \frac{2\beta_C^* E^*}{(1-\sigma^*)} + 2\kappa_2 w^2\right] \tag{8.81b}$$

The stability limit is given by using an equality sign in Eq. (8.81a).

During the initial stages of decomposition, $A(w,t)$ may be written as

$$A(w,t) = A(w,0)\exp[\mathcal{I}(w)t] \tag{8.82a}$$

where the kinetic amplification factor, $\mathcal{I}(w)$, is given by

$$\mathcal{I}(w) = -\hat{M}w^2\left[\left(\frac{\partial^2 f}{\partial C^2}\right)_{C_\infty} + \frac{2\beta_C^{*2} E^*}{(1-\sigma^*)} + 2\kappa_2 w^2\right] \tag{8.82b}$$

with $\hat{M}$ as the atomic mobility.

It should be noted that $\mathcal{I}(w)$ is negative when the solution is stable against fluctuations so that such fluctuations will be damped out. When the solution is unstable against fluctuations of wavelength greater than λ_c, such fluctuations will grow with time and $\mathcal{I}(w)$ is greatest for $\lambda = \sqrt{2}\lambda_c$ so such components should grow the fastest and come to dominate the decomposition process.

Outside the spinodal, the critical nucleus for new phase formation conforms to the classical picture of a fluctuation large in degree and small in extent. As the spinodal is approached, the critical nucleus begins to take on the appearance of a fluctuation small in degree and large in extent. Nucleation via such a process begins to resemble a process of continuous decomposition. For C_∞ just inside the spinodal field, the solution is stable against short wavelength fluctuations and unstable only against fluctuations sufficiently large in extent that the increase in gradient energy is smaller than the degree in volume energy. The scale of any such decomposition implies the necessity for long-range diffusive processes so that the transformation, for kinetic reasons, may take place via a nucleation and growth mechanism even though this is not via the critical nucleus path. For C_∞ well within the spinodal field, the solution is unstable even against fairly short wavelength fluctuations so the phase change should take place by spinodal decomposition.

References

Chapter 1

1. U. Nakaya, *Compendium on Meteorology* (Amer. Math. Soc.) Boston, p.207, (1954); *Snow Crystals,* (Harvard University Press, Cambridge, Mass., 1955).
2. B. J. Mason in *Art and Science of Crystal Growing* (Wiley, New York, 1963), pp. 119-50.
3. W. A. Tiller, *Science*, **146**, 871 (1964).
4. W. A. Tiller, S. Hahn and F. A. Ponce, *J. Appl. Phys.*, **59**, 3255 (1986).
5. B. Deal and A. S. Grove, *J. Appl. Phys.*, **36**, 3770 (1965).
6. W. A. Tiller, *J. Electrochem. Soc.*, **127**, 625 (1980); **127**, 619 (1980); **130**, 501 (1983).
7. W. A. Tiller, *The Science of Crystallization: Macroscopic Phenomena and Defect Generation* (Cambridge University Press, Cambridge, 1991).
8. E. M. Young and W. A. Tiller, *Appl. Phys. Lett.*, **42**, 63 (1983); **50**, 46 (1987); **50**, 80 (1987).

Chapter 2

1. E. H. Pearson, T. Halicioglu and W. A. Tiller, *Surface Science*, **184**, 401 (1987).
2. D. K. Choi, T. Takai, S. Erkoc, T. Halicioglu and W. A. Tiller, *J. Crystal Growth*, **85**, 9 (1987).
3. W. W. Mullins in *Metal Surfaces*, Eds. W. D. Robertson and N. A. Gjostein (Am. Soc. Metals, Metals Park, Ohio, 1963) Ch. 2.
4. W. A. Tiller, "On the Energetics, Kinetics and Topography of Interfaces", in *Treatise on Materials Science and Technology*, Vol. 1, Ed. H. Herman (Academic Press, New York, 1972), p.1.

5. K. A. Jackson, in *Solidification* (Am. Soc. Metals, Metals Park, Ohio, 1958), p. 180.
6. D. Turnbull, in *Thermodynamics in Physical Metallurgy* (Am. Soc. Metals, Metals Park, Ohio, 1950) p. 293.
7. J. R. Bourne and R. J. Davey, *J. Crystal Growth*, **36**, 278 (1976).
8. J. P. Hirth, in *Energetics in Metallurgical Phenomena, Vol. 11*, Ed. W. M. Mueller (Gordon and Breach, New York, 1965). p. 1.
9. C. Herring, in *Structure and Properties of Solid Surfaces*, Eds. R. Gomer and C. S. Smith (University of Chicago Press, Chicago, 1952) Ch. 1.
10. G. F. Bolling and W. A. Tiller, *J. Appl. Phys.*, **31**, 1345 (1960).
11. J. W. Cahn, *Acta Met.* **8**, 554 (1960).
12. B. Chalmers, *Principles of Solidification* (John Wiley and Sons, New York, 1964) p. 35.
13. D. Turnbull, *J. Phys. Chem.*, **66**, 609 (1962).
14. J. Cahn, W. B. Hillig and G. W. Sears, *Acta Met.*, **12**, 1421 (1964).

Chapter 3

1. W. A.Tiller and B. K. Jindal, *Acta Met.*, **20**, 543 (1972).
2. W. Lin and D. W. Hill, *Silicon Processing*, ASTM STP804, Ed. D. C. Gupta (Am. Soc. for Testing and Materials, Philadelphia, Pa., 1983).
3. F. C. Frank, in *Growth and Perfection of Crystals*, Eds. R. H. Doremus, B. W. Roberts and D. Turnbull (J. Wiley and Sons, New York, 1958) p. 3.
4. R. J. Davey, *J. Crystal Growth*, **34**, 109 (1976).
5. N. Albon and W. J. Dunning, *Acta Cryst.*, **15**, 475 (1962).
6. N. Cabrera and D. A. Vermilyea, in *Growth and Perfection of Crystals*, Eds. R. H. Doremus, B. W. Roberts and D. Turnbull (J. Wiley and Sons, New York, 1958) p. 441.

Chapter 4

1. M. B. Panish and M. Ilegems, *Progress in Solid State Chemistry*, Vol. 7, Eds H. Reiss and J. O. McCaldin (Pergamon, New York, 1972) p. 39.
2. J. A. Burton, R. C. Prim and W. P. Slichter, *J. Chem. Phys.*, **21**, 1987 (1953).
3. R. N. Hall, *Phys. Rev.*, **88** (1952) 139; *J. Phys. Chem.*, **57**, 836 (1953).
4. W. A. Tiller and K. S. Ahn, *J. Crystal Growth*, **49**, 483 (1980).
5. W. A. Tiller, *The Science of Crystallization: Macroscopic Phenomena and Defect Generation* (Cambridge University Press, Cambridge, 1991).
6. J. C. Baker and J. W. Cahn in *Solidification* (Am. Soc. Metals, Metals Park, Oh, 1970) pp. 23–58.
7. K. S. Ahn, "Interface Field Effects on Solute Redistribution and Interface Instability During Steady State Planar Front Crystallization," Ph.D. Thesis, Materials Science and Engineering Department, Stanford University, Stanford, CA (1981).
8. B. K. Jindal and W. A. Tiller, *J. Chem. Phys.*, **49**, 4632 (1968).
9. L. M. Goldman and M. J. Aziz, *J. Mater. Res.*, **2**, 524 (1987).

Chapter 5

1. F. C. Tompkins, *Chemisorption of Gases on Metals*, (Academic Press, New York, 1978) Ch. 6.
2. B. K. Chakraverty, in *Crystal Growth: An Introduction*, Ed. P. Hartman (North Holland, Amsterdam, 1973).
3. G. E. Rhead, *Surface Science*, **47**, 207 (1978).
4. W. A. Tiller, *J. Vac. Sci. Technol.*, **A7**, 1353 (1989).
5. H. J. Gossman and L. C. Feldman, *Phys. Rev.*, **B32**, 6 (1985); *Surface Sci* **155**, 413 (1985).
6. B. J. Garrison, M. T. Miller and D. W. Brenner, private communication.
7. S. S. Iyer, T. F. Heinz and M. M. T. Loy, *J. Vac. Sci. Technol.*, **B5**, 709 (1987).
8. J. Knall, J. E. Sundgren, G. V. Hansson and J. E. Greene, *Surface Sci.* **166**, 512 (1986).
9. W. A. Tiller, *The Science of Crystallization: Macroscopic Phenomena and Defect Generation*, (Cambridge University Press, Cambridge, 1991).
10. J. Bloem and L. J. Giling in *VLSI Electronics: Microstructure Science*, Vol. 12, Ch. 3; Eds. N. G. Einspruch and H. Huff (Academic Press, New York, 1985).
11. J. Nishigawa in *Crystal Growth: Theory and Techniques*, Vol. 2, Ch. 2; Ed. C. H. L. Goodman (Plenum Press, New York, 1978).
12. A. A. Chernov, *J. Crystal Growth*, **47**, 207 (1978).
13. J. R. Arthur, *J. Appl. Phys.*, **39**, 4032 (1968); *The Structure and Chemistry of Solid Surfaces*, Ed. G. A. Somerjai (Wiley, New York, 1969) p. 46.

Chapter 6

1. M. V. Rao, R. Hiskes and W.A. Tiller, *Acta Met.*, **21**, 733 (1973).
2. C. D. Thurmond and M. Kowalchik, *Syst. Tech. J.*, **39**, 169 (1960).
3. M. V. Rao and W. A.Tiller, *Materials Science and Engineering*, **11**, 61 (1973).
4. J. W. Gibbs, '*The Collected Works*, Vol. 1 (Yale University Press, New Haven, 1957).
5. C. D. Thurmond, Ch. 4 in *Semiconductors*, Ed. N. B. Hannay (Reinhold Publishing Corp., New York, 1959).
6. F. A. Trumbore, *Bell Syst. Tech. J.*, **39**, 205 (1960).
7. R. A. Swalin, *Thermodynamics of Solids*, (John Wiley & Sons, Inc., New York, 1962).
8. J. A. Van Vechten, *Handbook on Semiconductors*: Vol. 3, *Materials, Properties and Preparation* Vol. ed., Seymour P. Keller; Series ed. T. S. Moss (North Holland, New York, 1980) p.82.

Chapter 7

1. J. W. Gibbs, *The Collected Works*, Vol. 1 (Yale University Press, New Haven, 1957).
2. W. W. Mullins in *Metal Surfaces: Structure, Energetics and Kinetics*, (Am. Soc. Metals, Metals Park, Ohio, 1963).
3. A. E. Carlsson, in *Solid State Physics: Advances in Research and Applications*,

Eds. H. Ehrenreich and D. Turnbull, Vol. 43 (Academic Press, New York, 1989).

4. J. F. Nickolas, *Aust. J. Phys.* **21**, 21 (1968).
5. E. M. Lifshitz, *Soviet Physics, JETP*, **2**, 73 (1956).
6. J. H. Rose, J. R. Smith and J. Ferrante, *Phys. Rev.* B **28**, 1835 (1983).
7. H. Brooks in *Metal Interfaces* (Am. Soc. Metals, Metals Park, Ohio, 1952).
8. J. W. Cahn and J. E. Hilliard, *J. Chem. Phys.*, **28**, 258 (1958).

Chapter 8

1. C. Wert and C. Zener, *Phys. Rev.*, **76**, 1169 (1949).
2. S. Glasstone, K. Laidler and H. Eyring, *The Theory of Rate Processes* (McGraw-Hill, New York, 1941).
3. J. Feder, K. C. Russel, J. Lothe and G. M. Pound, *Advances in Physics*, **15**, 111 (1966).
4. D. Turnbull, "Principle of Solidification" in *Thermodynamics in Physical Metallurgy* (ASM, Cleveland, Ohio, 1950).
5. D. Turnbull and R. E. Cech, *J. Appl. Phys.*, **21**, 804 (1950).
6. G. M. Pound and V. K. LaMer, *J. Am. Chem. Soc.*, **74**, 2323 (1952).
7. D. Turnbull, *J. Chem. Phys.*, **20**, 411 (1952).
8. J. H. Perepezko and D. H. Rasmussen, *TMS-AIME*, **9A**, 1490 (1978). J. H. Perepezko and E. I. Anderson, TMS-AIME Symposium on *Synthesis and Properties of Metastable Phases*, Eds. T. J. Rowland and E. S. Machlin (Warrendale, Pennsylvania, 1980) p. 31.
9. M. A. Sharaf and R. A. Dobbins, *J. Chem. Phys.*, **77**, 1517 (1982).
10. G. M. Pound, *Metall. Trans.* A, **16A**, 487 (1985).
11. D. Turnbull and B. Vonnegut, *Ind. Engng. Chem.*, **44**, 1291 (1952).
12. B. E. Sundquist and L. F. Mondolfo, *Trans. Am. Inst. Min. Metall. Engrs.*, **221**, 157 (1961).
13. W. A. Tiller and T. Takahashi, *Acta Met.*, **17**, 483 (1969).
14. E. M. Pearson, "Computer Modeling of Atomic Interactions with Application to Silicon," Ph.D. Thesis, Materials Science and Engineering Department, Stanford University, Stanford, CA (1985).
15. J. W. Cahn and J. E. Hilliard, *J. Chem. Phys.*, **28**, 258 (1958) and **31**, 688 (1959).

Index

Printed in the United States
By Bookmasters